Sustainable Development

This groundbreaking volume offers an innovative exploration of sustainability challenges through complementary lenses that traditional approaches have often kept separate. It bridges the crucial gap between legal frameworks and technological innovations, providing essential insights into how governance, policy, and emerging technologies collectively drive sustainable development initiatives. The volume's distinctive interdisciplinary approach encompasses diverse themes, including renewable energy law, indigenous knowledge systems, gender justice, food security, and digital democracy. Featuring contributions from leading experts across the globe, the book balances theoretical discourse with practical application through illuminating case studies, judicial analyses, and policy evaluations.

By combining scholarly rigor with accessible insights, this book is both an intellectual contribution to academic discourse and a practical guide for those working to shape a more sustainable future through the powerful integration of legal frameworks and technological innovation. It will be indispensable for students and academics seeking cutting-edge research; legal professionals navigating evolving regulatory landscapes; policymakers crafting effective sustainability frameworks; environmental activists pursuing systemic change; and corporate leaders implementing ESG compliance strategies.

Usha Tandon, Senior Professor and Vice Chancellor, Dr. Rajendra Prasad National Law University Prayagraj, Prayagraj, India.

Sustainable Development
A Techno-Legal Perspective

Edited by Usha Tandon

Routledge
Taylor & Francis Group
LONDON AND NEW YORK

Designed cover image: Getty Images

First published 2027
by Routledge
4 Park Square, Milton Park, Abingdon, Oxon OX14 4RN

and by Routledge
605 Third Avenue, New York, NY 10158

Routledge is an imprint of the Taylor & Francis Group, an informa business

British Library Cataloguing-in-Publication Data
A catalogue record for this book is available from the British Library

ISBN: 978-1-041-22145-6 (hbk)
ISBN: 978-1-041-22737-3 (pbk)
ISBN: 978-1-003-73375-1 (ebk)

DOI: 10.4324/9781003733751

Typeset in Sabon
by Apex CoVantage, LLC

Contents

Contributors

Usha Tandaon
Senior Professor and Vice Chancellor
Dr. Rajendra Prasad National Law University
Prayagraj, Uttar Pradesh, India

Narender Bishnoi
Assistant Professor
Campus Law Centre, University of Delhi
India

Klaus Bosselmann
Professor of Law and Founding Director
New Zealand Centre for Environmental Law, University of Auckland
New Zealand

Philippe Cullet
Professor of International and Environmental Law
SOAS University of London
UK

Neha Dubey
Assistant Professor, English
Dr. Rajendra Prasad National Law University Prayagraj
Prayagraj, India

Neeraj Kumar Gupta
Assistant Professor, Grade-I
Rajiv Gandhi School of Intellectual Property Law
IIT Kharagpur, West Bengal, India
Assistant Professor

Pallavi Gupta
Assistant Professor
Marwadi University
Rajkot, India

Moon-hyun Koh
Professor
Soongsil University
Seoul

P. Leelakrishnan
Professor Emeritus
Cochin University School of Legal Studies
Kochi, India

Swati Kumari Mawandiya
Assistant Professor
Institute of Law, Nirma University
Ahmedabad, India

Erimma Gloria Orie
Head of Department of Private and Property Law
Faculty of Law, National Open University of Nigeria
Nigeria

Rinzin Penjor
Former Judge
Supreme Court of Bhutan

Robert Russo
Professor
Peter A. Allard School of Law
University of British Columbia
Canada

Shaiwal Satyarthi
Professor
Faculty of Law
University of Delhi
India

Sonika
Assistant Professor of Law
Dr. Rajendra Prasad National Law University Prayagraj
Prayagraj, India

Amber Prasad Pant
Professor & Former Head & Dean
Faculty of Law
Tribhuvan University, Nepal

Prakash Tripathi
Assistant Professor
Dr. Rajendra Prasad National Law University Prayagraj
Prayagraj, India

Shivang Tripathi
Research Scholar
Banaras Hindu University
India

Arun Bhansali
Chief Justice
High Court of Judicature at Allahabad
November 25, 2025

Foreword

The global conversation on sustainable development has undergone a profound transformation in recent decades. The accelerating pace of technological advancement, coupled with the unprecedented ecological pressures on our planet, has made it imperative for policymakers, jurists, researchers, and civil society to rethink conventional paradigms of governance. Law, today, stands at a decisive crossroads, called upon not merely to regulate technological progress but to actively harness it for securing a just, equitable, and ecologically responsible future.

In this context, the edited volume *Sustainable Development: A Techno-Legal Perspective* makes a timely and significant contribution. The work brings together eminent scholars and practitioners from around the world to examine the intersections of law, technology, and sustainability, offering a composite and interdisciplinary understanding of the challenges that lie ahead.

The chapters in this volume traverse a wide thematic landscape, ranging from the Earth System approach to international environmental governance, energy efficiency and renewable energy transitions, tribal ecological knowledge, gendered dimensions of digital technologies, food security, digital access to justice, ethical dimensions of AI-assisted decision-making, and judicial responses to climate change. Despite their diversity, these themes converge on a common insight: that technology, when guided by sound legal principles and informed by social and ecological realities, can be a transformative tool for sustainable development.

As the world strives to achieve the United Nations' 2030 Agenda, it becomes increasingly clear that sustainable development cannot be secured through isolated interventions. It requires a synergistic approach, integrating the strengths of law, science, technology, and community action. This volume stands as a thoughtful and compelling contribution to that endeavour.

I congratulate the editors and contributors for their scholarly commitment and intellectual rigour. I am confident that this work will serve as a valuable resource for students, researchers, policymakers, and all those engaged in advancing a more sustainable and harmonious future.

Arun Bhansali
Chancellor
Dr. Rajendra Prasad National Law University
Prayagraj

Manoj Kumar Gupta

Judge, High Court of
Judicature at Allahabad
Allahabad

Foreword

Sustainable development is our constitutional and ethical obligation to the present and future generations. Today, it stands at a critical crossroads where rapid technological innovation intersects with the normative discipline of law, global challenges like climate change, environmental degradation, energy insecurity, and governance deficits, transcending territorial boundaries, and sectoral silos. These challenges demand integrated responses rooted in constitutional values, international cooperation, scientific advancement, and legal accountability. It is within this evolving global context that the present theme, *Sustainable Development: A Techno-Legal Perspective*, acquires relevance, offering a reflective and rigorous inquiry into how law may regulate and harness technology in furtherance of sustainability.

The present book is a scholarly outcome of the three-day International Conference on Law, Technology and Sustainable Development organized by the Dr. Rajendra Prasad National Law University Prayagraj (RPNLUP), which reflects its academic policy of fostering interdisciplinary research that is socially responsive, globally engaged, and attuned to global concerns. Bringing together jurists, scholars, and practitioners from all over the world, this book examines sustainable development through a techno-legal lens, which is now indispensable.

A defining strength of this work lies in its multidisciplinary and integrative approach. The book engages with international environmental governance and expands to examine renewable energy, indigenous ecological knowledge, gender justice in digital spaces, food security, technology-enabled access to justice, ethical dimensions of AI-assisted decision-making, and evolving judicial responses to environmental and climate challenges. Comparative perspectives from jurisdictions such as Bhutan, Nepal, and South Korea further enrich the analysis and underscore the transnational nature of sustainability concerns.

The book offers a nuanced engagement with the techno-legal interface as a site of both opportunity and caution. While technology can address climate challenges, the absence of legal foresight may exacerbate inequality and weaken accountability. The contributors emphasize that sustainable development requires coherence between law, technology, social equity, and ecological limits, in line with constitutional jurisprudence, which regards environmental protection and climate justice as integral to the right to life and human dignity.

I commend Professor Usha Tandon for her academic leadership and editorial rigour in curating these thoughtful scholarly collections. The book reflects RPNLUP's commitment to interdisciplinary legal research and scholarship, and I am confident that this work will be of enduring value to scholars, judges, policymakers, and students while contributing meaningfully to the pursuit of a just, inclusive, and sustainable future.

Manoj Kumar Gupta
Judge, High Court of Judicature at Allahabad
Allahabad

Preface

This book emerges out of a three-day international conference titled "Law, Technology and Sustainable Development", organized by the Dr. Rajendra Prasad National Law University Prayagraj, Prayagraj, March 7–9, 2025. A few selected papers, after blind peer review, have been meticulously edited to form the content of this book. The book explores the multidimensional nature of sustainability through a legal and technological lens. This compilation seeks to provide a comprehensive understanding of how legal frameworks and technological advancements can jointly facilitate the realization of sustainability goals. This book presents an interdisciplinary approach to sustainable development, integrating legal scholarship with technological innovations.

The full-length papers from all participants were invited by December 2024. The initial intention was to release the edited volume during the conference itself, a practice I had observed at academic conferences elsewhere. However, upon gaining a deeper understanding of the processes and compromises often underlying such expedited publications, I consciously chose to abandon this approach in favour of ensuring academic rigour, thorough peer review, and editorial integrity. The manuscript underwent further scrutiny by the publisher, and the agreement for publication of the book was ultimately signed on September 22, 2025, with Routledge, an imprint of Informa UK Ltd., trading as the Taylor & Francis Group. I hugely appreciate the understanding, cooperation, and patience of the contributors, who had been longing to see the publication in March 2025. I am also thankful to all the contributors for seriously attending to the editorial comments and graciously meeting the timeline. Various persons at different stages have helped with the onerous task of reviewing, editing, and proof-reading the manuscript. Dr. Prakash Tripathi deserves special appreciation in this regard.

I am deeply grateful to Honourable Chief Justice Arun Bhansali, Allahabad High Court, for graciously taking time from his demanding judicial responsibilities to write a scholarly and deeply enriching foreword for this book. Thanks are also due to Honourable Justice Manoj Kumar Gupta, Senior Judge, Allahabad High Court, for writing a foreword to this volume, which lends enduring intellectual value to the book.

I thank the Taylor & Francis Group, and particularly Dr. Shashank S. Sinha, for undertaking the task of publishing this book.

Given the interdisciplinary approach of the book, addressing both legal and technological aspects of sustainability, the target audience that may draw benefit from the book includes academics, researchers, students, legal and policy professionals, and international organizations, as well as industry and technology professionals.

Usha Tandon

Part I

Earth Approach

1 The Earth System Approach to International Environmental Law*

Klaus Bosselmann

1.1 Introduction

The Earth system approach allows us to see the environment as an integrated and interconnected whole rather than as fragmented elements of Earth, which are supposed to work together in order to regulate life on the planet.[1] The notion of "planetary boundaries" aptly describes the challenge for global environmental law. This science-based concept sets nine global biophysical indicators that measure the degree to which human activities have degraded Earth's life-support system. The idea is that we can – and must – limit the impacts of our activities so that these impacts stay within biophysical thresholds. Only then can Earth continue to provide the natural processes that sustain life.[2] However, humans have already crossed six of the nine boundaries.[3]

This chapter offers some ideas about a concept of environmental governance that respects the Earth system and its planetary boundaries. Living safely within the boundaries of our planet cannot be achieved by relying on governance practices around traditional interpretations of state sovereignty. For too long, states have been resistive to obligations towards areas outside their national jurisdictions. Essentially, national interests can only be met through safeguarding the planetary commons,[4] that is the polar regions, atmosphere, oceans, biosphere, in a word, Earth as a whole.[5]

The Earth is a complex system formed by a large variety of sub-systems (biosphere, atmosphere, lithosphere as well as social and economic systems, etc.), which interact by exchange of matter, energy, and information. Because of these inter-relations, the Earth is a complex and evolving network. We may choose to consider each subsystem separately, but the growing understanding of the whole system Earth suggests that one should consider the interactions between these subsystems.[6]

Over the past two decades, evidence has been mounting that planetary-scale changes are occurring rapidly in response to the forcing and feedback that characterize the internal dynamics of Earth System.[7] Scientific revelations have informed us about four aspects of the Earth systems that illustrate how human enterprises are pushing towards "*planetary terra incognita*."[8] This phenomenon of global change represents a profound shift in the relationship

DOI: 10.4324/9781003733751-2

between humans and the rest of nature. They are non-linear, catastrophic shifts: tipping points. To begin with, complex social-ecological systems which underpin human and non-human well-being such as coral reef ecosystems, agro-ecological landscapes, forests, and freshwaters can shift with irreversible damage.[9] The next aspect is the scale. The non-linear properties of vital social-ecological systems are not limited to regional or local scale examples. The potential of irreversible shift in the Earth systems, abrupt climate change,[10] tipping elements in Earth system,[11] planetary boundaries,[12] a proposed possible state-shift in Earth's biosphere[13] are all examples of attempts to explore the possibility of rapid, aggregated, and destructive change on global scale.

These aspects reveal that we are no longer able to predict how deterioration to the ecosystem could affect processes and functionality at the planetary scale and how ecosystems themselves may react to disturbance. What we do know, however, is we must bring to scale human interference beyond the planetary boundaries and ensure we don't cross the safe operating space for humans. One important conclusion is that the "boundaries" presently perceived as "safe" could move over time as the Earth system or our understanding of it evolves. This dynamic interplay between systems behaviours, values, and politics should be considered in the environmental decision-making dimensions.

The Earth system approach to international environmental law requires a shift from state-centred governance to new forms of multi-actors governance. The shift involves states, but not exclusively. The new approach emphasizes the role of the citizen rather than the nation-state as the source of governance. In a democracy, government is empowered by citizens and accountable to them. It is appropriate, therefore, to perceive of governments as fiduciaries or trustees acting on behalf of its citizens. In the Anthropocene, however, citizenship is no longer defined purely through national identity; it is also defined through transnational, global identity. This calls for transnational processes of forming the collective will, at least potentially. In this way, we can perceive areas outside national jurisdiction, that is the global commons, as an additional point of reference and the subject of stewardship or trusteeship.

The chapter aims to show how states can, and must, be perceived as institutions of trusteeship for the global commons and the Earth system.

1.2 The Challenge of Earth System Science

The recognition of Earth as a system has roots in early scientific inquiries into geology, geography and ecology. Early pioneers such as geologist James Hutton and geographer Alexander von Humboldt described the Earth as being governed by consistent natural processes over time. In 1788, Hutton posited that the same geological forces observed today – erosion, sedimentation, and volcanic activity – had operated in the past, maintaining Earth's equilibrium.[14] Similarly, Alexander von Humboldt described the course of nature as dynamic, interrelated, and self-generating that all life (including human life) is part of.[15] In 1926, Vladimir Vernadsky established the biosphere concept and emphasized the impact of living organisms on the Earth's chemical composition and energy flows. Vernadsky argued that life is a geological force, profoundly influencing processes such as the carbon and nitrogen cycles. His influential work, *The Biosphere*, established the intellectual basis for seeing the Earth as a system shaped by biological and physical processes.[16]

The 20th century saw the development of systems theory, which stressed the interdependence of Earth's components. Ludvig von Bertalanffy's General System Theory introduced a comprehensive approach, emphasizing feedback loops and interactions in complex systems.[17] These principles were applied to Earth sciences, integrating climatology, oceanography, and biology disciplines, and suggested that Earth functions as a self-regulating system where biotic and abiotic components interact to maintain conditions conducive to life.[18]

Not only is Earth system science the study of the physical processes of atmospheric chemistry, oceanic circulation, biodiversity, ecosystem physiology, and so forth; it is also the study of the complex interaction of human beings with the environment. Its core function is to work out how to retain the Eden-like Holocene state that has fostered human and ecological health. There is a strong focus on understanding the deeply embedded connections between the subsystems and how these affect the global system. Earth system science contains a level of complexity well beyond the scope or understanding of any one discipline and reflects most accurately the true complexity of modern life. Undaunted, it attempts to uncover the socio-ecological system that already exists and is a moving target, constantly mutating in unexpected ways. It calls for a community of human effort from all directions and a flexibility in law and in democratic governance structures that can adapt and keep pace with the new knowledge and changing aspects of the earth system.

As human impact on the planet increases, the chances of achieving well-being for everyone and sustainability of natural resource use at global, national, and local levels, grows dimmer. We only have one planet and yet, the Global Footprint Network calculates that we already use the equivalent of 1.75 Earths to provide for current annual resource use, with this measure set to rise to 2 Earths by the 2030s.[19] The calculation is based on how quickly the planet can regenerate what we actually use in a year. Whatever measure is used to determine Earth system resilience, it is clear that we are living on borrowed natural capital, which at some point must be repaid. The situation causes an ecological overshoot that over time steadily erodes the planet's resilience levels and ability to support life, including our lives and that of future generations.[20]

International environmental law has attempted to stem the damage but has been largely unsuccessful except at local and problem-specific levels. Vital questions environmental law scholars grapple with are as follows: How can the law protect the natural environment both *from* humanity and *for* humanity? Has the time finally arrived to consider protecting the environment simply for its own sake and beyond our own human interests? And, why is it that the body of law so far created does not actually protect the environment except in a sporadic and localized fashion?

These questions raise issues that range far outside the traditional purview of the law. Who are we as human beings? What is our place and how do we fit into the natural world? Since Alexander von Humboldt and Charles Darwin, we have come to understand how closely embedded humans are with all other life and yet Western legal systems (including international law) continue to ignore this reality. Humanity is now facing the consequences as we enter a whole new era of geology, that is the Anthropocene.[21]

Considering the ongoing dominance of short-sighted national interests and misguided state sovereignty, the challenges for international environmental law are enormous. However, unlike other branches of international law, which do not necessarily need to look outside the boundaries of traditional primary and secondary legal sources, the subject of international environmental law is the global environmental system of interactions between humans and the natural environment. This requires a system theoretical approach to environmental law, an external birds-eye view that sees the human and physical world as one,[22] in other words as a single Earth system. The Earth system approach immediately exposes the "gaps" between ecological and social realities and between legal constructs and prevailing economic interests.

Like other disciplines, the complexity of the modern world now requires the law to accommodate pluralistic (anthropocentric and ecocentric) concepts within its framework. An underlying *grundnorm* of ecological integrity is ecocentric in nature, which is why it has been rejected by the anthropocentrically focused legal system. An interesting example of how the law's anthropocentric focus has developed internal distortions is that along with humans, corporations are recognized as "legal persons" protecting economic interests whereas the natural environment has no such legal status. This has created the rather surreal situation where the law is focused on protecting economic interests over and above the very life support system of the people conducting those economic activities. This surely must cause us to question and re-evaluate the core values underlying the law.[23]

If "ecological integrity" (carefully defined to be legally effective) can be shown to be fundamental in terms of law and governance, this automatically calls for guardianship or trusteeship concepts to form part of international and state institutional arrangements.[24]

The idea that all systems, whether natural, social, or legal, function best with a certain level of underlying integrity[25] is one that is worth exploring in this report. "Integrity" here is used to mean that which a particular system needs to maintain coherent order between stability and change and to function in a resilient manner without collapse.[26]

The concept of "ecological integrity" is already well known in international environmental law. At least 25 international soft and hard law agreements, including the 2015 Paris Agreement on Climate Change, contain specific reference to it.[27] Yet states continue to shy away from associated legal obligations. In fact, environmental law as a legal discipline has failed (as explained by Staffan Westerlund)[28] because it has looked at sustainability from the perspective of the law instead of looking at the law from the perspective of sustainability. Essentially, sustainability requires the preservation and restoration of ecological integrity.[29]

Douglas Fisher's analysis of the methodology of judges, administrators, and politicians, who consider environmental matters, has shown that decisions are reached without specific environmental "points of commencement." Instead, traditional assumptions concerning human rights and economic imperatives play the dominant (albeit sometimes unconscious) role in the decision-making process. Fisher concludes that the processes of legal reasoning need to reflect fundamental *grundnorms* in the form of the Rule of Law and of ecological sustainability in the context of decision-making.[30]

Victor Galaz aptly describes the challenges ahead as being threefold[31]:

1. Our mental models and causal beliefs are being seriously challenged by the complexity, scale and speed of global change.
2. Our analytical approaches [meaning political approaches] are increasingly failing us as we gain increasing insights about the anatomy of earth system change.
3. Our political institutions at multiple levels of social organisation are unable to effectively respond to novel risks and opportunities induced by interacting environmental, political and technological change.

Discussions about altering the existing structure of environmental law and governance must also deal with the historical and existing recognition of state sovereignty. The success of states advancing their own interests – in the name of protecting their citizens – achieves nothing if the integrity of the Earth's ecological system is at stake. This will not change until the protection of the global commons is seen as a prerequisite for the protection of citizens.

1.3 The Global Commons and International Law

One prominent example of the global commons is the atmosphere. Intuitively, the atmosphere belongs to all living things including humans as our well-being and existence depends on a relatively stable climate system. The problem here is that, currently, the atmosphere is being treated as an open access resource without any legal status; it is widely regarded as *ius nullius*, that is a legal nullity or vacuum. This has worked to the advantage of property owners who filled the vacuum by exercising their property rights. Property rights may not include a right to pollute, but the absence of someone who could claim violation of own's rights means that actual pollution goes without any sanction.

In fact, it is free. You and I or the entire fossil fuel industry, for that matter, can freely emit greenhouse gases into the atmosphere as nobody's property rights are affected. This is the legal status quo. It will only be qualified if, and in so far, the law sets rights-limiting emissions standards. To date, this has been an uphill battle, which has not been made any easier merely by having the 2015 Paris Agreement on Climate Change (requiring additional binding commitments, targets, and timelines to make it work).

But we can win the battle, and it is quite simple. By asserting that all humanity, living now and in the future, own the atmosphere, we begin to use the institutions of law working in our favour. As legal owners we can charge for individual damage of our common property, provide rewards to those who protect it (e.g., producers and users of renewable energy), and in this way eliminate greenhouse gases. All we have to do is to declare the atmosphere a public common good, owned by all humanity and protected by law. This would provide an effective legal shield against the fossil fuel industry and its supporters (states, banks, corporations). The concepts of private property and state sovereignty would stay unchanged, but they would end where common property begins. The new rule for the Anthropocene is that common ownership prevails over private property.

This is a fairly simple legal construct. It could, for example, be supported by the well-established public trust doctrine. The public trust doctrine says that natural commons should be held in trust as assets serve the public good. It is the responsibility of the government, as trustee, to protect these assets from harm and ensure their use for the public and future generations. So nationally, the government would act as an environmental trustee. Internationally, states would jointly act as trustees for the global commons such as the atmosphere. Considering that only about 90 companies are responsible for two-thirds of carbons emitted into the atmosphere, a global trusteeship institution could quickly fix the problem of climate change.[32]

The idea of global nature's trusts has been promoted for some time, by environmental lawyers such as Mary Wood[33] or Peter Sand[34] and economists such as Peter Barnes[35] or Robert Costanza. Trusteeship governance is also advocated by the rich literature on the commons.[36] The "Reclaiming the Commons" movement has certainly found a momentum in recent times.

A key driver for change is of course the ever increasing deterioration of the global commons: the atmosphere (climate change), the hydrosphere (oceans, freshwater systems), the lithosphere (soils), and the biosphere (biodiversity loss). All four spheres are interrelated and hence require a much stronger focus on the Earth system. Climate change, for example, is closely connected with ocean acidification and biodiversity loss and none of these problems can ever be solved in isolation from the others. The recent global pandemic was a poignant reminder that nature is in charge, not governments merely exercising crisis management. Now more than ever before, states need to acknowledge the urgency of an Earth system approach to decision-making and define obligations accordingly.

Several recent declarations and agreements have a greater focus on fundamental obligations towards Nature and Earth as a whole. One is the 2016 IUCN World Declaration on the Environmental Rule of Law. Principle 1 postulates the "*Obligation to Protect Nature. Each State, public or private entity, and individual has the obligation to care for and promote the well-being of nature, regardless of its worth to humans, and to place limits on its use and exploitation.*"[37]

Further, the 2017 Draft Global Pact for the Environment says in Article 2: "*Every State or international institution, every person, natural or legal, public or private, has the duty to take care for the environment. To this end, everyone contributes at their own levels to the conservation and restoration of the integrity of the Earth's ecosystem.*"[38]

And the 2018 Hague Principles for a Universal Declaration on Responsibilities for Human Rights and Earth Trusteeship states in Principle 1: "*All human beings, individually and collectively, share responsibility to protect Nature, of which we are an integral part, the integrity of Earth's ecological systems and Earth as a whole, home of all living beings. Each State individually, and the international community of States collectively, acknowledge that they have, and share, responsibilities for Nature, in cooperation and in alliance with their citizens as equal trustees of Earth and the integrity of Earth's ecological systems.*"[39]

The Hague Principles represent a consensus derived from the Earth Charter, the World Declaration on the Environmental Rule of Law, the Draft Global Pact for the Environment, and further 15 international agreements defining responsibilities that humans have for each other and for the natural environment that they are part of.[40]

A strong case can indeed be made that international law and the United Nations are not only in need of but practically ready to develop institutions of trusteeship governance. There is, for example, a tradition of UN institutions with a trusteeship mandate including the (now defunct) UN Trusteeship Council, the World Health Organization (WHO) with respect to public health, and ironically also the World Trade Organization (WTO) with respect to free trade.[41] A number of other UN or UN-related institutions with weaker trusteeship functions exist as well. Quite obviously, states are capable of creating international trusteeship institutions, either expressively or implicitly. These developments prove that sovereignty of states can be transferred to international levels (as also the European Union demonstrates).

The underpinning motives are not so much of a moral nature, but driven by political interests. More often than not, political interests are short-sighted and selfish, but they can change as morality can change to include responsibility for all people (human rights) and the entire planet (sustainability). It is important to note that citizens and social institutions are the agents of moral change, not the state per se. At present, states follow the outdated morality of national interests over global interests. What does it take for states to accept trusteeship responsibilities for the global commons?

We cannot expect the trusteeship governance to initiate from the "top," that is by the UN and its member states themselves, but rather by forces outside the system, in particular global civil society. To this end, we can build on many years of activism and proposals for institutional change. Nor should we envisage states to be in sole charge of running and controlling global trusteeship institutions such as a World Environment Organization or a Global Atmospheric Trust. Rather, trusteeship governance should be seen as a joint effort of the UN, states, and civil society organizations with an equal say in decision-making.

1.4 Sovereignty and Trusteeship

There is, at present, an unholy alliance between politics (sovereignty) and private interests (property), severely undermining democratic process and the public concern for safeguarding the global commons. Furthermore, as Barnes points out, "not even seated at democracy's table – not organized, not propertied, and not enfranchised – are future generations, ecosystems, and non-human species."[42]

The practice of state governance in times of economic neoliberalism has affected the way how environmental policies and laws are being conceived, namely as entirely discretionary and located somewhere between resource exploitation and anthropocentric environmental protection. Mary Wood calls this a "discretionary frame,"[43] meaning that governments see themselves as perfectly entitled to favour short-term resource exploitation over long-term conservation.

We can clearly see that state governance today is about a quid pro quo, symbiotic relationship between political institutions and corporations. The rewards include property rights, friendly regulators, subsidies, tax breaks, and free or cheap use of the commons. This leaves very little for the "common" good. In the words of Peter Barnes, "we face a disheartening quandary here. Profit-maximizing corporations dominate our economy. Their programming makes them enclose and diminish common wealth. The only obvious counterweight is government, yet government is dominated by these same corporations."[44]

Fundamentally, the legitimacy of the state rests on its function to act for, and on behalf of, its citizens. This requires consent with the governed as, for example, John Locke stated: "(G)overnment is not legitimate unless it is carried on with the consent of the governed."[45] Governmental duties can therefore be understood as fiduciary obligations toward its citizens.[46] Such fiduciary obligations are recognized typically in public law, but generally exist in common law and civil law (although in varying forms and degrees) and are also known in international law. The fiduciary function of the state can also be described as a trusteeship function.[47]

In the following, we will examine how state sovereignty can be reconciled with trusteeship. Prime facie both seem to have different purposes, yet as we will see they are part of the same basic function of the state, that is to serve the citizens it depends on and is accountable to.

The environmental crisis and the state of the global commons give rise to the need for revisiting the relationship between sovereignty and trusteeship. Trusteeship must be pursued at both the international level and the domestic "internal" level. As Eyal Benvenisti notes, the present self-contained concept of sovereignty is less compelling than it was in the past because of the "glaring misfit between the scope of the sovereign's authority and the sphere of the affected stakeholders."[48] This misfit engenders inefficient, undemocratic, and unjust outcomes for under- or unrepresented affected stakeholders. Non-citizens, future generations, and the natural environment all fall into such a category of affected stakeholders.

There are two challenges to advancing the idea of trusteeship and both boil down to an outmoded understanding of sovereignty. On the one hand, to propose a system of international trusteeship is to directly challenge the principle of non-interference in states' domestic affairs. To propose that states become trustees themselves, in addition to an international system of trusteeship, is again, an intrusion into state's sovereign right to determine their approach to the environment. However, without the latter we will not achieve the former. Regardless of what one thinks about the legitimacy of sovereignty and the entire makeup of international relations, the reality is that states are calling the shots. Leaving aside the theoretical possibility of radically reorganizing global politics, we need to work within the context of nation-states.

Trusteeship is an idea, which softens the blow of what would otherwise be seen as an unprecedented intrusion into sovereign state affairs. In the context of global governance, trusteeship serves the common interests of states. Yet trustees are not states; a trust council might not even be an intergovernmental institution if it were comprised of individuals rather than drawn exclusively from "states." Arguably this represents a less threatening intrusion into sovereignty. As Catherine Redgwell explains, "trust arrangements do not challenge sovereignty directly, for one of the advantages of trusteeship arrangements is the absence of sovereignty in the exercise of trusteeship functions – there is no transfer of sovereignty to the trust authority."[49]

Nevertheless, the many proposals of trusteeship arrangements at the level of the United Nations have been, more often than not, greeted with hostility. States seem too attached to the principle of non-interference to appreciate co-operation of this kind. Yet the very origins of the concept of state sovereignty are closely linked with humanitarian concerns. The Peace of Westphalia, as the foundation of state sovereignty, was a key instrument for upholding humanitarian precepts related to freedom of conscience and religion. To the extent that it resolved a crisis of freedom of conscience and equality before the law and many pre-existing institutions had lost their legitimacy and ultimately collapsed, sovereignty has been and can be justified. But it should also be remembered that humanitarian concerns were at the root of the crisis that the new order resolved. Where new crises emerge that can only be solved through full cooperation, the principle of non-interference can be counterproductive and must be interpreted within a given context.

Likewise, with regard to the state itself as environmental trustee for those over whom it governs, it could hardly be refuted that a democratically elected government does not owe its citizens a duty to govern their natural wealth and resources in a sustainable way. The first step, then, is reminding ourselves as citizens and society that these rights and responsibilities rest with us, despite the state acting as our representative. The second step is convincing the consumer society of what these rights and responsibilities entail. This is no small feat.

Although there is a dedicated green movement, and ever more "lite" green sentiment, convincing people to alter their engrained, even unconscious neoliberal proclivities, in favour of an ethic of stewardship and trusteeship is a very trying task. Yet, without a mobilized civil society – a demos – that is willing to hold governments to account, and demand that they represent their ecological interests internationally, states will continue to behave in the way they always have – reacting to the global environmental crisis according to the conflict model of international law that they are so used to.

1.5 Trusteeship Functions of the State

The only way to turn things around and move international law from the Westphalian conflict model to a 21st-century cooperation model is to redefine states as trusteeship organizations. Sovereignty and trusteeship must be seen as complementary and not mutually exclusive. The argument in favour of states as trustees goes as follows.

The state gains its legitimacy exclusively from the people who created it. While the legality of a state depends on recognition by other states, once in existence a state can only ever legitimize its continued existence through ongoing trust by its people. The core idea of the modern democratic state is that it acts through its people, by its people, and for its people. This implies a fiduciary relationship between people and state and was arguably the only legitimate basis for political authority in the English civil war, American Revolution, and then again confirmed in the French Revolution. It is echoed in constitutional documents such as in Article IV of the Pennsylvanian Constitution of 1776: "[A]ll power being . . . derived from the people; therefore all officers of government, whether legislative or executive, are their trustees and servants, and at all times accountable to them." Likewise, John Locke had asserted that legislative power is "only a fiduciary power to act for certain ends" and that "there remains still in the people a supreme power to remove or alter the legislative, when they find the legislative act contrary to the trust reposed in them" (Locke, 1690, Sect. 149).

In the same vein, Immanuel Kant drew the moral basis of fiduciary obligations from the duty-bound relationship between parents and children. Kant claimed that children have an innate and legal right to their parents' care. In a similar sense, he believed that state legitimacy was the result of a contract that is necessarily created between people to form a Rousseauian "general will." Through this process, Kant claimed, we jointly authorize the state to announce and enforce law.

The idea that state sovereignty entails trusteeship obligations can be traced back not only to Middle Eastern origins, but also to Roman and Germanic law and equally to religious teachings. The idea is perhaps even more prevalent in non-Western societies than present-day Western societies because the former emphasize collective identities (e.g., family, clan, nation, religion) over individual freedom and dignity, imbuing implied fiduciary obligations into the structure of public and private legal institutions.

Article 21(3) of the Universal Declaration of Human Rights states that "the will of the people shall be the basis of the authority of government." But as Ron Engel has pointed out, "democracy" can have differing interpretations. There is the "thin" interpretation that includes procedural democracy, liberal democracy, representative democracy, or simply put "the democratic process." And there is "thick" or material democracy where the government acts in trusteeship of the common good.[50] As the COVID-19 pandemic and many other unexpected or unforeseeable events have shown, existing means of democratic decision-making have failed.

So although we may have democracy (and many places do not) in a technical sense, we have lost sight of what duty the state owes to those it governs. At its most simplistic, the state's legitimacy to govern is based on its ability to serve the common interest. Benvenisti conceives of two normative bases according to which we should ascribe a trusteeship function to states' mandate to govern.[51] Firstly, sovereignty should be viewed as a vehicle for the exercise of personal and collective self-determination. Collective self-determination embodies the freedom of a group to pursue its interests, further its political status, and "freely dispose of [its] natural wealth and resources." An outdated conception of sovereignty, which equates the voting constituency with the only affected stakeholders, can undermine communities' ability to exercise their right to self-determination. Secondly, Benvenisti refers to a conception of sovereign states as agents of humanity as a whole. He bases this conception largely on the equal moral worth of all human beings and the corresponding foundation of international law in human rights. He argues that it is humanity at large that assigns to certain groups of citizens the power to form national governments. Accordingly, states can and should be viewed as agents of a global system that allocates competences and responsibilities for the promotion of the rights of all human beings and their interest in the sustainable utilization of global resources. As such, the corollary of states' authority to manage public affairs within their domestic jurisdictions is an obligation to take account of external interests and balance internal against external interests.

Similarly we could say that the privilege of territorial sovereignty can be legitimized only in so far as universal interests of humanity as a whole are not severely affected. This argument is based on ecological realities defying national state boundaries, but also on the observation that boundaries of states not necessarily coincide with boundaries of nationalities, or more generally, with the boundaries of the groups whose members commonly share a distinct interest or conception of the good.

Benvenisti also refers to a conception of sovereignty as the power to exclude portions of global resources. He notes that both ownership and sovereignty are claims for the intervention in the state of nature by carving out valuable space for exclusive use. Such a perception of states as power-wielding property owners provides a solid normative foundation for the imposition of a positive obligation on states to take other-regarding considerations into account when managing the resources assigned to them. Property law theory can thus provide us a framework within which we can translate these moral grounds into legal obligations. Thus, we can and should conceptualize ownership of global resources as originating from a collective regulatory decision at the global level, rather than as an entitlement of sovereign states.

Although we might all agree that a government has fiduciary duties, there is seemingly little to establish a precedent of state environmental trusteeship.[52] However, this may be due to a lack of appreciation of just how closely connected rights and responsibilities have been in the past. The English "commons" or German "Allmende" were governance systems based on common property. States or local principalities exercised their powers in a way that access of the commoners to land and natural resources was assured. It is often forgotten, for example, that the Magna Charta of 1215 was accompanied by the Carta de Foresta (Charter of the Forest) of 1217 in an attempt to restrict exploitation of the forests by the King at the time. The Charter provided a right of common access to (royal) private lands. Some of its clauses remained in force until the 1970s.[53]

For centuries, environmental duties and environmental trusteeship were part of ways of life around the world. Modern capitalism has dramatically turned things around. It replaced nature – as the basis of everything – with money. In the course of unfolding – and now unfettered – capitalism, governments, and business had little regard to safeguarding natural cycles and ecological integrity. More than ever before we need to revive and develop environmental trusteeship. In the context of international environmental law, this has been advanced to the concept of Earth trusteeship.[54]

1.6 Conclusion

The Earth system approach reverses the traditional rule that international law and governance end where national borders begin. The dichotomy between national law and international law defies ecological realities and is counterproductive to protecting the Earth system. States need to internalize transnational ecological aspects into the interpretation of territorial sovereignty and in this way incorporate trusteeship obligations into the concept and practice of state sovereignty. The requirement for states, therefore, is to accept Earth trusteeship by restricting territorial sovereignty to environmental uses and impacts that are limited to their own territory.[55] This legal correction would remove the present paradox: instead of state sovereignty setting limits to environmental protection, environmental protection would set the limits to state sovereignty. This insight would help to overcome the design flaw of international law: "Limiting the self-interest of states by taking into account global concerns of humanity has become a fundamental aspect of international law."[56]

The 2024 UN Summit of the Future has addressed these issues in its 2024 Pact for the Future.[57] The Pact calls for "bold, ambitious, accelerated, just and transformative actions to implement the 2030 Agenda" (Action 1) by, *inter alia*, giving future generations a voice (Actions 34–37) and transforming global governance (Actions 38–56). As the preparatory UN Report "A Breakthrough for the Planet and the People" pointed out: "Our responsibilities to current and future generations can only be met if we act in trusteeship for the planet."[58] The Earth system approach to international environmental law shows how this can be achieved: Along with us as individual citizens, states need to incorporate and practice Earth trusteeship.

Notes

* This chapter is based on K. Bosselmann, "Global Governance in the Anthropocene", in: D. DellaSala and M.Goldstein (eds.), *Encyclopedia of the Anthropocene*, vol. 4, Elsevier Publishing, 2018, 265–269, and has been substantially changed and updated following subsequent developments and research.

1 W. Steffen, P. Tyson, et al., "Earth System Science: An Integrated Approach", *Environment: Science and Policy for Sustainable Development* 43:8 (2001) at 22.

2 J. Rockström, W. Steffen, et al., "Planetary Boundaries: Exploring the Safe Operating Space for Humanity", *Ecology and Society* 14:1 (2009) at 472–475.

3 https://www.scientificamerican.com/article/humans-have-crossed-6-of-9-planetary-boundaries/?v3=true (2023).

4 J. Rockström, L. Kotzé, et al., "Planetary Commons for the Anthropocene", *Proceedings of the National Academy of Sciences* (2024), https://doi.org/10.1073/pnas.2301531121.

5 K. Bosselmann, *Earth Governance: Trusteeship for the Global Commons* (Edward Elgar, Cheltenham, 2015).

6 R. Donner, S. Barbosa, J. Kurths and N. Marwan, "Understanding the Earth as a Complex System – Recent Advances in Data Analysis and Modelling in Earth Sciences", *European Physical Journal Special Topics* (2009) at 1–9.

7 W. Steffen, A. Sanderson and P. Tyson, *Global Change and the Earth System – A Planet under Pressure* (Springer, Berlin, 2004) at 4.

8 W. Steffen, P. Crutzen and J. McNeil, "The Anthropocene: Are Humans Now Overwhelming the Great Forces of Nature?", *Ambio* 36:8 (2007), 614–621; See further V. Galaz, *Global Environmental Governance, Technology and Politics – The Anthropocene Gap* (Edward Elgar, 2014), 1–15.

9 See S. Diaz, J. Settele, et al., "Summary for Policy Makers of the Global Assessment Report on Biodiversity and Ecosystem Services of the Intergovernmental Science-Policy Platform on Biodiversity and Ecosystem Services" (2019) at 11, https://www.ipbes.net/global-assessment.

10 S. Salomon, D. Qin, et al (eds.), *Climate Change 2007: The Physical Science Basis*, Contribution of Working Group I to the International Panel on Climate Change (Cambridge University Press, Cambridge, 2007).
11 T. Lenton, H. Held, et al., "Tipping Elements in the Earth's Climate System", *Proceedings of the National Academy of Sciences* 105:6 (2008), 1786–1793.
12 J. Rockström, W. Steffen, et al., above n 1.
13 A. Barnosky, E. Hadley, et al., "Approaching a State Shift in Earth's Biosphere", *Nature* 486 (2012), 52–58.
14 J. Hutton and A. Geikie, *Theory of the Earth, with Proofs and Illustrations. In Four Parts* (Edinburgh, 1795), ch. 1.
15 A. von Humboldt, *Views of Nature* (Henry G. Bohn, London, 1850).
16 V. Vernadskiiĭ and M. McMenamin, *The Biosphere* (Copernicus, New York, 1998) at 12.
17 E. Edmonds, "General System Theory: Foundations, Development, Applications by Ludwig von Bertalanffy", *Leonardo* 10:3 (1977) 248 at 248.
18 J. Lovelock, *Gaia: A New Look at Life on Earth* (Oxford University Press, Oxford, 2000) at 10.
19 Global Footprint Network, "Ecological Footprint", www.footprintnetwork.org.
20 T. F. Stocker, D. Qin, et al. (eds.), *Climate Change 2013: The Physical Science Basis*. Contribution of Working Group I to the Fifth Assessment Report of the Intergovernmental Panel on Climate Change (Cambridge University Press, Cambridge, 2013).
21 The Working Group on the Anthropocene voted in favour of designating the Anthropocene as a new geological epoch and plans to submit a formal proposal to the International Commission on Stratigraphy in 2021. Working Group on the Anthropocene "Results of Binding Vote on AWG" (2019) Subcommission on Quaternary Stratigraphy http://quaternary.stratigraphy.org/working-groups/anthropocene/.
22 S. Westerlund, "Theory of Sustainable Development, Towards or Against?", in H. C. Bugge and C. Voigt (eds.), *Sustainable Development in International and National Law* (Europa Law, Groningen, 2008), 49 at 55–58.
23 R. Kim and K. Bosselmann, "Operationalizing Sustainable Development: Ecological Integrity as a *Grundnorm* in International Law", *Review of European, Comparative and International Environmental Law* 24:2 (2015), 194–208.
24 K. Bosselmann, "Mind the Gap: State Governance and Ecological Integrity", in L. Westra, J. Gray and V. Karageorgou (eds.), *Ecological Systems Integrity: Governance, Law and Human Rights* (Routledge, London, 2015) at 272–273.
25 Bosselmann, above n 24, at 276.
26 Bosselmann, above n 24, at 273.
27 Bosselmann, above n 24, at 274.
28 Westerlund, above n 22, at 65.
29 See K. Bosselmann, *The Principle of Sustainability: Transforming Law and Governance* (2nd ed., Routledge, London, 2017).
30 D. Fisher, *Legal Reasoning in Environmental Law: A Study of Structure, Form and Language* (Edward Elgar, Cheltenham, 2013) at 433. See also Bosselmann, above n 29, 93–101.
31 V. Galaz, *Global Environmental Governance, Technology and Politics – The Anthropocene Gap* (Edward Elgar Publ., 2014) at viii.
32 R. Costanza, "Claim the Sky!", *Solutions* 6:1 (2015), 18–21.
33 M. Wood, *Nature's Trust: Environmental Law for a New Ecological Age* (Carolina University Press, New York, 2013).
34 P. Sand, "The Rise of Public Trusteeship in International Law", *Global Trust Working Paper Series* 4 (2013), 21.
35 P. Barnes, *Capitalism 3.0: A Guide to Reclaiming the Commons* (Berret-Koehler Publ., San Francisco, 2006).

36 For example D. Bollier, *Think Like a Commoner: A Short Introduction to the Life of the Commons* (New Society Publ., Gabriola Island, 2014); S. Helfrich and J. Haas (eds.), *The Commons: A New Narrative of Our Time* (Heinrich-Böll-Stiftung, Berlin, 2009).
37 https://iucn.org/sites/default/files/2022-10/world_declaration_on_the_environmental_rule_of_law_final_2017-3-17.pdf.
38 https://globalpactenvironment.org/uploads/EN.pdf.
39 https://www.earthtrusteeship.world/the-hague-principles-for-a-universal-declaration-on-human-responsibilities-and-earth-trusteeship/.
40 Listed as Annex to the Hague Principles. See also K. Bosselmann, "The Hague Principles: Responsibilities and Rights concerning Humans and the Earth", in J. Sobion and H. V. Willenswaard (eds.), *Reflections on Earth Trusteeship: Mother Earth and a New 21st-Century Governance Paradigm* (INI Books, Nonthaburi/Thailand, 2023), 115–127.
41 K. Bosselmann, above n 5, 198–232.
42 Barnes, above n 35, 38.
43 Wood, above n 33, 592.
44 Barnes, above n 35, 45.
45 R. Ashcraft (ed.), *John Locke: Critical Assessments* (Routledge, London, 1991), 524.
46 E. Fox-Decent, *Sovereignty's Promise: The State as a Fiduciary* (Oxford University Press, Oxford, 2012).
47 P. Finn, "The Forgotten 'Trust': The People and the State", in M. Cope (ed.), *Equity: Issues and Trends* (The Federation Press, Sydney, 1995), 131–151.
48 E. Benvenisti, "Sovereigns as Trustees of Humanity: On the Accountability of States to Foreign Stakeholders", *American Journal of International Law (AJIL)* 107:2 (2013), 295–333.
49 C. Redgwell, "Reforming the UN Trusteeship Council", in W. Chambers and J. Green (eds.), *Reforming International Environmental Governance: From Institutional Limits to Innovative* (United Nations University Press, New York, 2005), 179.
50 R. Engel, "Contesting Democracy", in R. Engel, L. Westra and K. Bosselmann (eds.), *Democracy, Ecological Integrity and International Law* (Cambridge Scholars Publications, London, 2010), 28.
51 Benvenisti, above n 48, 301–309.
52 Bosselmann, above n 5, 155–197; Bosselmann, above n 29, 145–174.
53 N. Robinson, "The Charter of the Forest: Evolving Human Rights in Nature", in D. Magraw et al. (eds.), *Magna Carta and the Rule of Law* (Pace Law Faculty Publications, 2014).
54 Above n 39 and 40. Further, K. Bosselmann, *Earth Trusteeship and the Sovereign State: Transforming International Environmental Law* (Routledge, London, 2025).
55 K. Bosselmann, "Environmental Governance: A New Approach to Territorial Sovereignty", in R. Goldstein (ed.), *Environmental Ethics and Law* (Ashgate Publishing, Burlington, 2004), 293–313.
56 S. Stec, "Humanitarian Limits to Sovereignty: Common Concern and Common Heritage Approaches to Natural Resources and Environment", *International and Comparative Law Review* 12:361 (2010) at 364.
57 https://www.un.org/sites/un2.un.org/files/sotf-pact_for_the_future_adopted.pdf.
58 High-Level Advisory Board on Effective Multilateralism (HLAB), *A Breakthrough for People and the Planet: Effective and Inclusive Global Governance for the Today and the Future* (United Nations University Press, New York, 2023) at 23.

Part II

Energy Efficiency

2 Optimizing Technology to Enhance RE-Development in India

A Reconnaissance Legal Study*

Usha Tandon

2.1 Introduction

With increasing awareness of environmental challenges and the crying need to combat "climate change," Sustainable Development (SD) has become a guiding principle for global progress.[1] In this regard, the advancement of renewable energy technologies is most promising strategy to achieve SD.[2] Renewable Energy (RE) and SD are inextricably linked, forming a symbiotic relationship that drives global efforts to achieve environmental, economic, and social goals.[3] As the world, including India, grapples with the challenges of "climate change," resource depletion, and energy insecurity, the role of RE becomes increasingly critical in creating a sustainable future.[4]

At COP 21, in the Paris Agreement, 2015, India had intended to achieve the target of 40% of its "installed electricity capacity from non-fossil energy sources" by 2030. This was mentioned in its nationally determined contributions (NDCs). It is gratifying to mention that India has already achieved this target nine years in advance, in November 2021 itself.[5] India now has set a target of 50% "cumulative electric power installed capacity from non-fossil fuel-based energy resources" by 2030. This goal is part of India's larger plan to reduce its carbon footprint towards cleaner energy sources. India has also announced plans to reach net zero emissions by 2070.[6] In this context, adopting innovative technologies to enhance RE development is crucial for achieving India's ambitious RE targets and transitioning to a sustainable energy future. The Indian Government, as a matter of fact, has been promoting the use of technology to enhance RE development in sectors like solar energy, wind energy, hydro energy, and so forth.

Indian laws and policies can play a crucial role in promoting the use of technology to advance RE development. For instance, India's policy framework, including the "National Solar Mission" (NSM), 2010,[7] and the "National Wind-Solar Hybrid Policy" (NWSHP), 2018,[8] encourages the adoption of technology in RE development, "The Electricity Act" (EA), 2003,[9] and the "National Tariff Policy" (NTP), 2016,[10] mandate Renewable Purchase Obligations (RPOs), all promoting RE adoption and technology integration. The Indian judiciary has recognized the integration of technology for renewable

DOI: 10.4324/9781003733751-4

energy to address environmental concerns. Cases like *Hindustan Zinc Ltd* v. *Rajasthan Electricity Regulatory Commission*[11] and *Juniper Hotels Private Limited* v. *Delhi Electricity Regulatory Commission*[12] demonstrate the Indian judiciary's concerns for promoting renewable energy policies and using technology to achieve SD goals.

In the light of the aforementioned, this chapter undertakes a broad assessment of how effectively Indian legal framework facilitates the integration of technology in advancing RE. It elicits through this preliminary study that the use of technology in promoting RE is accompanied by a range of legal challenges, such as intellectual property rights, data privacy, cybersecurity threats, and environmental impacts. The chapter concludes by emphasizing the necessity of addressing these challenges through enacting specific comprehensive law on RE and amending existing laws to establish a sustainable, technology-driven RE ecosystem in India.

2.2 How Renewable Energy Is Crucial for Sustainable Development

SD is about meeting the "needs of the present without compromising the ability of future generations to meet their own needs."[13] It's a holistic approach that recognizes the interconnectedness of "economic growth," "social progress," and "environmental protection." The goal is to create a world where everyone has access to the resources they need for a good quality of life, while ensuring the planet's resources are not depleted and the environment is not damaged beyond repair. This requires a shift away from the traditional model of development that often prioritizes economic growth at the expense of the environment and social equity.

RE sources, such as solar, wind, hydro, geothermal, and biomass, are crucial for achieving "sustainable development." Fossil fuels (coal, oil, and natural gas) are finite and lead to pollution, whereas RE sources are infinite and have a lesser environmental impact. Greenhouse gases are released in the atmosphere when fossil fuels are burnt causing global warming.[14] RE sources are essential to combat climate change as they release little greenhouse gas emissions.[15]

Many countries rely on imports of fossil fuels, making them vulnerable to price fluctuations and geopolitical instability. RE sources, on the other hand, are often locally available, reducing dependence on foreign energy sources and increasing energy security.[16] This is particularly important for developing countries that may lack access to fossil fuel resources. The RE sector is a growing industry that creates jobs in manufacturing, installation, maintenance, and research. Investing in RE can stimulate economic growth and create new opportunities for communities.

In many parts of the world, particularly in rural areas, access to electricity is limited or non-existent. RE technologies, such as solar panels and small-scale wind turbines, can provide decentralized power solutions, bringing electricity to remote communities and improving quality of life. This is highlighted in the example of the Indian government's efforts to bring electricity to all areas through RE projects.[17] The extraction, processing, and burning of fossil fuels can have significant environmental impacts, including air and water pollution, habitat destruction, and oil spills. RE sources have a much lower environmental footprint, reducing pollution and protecting ecosystems.

2.3 How Emerging Technologies Enhance RE Growth

The field of RE is changing quickly because of new technologies and the pressing need for cleaner energy. These cutting-edge technologies are key to making future RE projects better – more efficient, affordable, and able to be scaled up. While RE technologies have come a long way, research and development continue to boost their innovation and effectiveness. These new technologies can solve some of the problems with clean energy, like its inconsistent supply, storage difficulties, and how to connect it to the power grid. They provide ways to make RE projects more dependable, easier to expand, and more cost effective, which helps them to be adopted more widely and have a bigger impact. Some of the key emerging technologies in the field of RE include photovoltaic solar cells (PV),[18] concentrated solar power (CSP),[19] perovskite solar cells,[20] offshore wind turbines,[21] floating wind farms,[22] vertical axis wind turbines,[23] flow batteries,[24] and hydrogen storage.[25]

India has made remarkable progress in advancing its solar energy technologies, with a focus on enhancing the efficiency of photovoltaic cells, bifacial solar cells,[26] solar-integrated building materials such as solar roof tiles, and solar facades. It has seen the introduction of higher-capacity wind turbines with larger rotor diameters, enabling them to capture more wind energy.[27] India has witnessed the establishment of biomass power plants and[28] waste-to-energy plants.[29]

India's vast coastline offers significant potential for harnessing ocean energy, including tidal and wave energy. The potential for tidal energy in India is estimated to be around 8000 MW.

India's long coastline also provides ample opportunities for wave energy.[30] Emerging technologies, thus, have huge role to play for harnessing wave energy in India.

2.4 Do Indian Laws and Policies Facilitate Integration of Technology for RE Promotion?

As stated, India has set ambitious RE targets, aiming to achieve 50% of its installed power capacity from non-fossil fuels by 2030. Various efforts have been made by the Government of India to achieve this goal. The role of Indian laws and policies in promoting technology for RE upgradation can be examined by referring to key legislations, regulations, and schemes.

2.4.1 *Key Policies Relating to RE*

India's policy framework on RE, *inter alia*, includes the National Solar Mission (NSM), 2010, the National Tariff Policy (NTF), 2016, and the National Wind-Solar Hybrid Policy (NWSHP), 2018. A brief account of these policies, examining the promotion of technology therein, is presented herein.

2.4.1.1 *The National Solar Mission, 2010*

The National Solar Mission, launched in 2010[31] under the National Action Plan on Climate Change (NAPCC), 2008,[32] is a flagship initiative by the Government of India to promote SD through the widespread adoption of RE. This ambitious mission, revised twice, aims to make India the leader in solar energy[33] by fostering its deployment across residential, commercial, and industrial sectors, thereby reducing the dependency on fossil fuels and mitigating climate change. The primary mechanism for advancing solar power is the implementation of an RPO, which mandates power utilities to include a designated solar component.[34] One of the primary objectives of the mission is to achieve significant capacity additions in grid-connected and off-grid solar power.[35] To meet these objectives, the mission has implemented various policies, including financial incentives[36] like capital subsidies, viability gap funding, and feed-in tariffs, making solar energy cost-competitive and accessible to a larger population.[37]

The mission actively promotes the use of technology to advance RE, particularly solar power. By fostering innovation in photovoltaic (PV) and concentrated solar power (CSP) technologies, the mission aims to enhance the efficiency and affordability of solar energy systems. It provides funding for research and development (R&D)[38] in solar energy technologies, encouraging collaboration between public and private sectors to develop cutting-edge solar technologies suited to India's unique climatic conditions.[39] The mission also supports the deployment of advanced manufacturing technologies for solar panels, energy storage, and grid integration to address intermittency issues. By incentivizing large-scale solar installations and rooftop systems, it leverages smart grid technology and modern monitoring tools to optimize energy distribution.[40] NSM supports training and capacity-building programs for stakeholders, including manufacturers, developers, and policymakers to enhance their skills and knowledge in solar energy technologies. While it promotes the development of standards and certification procedures for solar energy equipment to secure quality and reliability, it also encourages the development and deployment of solar energy storage such as batteries, to enhance the efficiency and reliability of solar power systems. By promoting technology development, deployment, and adoption, the NSM has played an important role in the expansion of the solar industry in India and reducing the country's dependence on fossil fuels. In this way, NSM represents a commendable step towards promoting RE and addressing climate change in India.

However, the implementation of NSM faces several challenges that warrant critical evaluation. One key issue is the financial sustainability of the mission. While incentives like subsidies and viability gap funding have driven initial adoption, the long-term economic viability of solar projects depends on ensuring affordability without continued government support. The high cost of solar infrastructure, particularly for rooftop systems, remains a barrier for many households and small businesses, limiting widespread adoption. Another concern is the reliance on imported solar modules, despite efforts to promote domestic manufacturing. Grid integration of solar power is another significant challenge. The intermittent nature of solar energy requires advanced grid infrastructure and energy storage solutions, which remain underdeveloped in India. This leads to inefficiencies in managing power supply and limits the full utilization of solar energy. The land acquisition for large-scale solar parks has raised concerns over displacement, environmental degradation, and conflicts with agricultural priorities. Lastly, while the mission has set ambitious targets, execution often falls short due to bureaucratic delays, inconsistent policies across states, and limited awareness and participation from local stakeholders.[41] To address these challenges, the Government must focus on strengthening domestic manufacturing, investing in energy storage and grid infrastructure, ensuring equitable land-use policies, and fostering better collaboration between stakeholders. This will enhance the mission's effectiveness and ensure a sustainable transition to RE.

2.4.1.2 The National Tariff Policy, 2016

The National Tariff Policy, 2016,[42] establishes a framework for electricity tariffs that incentivizes clean energy production; the policy aims to balance economic growth with environmental sustainability. One of the key features of the NTP is its emphasis on RPOs.[43] RPOs mandate distribution companies (DISCOMs) to procure a specific percentage of their power from RE sources. This provision encourages the development and deployment of RE technologies and has been instrumental in creating demand for renewables like solar, wind, and biomass energy. Moreover, the policy promotes differential tariffs for RE projects, allowing developers to recover their higher capital costs while encouraging investment in the sector. The NTP also recognizes the importance of long-term Power Purchase Agreements (PPAs),[44] providing financial stability to developers and ensuring bankability for RE projects. The policy emphasizes the adoption of competitive bidding mechanisms to reduce the cost of RE.[45] This approach enables developers to adopt the technologies to reduce costs and increase efficiency. The NTP promotes Renewable Energy Certificates (RECs), which can be traded and represent the "environmental attributes of one megawatt-hour of renewable electricity."[46] RECs provide an additional revenue stream for RE generators, encouraging the adoption of RE technologies. The Policy provides for tariff incentives for RE projects[47] and accelerated depreciation benefits.[48] These incentives encourage the adoption of RE technologies and promote the growth of RE sector. The Policy allows for technology-agnostic bidding,[49] which enables developers to choose the most suitable technology for their RE projects. To motivate DISCOMs to source power from RE, the Policy promotes a competitive bid-based tariff framework, enabling a gradual increase in tariffs over time.[50] This approach promotes innovation and adoption of new RE technologies.

The NTP emphasizes the importance of grid integration for RE sources. There are certain areas where the grid is not accessible so far or where the supply to the grid is not adequate. For such places, the Policy encourages the development of micro-grids.[51] To further promote RE, the policy stipulates that electricity generated from "solar and wind sources transmitted through the inter-state transmission system" for sale will not be subject to inter-state transmission charges or losses.[52] It encourages State Electricity Regulatory Commissions (SERCs) to stipulate Regulations to facilitate the generation as well as sale of electricity from roof-top solar system.[53] By incorporating these provisions, the NTP promotes the use of technology for RE promotion, encouraging the growth of RE sector and the transition to a cleaner and more sustainable energy mix.

Despite these strengths, NTP faces several challenges in its implementation. One major concern is the inconsistent enforcement of RPOs across states. While the Policy provides for mandatory RE procurement, weak regulatory mechanisms and non-compliance by DISCOMs undermine its effectiveness. Financially distressed DISCOMs often prioritize cheaper fossil fuel-based power over costlier renewable options, despite the declining costs of renewables in recent years. Another issue is the Policy's limited focus on addressing technical and infrastructural challenges, such as "grid integration" and "energy storage." However, the Policy does not adequately address these requirements, leading to inefficiencies in RE utilization. Furthermore, while the NTP encourages competitive bidding to reduce costs, it has sometimes resulted in aggressive tariff reductions that threaten the financial sustainability of RE developers. Unrealistically low tariffs can lead to project delays, cancellations, or compromises in quality, ultimately hindering the growth of the sector.

The Policy also lacks sufficient emphasis on consumer-level incentives for adopting RE. While it promotes RE generation at the utility level, there is limited support for residential and commercial adoption, such as rooftop solar systems or energy-efficient appliances. Thus, it can be said that NTP has significantly contributed to creating a favourable environment for RE development in India by introducing mechanisms like RPOs, differential tariffs, and competitive bidding. However, its success is hindered by challenges such as weak enforcement, insufficient focus on grid infrastructure, and financial pressures on developers and DISCOMs. Addressing these issues through stronger regulatory frameworks, increased investment in infrastructure, and enhanced support for decentralized renewable adoption will be critical to achieving India's RE goals and transitioning to a sustainable energy future.

2.4.1.3 *The National Wind-Solar Hybrid Policy, 2018*

The NWSHP, introduced by the Government of India in 2018, aims to optimize the utilization of RE resources by combining wind and solar power generation systems. This Policy addresses key challenges in RE development, including land constraints, grid integration, and variability in power supply, providing for a more sustainable and efficient energy mix.[54] The Policy promotes the establishment of hybrid systems, where both wind and solar energy are harnessed at the same location.[55] Solar energy is in abundance during the day, while wind energy is often stronger during evenings and nights. This synergy minimizes fluctuations in power generation, leading to a more steady and secure RE supply for the grid. The policy seeks to promote innovative technologies and approaches that facilitate the joint operation of wind and solar PV plants.[56] Additionally, the Policy emphasizes the efficient utilization of existing infrastructure, such as land and transmission networks, by retrofitting wind or solar energy systems onto already operational RE projects.[57] This reduces the costs associated with new land acquisition and transmission development, making hybrid projects more economically viable. To incentivize adoption, the Policy provides a framework for tariff-based competitive bidding and financial support mechanisms.[58] NWSHP encourages the deployment of advanced technologies, such as energy storage systems, to further enhance the efficiency and reliability of hybrid projects. Moreover, clear regulatory guidelines under the Policy aim to facilitate smoother implementation by addressing grid compatibility and technical challenges. It also contributes to the reduction of greenhouse gas emissions and strengthens India's commitment to combat climate change to SD.

NWSHP is a progressive step towards optimizing RE generation in India, but it faces few challenges as well, that merit critical evaluation. A major concern is the technical complexity of integrating hybrid systems into the existing energy infrastructure. The Policy's reliance on existing land and transmission infrastructure can lead to operational challenges, such as increased wear and tear on older systems or limited capacity to accommodate new hybrid installations. This approach may also result in uneven development, as hybrid projects could be concentrated in areas with existing renewable infrastructure, bypassing regions with untapped potential. The Policy's implementation also lacks comprehensive mechanisms for addressing land-use conflicts and environmental impacts, particularly in ecologically sensitive areas. Without robust guidelines, hybrid projects risk exacerbating land-related disputes or causing harm to biodiversity. Lastly, the Policy's emphasis on tariff-based competitive bidding may marginalize smaller developers, favouring larger companies, which can afford the high costs of hybrid systems.[59] To strengthen NWSHP, India must develop clear guidelines for land use and environmental conservation.

2.4.2 *Key Legislations on RE*

Currently, India does not have a dedicated law specifically regulating RE. The provisions of the Electricity Act (EA), 2003, undergoing amendments after amendments, apply to RE. The EA mainly deals with conventional sources of energy and it includes provisions on RE as well.[60] This is the cornerstone legislation, outlining the regulatory framework for electricity generation, distribution, and transmission having provisions for promoting RE sources.

2.4.2.1 *The Electricity Act, 2003*

The Electricity Act, 2003, is a landmark legislation in India that aims to promote the development of the power sector,[61] including RE. An examination of EA demonstrates that there are some provisions in the Act that encourage the adoption of technology for RE development, which are discussed in the following pages:

2.4.2.1.1 MANDATORY RENEWABLE PURCHASE OBLIGATIONS

RPOs serve as a crucial mechanism under the Act,[62] mandating power purchasers across all states – including DISCOMs, captive power producers, and open access consumers – to procure a specified minimum percentage of their power from RE sources each year. The primary objective of setting RPO targets is to drive the adoption of RE, foster a competitive market for clean energy technologies, and support the transition to a sustainable energy ecosystem, even when such technologies are not yet cost-competitive. To ensure compliance, states that fail to meet their RPO targets have the option to use the REC mechanism, which provides an alternative means of fulfilling RE obligations with equivalent value to direct RE procurement.

SERCs are required to set yearly RPO targets, taking into account factors like RE availability, current and future electricity demand, and the effect on consumer electricity prices. The RPO target has steadily increased from 2.75% in 2016 to almost 23% in 2023. In 2023, the Government increased the RPO targets, aiming for 39% RE by 2028.[63] to reflect the growing contribution of renewables. However, an analysis of public data shows that nearly 25 out of 30 states[64] are failing to meet their yearly solar RPO targets.

2.4.2.1.2 RENEWABLE ENERGY CERTIFICATES

As per the Regulations of Central Electricity Regulatory Commission (CERC), RECs[65] are tradeable certificates issued to RE generators in India. Each REC counts one megawatt-hour (MWh) of RE generated and fed into the grid. RECs enable obligated entities, such as DISCOMs and open access consumers, to meet their RPOs by purchasing RECs from RE generators. The RECs are traded on power exchanges, such as the Indian Energy Exchange

(IEX) and the Power Exchange India Ltd (PXIL), and are settled through the clearing and settlement mechanism of the exchanges. RECs provide an additional revenue stream for RE generators, encouraging the adoption of RE technologies.[66]

2.4.2.1.3 GRID CONNECTIVITY

The EA provides for the grid connectivity of RE sources, enabling them to feed electricity into the grid. This provision promotes the integration of RE sources into the grid and encourages the development of RE technologies. Provisions regarding grid connectivity provide that any generating company can connect to the grid by complying with the technical standards specified in it,[67] which includes requirements for grid connectivity and operation. The Grid is defined as "the high voltage backbone system of interconnected transmission lines, substations, and generating plants."[68] CERC is responsible for setting the Grid Code,[69] which outlines the standards for grid operation and maintenance based on the Grid Standards[70] set by the Central Electricity Authority under EA.[71]

2.4.2.1.4 OPEN ACCESS

EA allows for open access to the grid, enabling RE generators to sell power to any consumer. This provision promotes competition and encourages the adoption of RE technologies. The Act grants eligible consumers to have non-discriminatory open access to a licensee's network[72] on payment of applicable charges.[73] Open access[74] under EA refers to the right of any eligible consumer to buy electricity from a power generator of their choice, using the existing transmission and distribution network, by paying appropriate charges determined by SERCs, essentially allowing them to access the grid without being limited to their designated electricity supplier. This is facilitated by Section 42 of EA, which mandates non-discriminatory access to the grid for eligible consumers.

This provision is central to open access, requiring distribution licensees to give access to consumers upon payment of applicable charges. Each SERC is responsible for determining the terms and conditions for open access within their jurisdiction, including charges for using the transmission network. Open access promotes competition in the power market, enabling consumers to choose cheaper electricity sources and potentially access RE options.

2.4.2.1.5 TARIFF DETERMINATION

EA provides for the determination of tariffs for RE projects based on competitive bidding, which encourages efficiency and innovation in this sector. According to EA, tariff determination is primarily handled by the Appropriate Commission (either SERC or CERC), which lays down tariff for generation, transmission, and distribution of electricity. The tariff is generally determined based on the cost of "generation, transmission, and distribution," aiming to recover the costs incurred by the licensee, considering factors like cost recovery, consumer load factor, and the nature of supply, ensuring no undue preference is given to any consumer.[75] The Commission can differentiate tariffs based on factors like consumer load factor, power factor, time of usage, and geographical location. Before finalizing a tariff, the Commission may hold public hearings and consultations with stakeholders. In certain cases, tariffs may be determined through a competitive bidding process, where the lowest bidder is awarded the contract.[76]

2.4.2.1.6 NATIONAL ELECTRICITY POLICY, 2005

EC mandated the Central Government to formulate a National Electricity Policy (NEP) and in 2005, the Central Government notified it. NEP promotes the integration of RE technology by establishing a framework to encourage private sector participation in the power sector, setting targets for RE capacity addition, facilitating grid connectivity for RE sources, and promoting research and development for advanced RE technologies, ultimately aiming to enlarge the portion of RE in the overall electricity mix. The NEP explicitly includes goals for increasing RE capacity and promoting clean energy sources as part of its overall objective to provide reliable and quality power at reasonable rates. The Policy encourages the development of a robust transmission network that can efficiently integrate power from geographically dispersed RE sources such as solar and wind farms. It establishes a market-based approach to incentivize RE development by providing mechanisms like feed-in tariffs and RECs to encourage investment in RE projects. The NEP encourages research and development efforts to improve the efficiency and cost-effectiveness of existing RE technologies and to explore new advancements in the field. While the 2005 NEP laid the foundation for RE integration, subsequent updates and revisions have further strengthened the focus on RE development.[77]

2.4.2.1.7 CENTRAL AND STATE ELECTRICITY REGULATORY COMMISSIONS

EA provides for the establishment of CERC[78] and SERCs,[79] which are responsible for promoting RE in their respective states by setting RPO targets, grid connectivity standards, and tariff determination for RE projects within their respective states. Each state can formulate its own RE policies, aligning with the national framework, to further incentivize RE development.[80] These Commissions encourage the adoption of RE technologies.

The CERC currently sets general tariffs for RE plants based on rules established in the 2017 regulations.[81] SERCs can create their own regulations and resolve disputes between energy producers, distributors, and consumers. Decisions made by the SERCs can be appealed to the Appellate Tribunal for

Electricity (APTEL), a specialized body that handles electricity-related disputes. APTEL decisions can then be challenged in the Supreme Court of India. To further support the growth of the RE sector, the Central Government has proposed amendments[82] in the Electricity (Amendment) Bill, 2020. This Bill, along with the Draft National Electricity Policy 2021 and other policies, are expected to drive significant changes that will improve energy efficiency, keep pace with technological advancements, and help achieve climate change goals.

2.4.2.1.8 NET METERING

The Act provides for net metering,[83] which allows consumers to generate their own electricity from RE sources and feed any excess into the grid. This provision promotes the adoption of RE technologies by encouraging individuals and businesses to install rooftop solar panels by providing a financial incentive for generating RE. Net metering refers to a billing mechanism where a consumer with a rooftop solar system can essentially "credit" their electricity grid by exporting excess power generated from their solar panels, effectively deducting this exported energy from the electricity they draw from the grid, resulting in a bill based only on their "net" energy consumption. They only pay for the electricity they use after accounting for the power they feed back into the grid. A bi-directional meter is installed at the consumer's premises, recording both the electricity drawn from the grid and the electricity fed back into the grid. When a consumer generates more electricity than they use, the excess is credited to their account, which can be used to offset their electricity consumption during periods when their solar system is not producing enough power. The consumer's electricity bill is calculated based on the "net" energy consumed, which is the difference between the total electricity drawn from the grid and the total electricity fed back into the grid. SERC determines the specific Regulations for Net Metering within their jurisdiction, including the eligibility criteria, tariff structure for exported electricity, and technical requirements.[84] Consumers can potentially lower their electricity bills by utilizing their self-generated solar power.

2.4.2.2 The Energy Conservation Act (ECA), 2001

The Energy Conservation Act (ECA), 2001,[85] by promoting Energy Efficiency[86] and indirectly incentivizing RE initiatives, integrates technology for RE development. While not solely focused on RE, this Act promotes energy efficiency measures which can indirectly support the integration of renewables into the grid. The Bureau of Energy Efficiency, constituted under the Act, implements energy efficiency.[87] Efficient energy use by allowing the grid penetration of more renewables reduces the overall demand. As per the definition of "energy," electricity from RE sources is included; however, other applications of RE are not included within the definition of "energy."[88]

It institutionalizes energy conservation measures through regulatory frameworks, standards, and policies.[89] The ECA has not been able to keep pace with the rapidly rising energy consumption and energy losses. Thus, there is a need to strengthen ECA or enact new separate law for RE[90] emphasizing on the deployment and development of new technologies.

India's NAPCC, launched in 2008,[91] details the country's strategy for tackling climate change with a strong emphasis on promoting RE. The NAPCC contains missions such as the National Solar Mission and the National Mission for Enhanced Energy Efficiency. These missions are designed to improve energy security and support SD.[92]

2.4.2.3 *The National Renewable Energy Bill, 2015*

A dedicated law on RE, The National Renewable Energy Bill, 2015,[93] was introduced in the Indian Parliament by the Ministry of New and Renewable Energy in July 2015 to provide a comprehensive framework for promoting RE in India.[94] The Bill included provisions for promoting technology development and deployment in the field of RE, specifically by allowing the Central Government to identify key focus technologies, components, and materials to be prioritized within the National RE policy and plan, thus encouraging innovation and investment in advanced RE technologies. The Bill empowered the Government to identify crucial RE technologies that need focused development and support. Provisions were included to foster entrepreneurship[95] and incubation of start-ups focused on RE technologies. The bill aimed to provide support for research and development activities related to new and improved RE technologies.[96] The Government could designate specific areas as Renewable Energy Investment Zones[97] to facilitate concentrated development of RE projects, potentially including advanced technology deployment.

Although the bill could not be enacted into law,[98] it attempted to represent a significant step towards achieving India's RE goals. The National Renewable Energy Bill, 2015, if passed, would have covered almost all aspects of the RE supply chain[99] and would have significant implications for India's RE sector, including formulation of National Renewable Energy Policy,[100] establishment of National Renewable Energy Committee,[101] National Renewable Energy Fund,[102] and National Renewable Energy Advisory Group,[103] and so forth.

2.4.2.4 *The National Renewable Energy Credit Support Bill, 2021*

The National Renewable Energy Credit Support Bill, 2021,[104] was built upon the existing framework for RE development in India, particularly the Electricity Act, 2003. While EA provides the overall framework for the power sector, including RE, the Bill of 2021 sought to introduce specific provisions to support the development of RE.

This proposed legislation included tax incentives for those involved in RE, from production and distribution to consumption.[105] It also sought to boost RE infrastructure in underserved areas[106] by creating a regionally funded public–private partnership (Regional Renewable Energy Fund)[107] and a Rural Renewable Energy Security Programme.[108] The bill aimed to address the needs of vulnerable groups, such as scheduled castes and tribes, and those in geographically isolated communities.[109] It encouraged the use of RE for electricity generation and self-consumption in various sectors, including public buildings, homes, businesses, industries, and agriculture.[110] Finally, it mandated the government to develop incentives and support systems to encourage businesses and industries to invest in RE projects.[111]

Thus it can be said that the National Renewable Energy Credit Support Bill, 2021, incorporated measures to encourage the use of RE technologies.

2.4.2.4.1 WHETHER THE USE OF TECHNOLOGY IN RE PROMOTION RAISES LEGAL ISSUES: BROAD OBSERVATIONS

We have seen how Indian laws can play a crucial role in promoting the use of technology to advance RE development. However, the deployment of technology in RE promotion also raises various legal issues, including IP rights, data protection, cybersecurity risks, and environmental concerns.

The development and deployment of RE technologies, such as solar panels and wind turbines, may infringe on existing patents. RE companies may need to enter into technology licensing agreements with other companies, raising IPR issues related to royalty payments, territorial restrictions, and technology transfer. Many RE companies possess trade secrets related to their technologies, which can be misappropriated by competitors.[112] The use of software and digital content in RE technologies, such as energy management systems, can raise copyright infringement issues.[113] The application of open-source technology to RE involves IPR disputes related to ownership, licensing, and modifications.[114] Moreover, IPR disputes related to RE technologies are complex and challenging requiring specialized expertise and dispute resolution mechanisms.

Further, the use of technology in RE promotion also raises concerns regarding data protection and privacy. The sharing and analysis of energy usage data between stakeholders, including utilities, grid operators, and energy service providers, pose risks to data protection and privacy. RE companies sharing data with third-party service providers, such as data analytics firms, may encounter similar risks. The increasing use of Internet of Things devices in RE systems, such as smart thermostats and solar panels, generates vast amounts of data that must be protected. Advanced metering infrastructure and smart grids collect vast amounts of consumer data, including energy usage patterns and personal information. RE systems are vulnerable to cyberattacks, such as hacking and data breaches, which can compromise consumer data and disrupt energy supply.[115] Thus RE companies must obtain explicit consumer consent for data collection, storage, and usage, which can be challenging.[116]

The large scale of deployment of technology in RE promotion may also create environmental challenges. The production of RE technologies, such as solar panels and wind turbines, requires the extraction of raw materials, like rare earth metals, which can harm ecosystems. The production of RE technologies can generate emissions, including greenhouse gases and air pollutants. The installation of large-scale RE projects, such as solar parks and wind farms, can lead to land use changes, habitat disruption, and fragmentation. Some RE technologies, like concentrated solar power systems, require significant water resources for cooling, which can strain local water resources. Wind turbines can generate noise pollution, potentially harming local wildlife and human populations. RE technologies, such as wind turbines and solar panels, can generate electromagnetic interference, potentially affecting local wildlife and human health.[117] Further, the production of RE technologies requires energy and resources, which can lead to resource depletion and greenhouse gas emissions. Moreover, the disposal of RE technologies at the end of their life cycle can pose environmental risks if not managed properly.

2.4.2.4.2 HOW CAN INDIAN LAWS FACILITATE OPTIMUM AND FAIR USE OF TECHNOLOGY TO PROMOTE RE

The rapid pace of emerging technologies in RE requires the legal approach to continually evolve. To address the IPR issues related to RE efficiently, the patent application process can be simplified and streamlined to encourage innovation and reduce the time and cost associated with obtaining patents. Model agreements for technology transfer, licensing, and collaboration to facilitate the sharing of RE technologies can be developed. Guidelines need to be established for reasonable technology transfer fees to prevent exploitation and ensure fair compensation.[118] To promote open-source RE technologies, the Patents Act, 1970, can be amended to provide explicit provisions for open-source patents, allowing innovators to dedicate their patents to the public domain or license them under open-source terms. The Copyright Act, 1957, can be amended to provide clarity on the application of open-source licenses to software and other copyrighted works related to RE technologies. The National Policy on Information Technology, 2012, can be updated to include provisions for promoting open-source RE technologies, encouraging collaboration and facilitating the development of open-source RE solutions. Moreover, the Government can also consider fund support and other incentives for open-source RE projects.[119] It can also create a national repository for open-source RE technologies and innovations, providing incentives for collaborative R&D efforts between industry, academia, and research institutions to drive innovation in RE.

The legal challenges relating to RE in India can be adequately addressed by amending existing laws and enacting new ones to provide a robust framework for data protection. The Information Technology Act, 2000, can be amended to include provisions specific to the RE sector. Data protection law in India, the Digital Data Protection Act, 2023 (DPDP Act), needs to ensure the confidentiality, integrity, and availability of energy usage data. It is gratifying that the DPDP Act provides for the establishment of Data Protection Board of India.[120] Additionally, the data protection law must require the RE developers and grid operators to implement robust cybersecurity measures to protect against data breaches and cyberattacks. Furthermore, specific guidelines can be issued for the sharing and use of energy usage data, ensuring that it is done in a secure and transparent manner. Overall, the law can play a crucial role in promoting RE by providing a robust framework for data protection, ensuring the secure and transparent use of energy usage data and promoting trust among consumers and stakeholders.

Laws can also deal with the challenges of environment like material extraction and e-waste management to promote RE by amending the Hazardous Waste (Management, Handling and Transboundary Movement) Rules, 2016, to include specific provisions for the management of e-waste generated by RE technologies; implementing extended producer responsibility policies, which make manufacturers responsible for the waste generated by their products, including RE technologies. A law may be enacted to regulate the extraction of rare earth metals and other critical materials used in RE technologies, ensuring that extraction is done in an environmentally responsible manner. The Environmental Protection Act, 1986, can be amended to include provisions that specifically address the environmental challenges to RE growth. The Wildlife (Protection) Act, 1972, can also be amended to provide provisions for the protection of wildlife habitats and corridors that may be affected by RE projects. Further, the Forest (Conservation) Act, 1980, can be modified to provide guidelines for the diversion of forest land for RE projects, ensuring that the projects are developed in a sustainable and environmentally friendly manner. Overall, by amending and updating these legislations, India can ensure that its RE growth is environmentally sustainable and responsible.

EA 2003, and the NTF 2016 can be strengthened to mandate the use of technology, such as smart grids and energy storage, to ensure RPO compliance and to require grid operators to adopt technologies that facilitate RE integration, such as advanced weather forecasting and predictive analytics. Laws can provide greater incentives, such as tax benefits, subsidies, or low-interest loans, to encourage the adoption of specific technologies, like solar rooftops or wind-solar hybrids. Laws can mandate the adoption of net metering and energy storage technologies to promote rooftop solar and stabilize the grid.

Last but not the least, there is a need to enact new law that provides a comprehensive framework for RE development such as a dedicated Renewable Energy Act.

2.4.2.4.3 WHAT DO INDIAN COURTS AND TRIBUNALS HOLD ON RE TECHNOLOGIES

When it comes to deciding energy disputes, the primary authority in India is the APTEL,[121] which often rules on issues related to RE projects, including issues with power purchase agreements, regulatory compliance, and grid access. Several orders have dealt with the enforcement of RPOs, where APTEL has upheld the validity of SERC's regulations mandating a certain percentage of power generation from renewable sources by distribution companies. A prominent APTEL order regarding RPO enforcement is the directive instructing all SERCs to strictly comply with RPOs as per their Regulations, essentially emphasizing the need for robust RPO enforcement across India.[122] This order also guided the SERCs to use the tariff policy when defining tariff terms for RE sources, facilitating RPO compliance.[123] By enforcing RPOs, this order significantly pushed for increased procurement of RE across Indian states. In another matter the APTEL[124] ruled in favour of Himachal Pradesh Electricity Regulatory Commission, essentially upholding its decision to promote the development of RE sources within the state, including measures to incentivize and facilitate the construction of new green energy projects in Himachal Pradesh.[125]

Some notable examples of Court judgements related to RE technologies include *Hindustan Zinc Ltd* v. *Rajasthan Electricity Regulatory Commission*,[126] wherein the Court upheld the validity of Rajasthan's Renewable Energy Obligation Regulations, setting a precedent for other states to implement similar policies. While dismissing the appeal and reiterating the decision of the Rajasthan High Court, the Supreme Court ruled that companies with their own power plants (Captive Generation Power Plants) in Rajasthan must buy a minimum amount of RE, as they are covered by state law.[127] The Court stated that regulations set by Electricity Regulatory Commissions are not restrictive or in violation of fundamental rights. This decision strongly supports the mandatory purchase of RE, even if it raises electricity costs for industries relying on captive power generation.

Recently, in November 2023, the Delhi High Court[128] dismissed a writ petition challenging the increase in RPOs and surcharge imposition.[129] Previously, open access consumers were completely exempt from certain charges. However, this exemption has been partially removed for wheeling charges, transmission charges, cross subsidy surcharge, and additional surcharge.[130] Juniper Hotels Private Limited, an open access consumer[131] buying electricity through short-term contracts, argued that these changes harm consumers and make procuring green energy too expensive. They challenged the removal of exemptions and increased RPO requirements, claiming these actions were unconstitutional and violated due process. The petitioner argued for a return to full exemptions. However, the Court dismissed the petition, stating that open access consumers do not have a guaranteed right to exemption from wheeling, transmission, and additional surcharges. The petitioner is obligated to meet minimum RPO thresholds.[132]

2.4.2.4.4 CONCLUSION: CALL FOR DISTINCT AND COMPREHENSIVE RE LEGISLATION

India has demonstrated a significant commitment to transitioning towards RE, setting an ambitious target of "achieving 50% of its installed power capacity from non-fossil fuel sources by 2030." This aligns with global sustainability goals, particularly the Paris Agreement and India's updated nationally determined contributions. However, while this ambition reflects a proactive stance on addressing climate change and energy security, the framework supporting RE integration reveals critical gaps, particularly in legislative and policy dimensions. Technology plays a pivotal role in advancing RE deployment, facilitating the integration of solar, wind, hydropower, and other renewable sources into India's energy mix. The rapid evolution of technology, driven by advancements in efficiency, storage capabilities, and grid management, has been essential for enhancing the feasibility of renewable projects. However, these advancements often come with increased costs and complexities, requiring robust legal and policy mechanisms to ensure their effective adoption and scalability.

A significant challenge in India's RE landscape is the absence of a dedicated law governing RE. Attempts to establish such legislation have been made in the past but have not materialized into actionable frameworks. This lack of a specific legal mandate creates a fragmented regulatory environment, where RE policies are addressed indirectly through laws primarily focused on electricity regulation, such as the Electricity Act of 2003 and the ECA, 2001. Furthermore, the absence of a comprehensive national policy specifically tailored for RE integration exacerbates these challenges. Existing policy measures, such as the NEP and the National Tariff Policy, include provisions for RE but fall short of providing a holistic vision or addressing the nuances of integrating emerging technologies. This often leads to inconsistent implementation across states, delays in project approvals, and financial uncertainties for developers and investors.

To bridge these gaps, India requires a cohesive legislative framework that directly addresses the promotion of RE while accommodating the rapid pace of technological innovation. Such a framework should include provisions for research and development incentives, streamlined regulatory processes, and financial mechanisms to offset the initial costs of advanced technologies. Moreover, it should promote public–private partnerships and international collaboration to leverage global expertise and funding. In conclusion, while India's RE targets are commendable, achieving them necessitates a rigorous legal and policy infrastructure that not only encourages technological integration but also ensures equitable, sustainable, and efficient energy transition pathways. By addressing these legislative and policy challenges, India can solidify its position as a global leader in RE enhancement and SD.

Notes

* Though some of the studies examining Indian law on renewable energy are available, the author could not find any study detailing the interplay of renewable energy, technology, and law; hence, the chapter is titled "Reconnaissance Legal Study," which attempts to explore the intricate interrelationship of these three areas of study.

1 United Nations, "The Sustainable Development Agenda". Available at https://www.un.org/sustainabledevelopment/development-agenda/. (Last visited on December 15, 2024).

2 Poul Alberg et al., "Advances in Renewable Energy for Sustainable Development", *Renewable Energy*, 219(Part 1), 119377 (December 2023), https://www.sciencedirect.com/science/article/abs/pii/S0960148123012922. (Last visited on December 15, 2024).
3 See generally, JRP Manso & NB Behmiri, "Renewable Energy and Sustainable Development", *Studies of Applied Economics*, 31(1), 7 (March 2020); VR Prakash & IK Bhat, "A Figure of Merit for Evaluating Sustainability of Renewable Energy Systems", *Renewable and Sustainable Energy Reviews*, 14(6), 1640 (2010); K Kaygusuz, "Energy for Sustainable Development: A Case of Developing Countries", *Renewable and Sustainable Energy Reviews*, 16, 1116 (2012).
4 CR Kumar & MA Majid, "Renewable Energy for Sustainable Development in India: Current Status, Future Prospects, Challenges, Employment, and Investment Opportunities", *Energy, Sustainability and Society*, 10, 2 (2020).
5 Ministry of Environment, Forest & Climate Change, Government of India, "India Achieves Two Targets of Nationally Determined Contribution Well Ahead of the Time" (18 Dec. 2023). Available at https://pib.gov.in/PressReleaseIframePage.aspx?PRID=1987752. (Last visited on December 16, 2024).
6 Ministry of Science & Technology, Government of India, "India Is Committed to Achieve the Net Zero Emissions Target by 2070" (28 Sep. 2023). Available at https://pib.gov.in/PressReleaseIframePage.aspx?PRID=1961797. (Last visited on December 16, 2024).
7 Ministry of New & Renewable Energy, Government of India, "The Jawaharlal Nehru National Solar Mission", 2010. The Mission Has Adopted a 3-Phase Approach. Available at http://www.indiaenvironmentportal.org.in/files/mission-document-JNNSM.pdf. (Last visited on December 20, 2024).
8 Ministry of New & Renewable Energy, Government of India, "National Wind-Solar Hybrid Policy". Available at https://cdnbbsr.s3waas.gov.in/s3716e1b8c6cd17b771da77391355749f3/uploads/2022/12/2022122036.pdf. (Last visited on December 20, 2024).
9 Act no (36 of 2003).
10 Ministry of Power, Government of India, "National Tariff Policy" (28 Jan. 2016). Available at https://www.cercind.gov.in/2018/whatsnew/Tariff_Policy-Resolution_Dated_28012016.pdf. (Last visited on December 16, 2024).
11 Decided on 13 May 2015 SC. MANU/SC/0641/2015. (Last visited on December 16, 2024).
12 Decided on 3 November 2023. MANU/DE/7411/2023.
13 Report of the World Commission on Environment and Development, "Our Common Future". Available at http://www.un-documents.net/our-common-future.pdf. (Last visited on December 15, 2024).
14 Kumar & Majid, above n. 4.
15 U Shanker & N Mondal, "Analysing Law and Policy on Renewable Energy in Midst of Energy-Environment Debate", in S Bhat (ed.) *Energy Law and Policy*, 359 at 363 (National Law School of India University, 2016).
16 F Gralla, et al., "Energy Transitions and National Development Indicators: A Global Review of Nuclear Energy Production", *Renewable and Sustainable Energy Reviews*, 70, 1251 (2017).
17 According to the International Energy Agency (IEA), "India Is Projected to Become the World's Largest Energy Consumer by 2040, Surpassing China (IEA, 2020). The Country's Energy Consumption Is Driven by Sectors Such as Industry, Transportation, and Residential Use, with Electricity Demand Being a Significant Component" (IEA, 2020). Available at https://www.researchgate.net/publication/374294172_Advancements_in_Renewable_Energy_Technologies_An_Indian. (Last visited on December 23, 2024).

18 "Photovoltaic (PV) solar cells have been the backbone of solar power generation. However, emerging technologies are revolutionizing PV technology by enhancing efficiency and reducing costs. Innovations such as multi-junction cells, tandem cells, and perovskite-silicon hybrid cells are enabling higher energy conversion efficiency and making solar power more affordable." Enerdatics, *Shaping the Future: Emerging Technologies in Renewable Energy Projects*, July 2023. Available at https://enerdatics.com/blog/shaping-the-future-emerging-technologies-in-renewable-energy-projects/. (Last visited on December 15, 2024).

19 "Concentrated Solar Power (CSP) systems utilize mirrors or lenses to focus sunlight onto a receiver, generating heat that drives a turbine for electricity production. Emerging technologies in CSP, such as molten salt storage and advanced heat transfer fluids, improve the efficiency and storage capabilities of these systems, enabling continuous power generation even when the sun is not shining." Enerdatics, above n 18.

20 "Perovskite solar cells are a rapidly emerging technology in the solar power sector. These cells offer high efficiency, low-cost manufacturing, and a wide range of applications. Ongoing research focuses on enhancing the stability and durability of perovskite materials to ensure their long-term performance and commercial viability." Enerdatics, above n 18.

21 "Offshore wind power is gaining significant momentum due to its abundant resource potential and higher wind speeds compared to onshore locations. Emerging technologies in offshore wind include larger and more efficient turbines, floating wind farms, and advanced turbine designs that capture wind from multiple directions. These innovations increase energy generation, open up new offshore areas for development, and reduce visual and environmental impacts." Enerdatics, above n 18.

22 Floating wind farms are an innovative solution for harnessing wind energy in deeper waters where fixed foundations are not feasible. This technology involves anchoring wind turbines to floating structures, allowing for flexible deployment and access to wind resources in previously untapped locations. It opens up vast offshore areas for wind power generation and expands the potential for clean energy production." Enerdatics, above n 18.

23 "Vertical axis wind turbines (VAWTs) are gaining attention as an alternative to traditional horizontal axis wind turbines (HAWTs). VAWTs have unique advantages, including compact size, omnidirectional wind capturing, and reduced noise levels. Ongoing research focuses on optimizing VAWT designs, enhancing their efficiency, and exploring new applications in urban and distributed energy systems." Enerdatics, above n 18.

24 "Flow batteries offer an alternative approach to energy storage, utilizing liquid electrolytes stored in external tanks. They provide scalability, long cycle life, and the ability to decouple power and energy capacity." Enerdatics, above n 18.

25 "Hydrogen Storage technologies enable the conversion of excess renewable energy into hydrogen, which can be stored and used to generate electricity or power fuel cell vehicles." Enerdatics, above n 18.

26 Capable of capturing sunlight from both sides of the panel (Indian Express, 2021).

27 "Several wind energy installations in India have made significant contributions to electricity generation and have become noteworthy examples of successful projects, namely, Jaisalmer Wind Park: Located in the Jaisalmer district of Rajasthan, the Jaisalmer Wind Park is one of the largest wind energy projects in India. It consists of multiple wind farms with a combined capacity of over 1,600 MW. This project has been instrumental in boosting Rajasthan's wind energy capacity and contributing to India's renewable energy goals (Economic Times, 2021) and Tehri Hydro Development Corporation (THDC) India Limited:

THDC India Limited has established wind energy projects with a total capacity of over 63 MW in various states, including Gujarat and Karnataka" (The Hindu Business Line, 2018).

28 "designed to generate electricity from organic materials, including crop residues, wood, and agro-industrial waste. Innovations in combustion and gasification technologies have improved the overall efficiency of these power plants while reducing their environmental impact" (Business Today, 2020).

29 "Utilizing agricultural and organic waste for renewable energy production, namely, Agricultural residues like rice straw, wheat straw, and sugarcane bagasse are valuable feedstocks for biomass power generation" (The Economic Times, 2020). "Waste-to-energy plants, like the one in Ghazipur, Delhi, are turning urban waste into renewable energy while addressing waste management challenges" (Livemint, 2020).

30 A Paliwal, "Gujarat Set to Develop India's First Tidal Energy Pant: But the Country Lacks a Tidal Energy Policy", *Down to Earth* (23 Apr. 2012). Available at https://www.downtoearth.org.in/energy/gujarat-set-to-develop-india-s-first-tidal-energy-plant-38042. (Last visited on December 20, 2024).

31 Ministry of New & Renewable Energy, Government of India, "The Jawaharlal Nehru National Solar Mission", 2010. The Mission Has Adopted a 3-Phase Approach. Available at http://www.indiaenvironmentportal.org.in/files/mission-document-JNNSM.pdf. (Last visited on December 20, 2024).

32 Prime Minister's Council on Climate Change, Government of India, "National Action Plan on Climate Change". Available at https://archivepmo.nic.in/drmanmohansingh/climate_change_english.pdf. (Last visited on December 20, 2024).

33 Ministry of New & Renewable Energy, Government of India, above n 31, para 3, pg. 2.

34 Ministry of New & Renewable Energy, Government of India, above n 31, para 4 (A), pg. 4.

35 Ministry of New & Renewable Energy, Government of India, above n 31, para 5, pg. 7.

36 Ministry of New & Renewable Energy, Government of India, above n 31, para 6, pg. 8.

37 Ministry of New & Renewable Energy, Government of India, above n 31, para2, pg 1.

38 Ministry of New & Renewable Energy, Government of India, above n 31, para 4 (E), pg. 6.

39 Ministry of New & Renewable Energy, Government of India, above n 31, para 7, pg. 12.

40 Ministry of New & Renewable Energy, Government of India, above n 31, para 6, pgs. 7 and 10.

41 See generally SP Upadhyay & U Singh, "Jawaharlal Nehru National Solar Mission: A Critical Analysis of Evolution and Challenges", in AK Agarwal, et al. (eds.), *New Research Directions in Solar Energy Technologies (Energy, Environment, and Sustainability)*, 11 (Springer, 2021); PK Yadaw & MM Umar, "Unleashing India's Solar Potential: A Review of the National Solar Mission and the Path to Sustainable Energy in India", *Journal of Technology Innovations and Energy*, 3(2), 49–63 (2024); G Shrimali & V Nekkalapudi, "How Effective Has India's Solar Mission Been in Reaching Its Deployment Targets?", *Economic and Political Weekly*, 49(42), 54–63 (18 Oct. 2014).

42 Under sec 3 (3) of the Electricity Act 2003, the Central Government notified the Tariff Policy on 6 January 2006. Further amendments to the Tariff Policy were notified on 31 March 2008, 20 January 2011, and 8 July, 2011. The Central Government notified the revised Tariff Policy in 2016. See The National Tariff Policy, 2016, Decided on 13 May 2015 SC, above n 11, para 1.1.

43 Decided on 13 May 2015 SC, above n 11, para 6.4 (1), (5).
44 Decided on 13 May 2015 SC, above n 11, para 5.5.
45 Decided on 13 May 2015 SC, above n 11, para 6.4 (2).
46 Decided on 13 May 2015 SC, above n 11, para 6.4 (1) (iii &iv).
47 Decided on 13 May 2015 SC, above n 11, para 5.8.
48 Decided on 13 May 2015 SC, above n 11, para 5.11 (C).
49 Decided on 13 May 2015 SC, above n 11, para 6.1; 6.4 (2).
50 Decided on 13 May 2015 SC, above n 11, para 6.4 (4).
51 Decided on 13 May 2015 SC, above n 11, para 8.0.
52 till such period as may be notified by the Central Government. Decided on 13 May 2015 SC, above n 11, para 6.4 (6).
53 Decided on 13 May 2015 SC, above n 11, para 6.4 (7).
54 Act no (36 of 2003), above n 9, para 2.
55 Act no (36 of 2003), above n 9, para 4.
56 Act no (36 of 2003), above n 9, para 2.2.
57 Act no (36 of 2003), above n 9, para 5.3.
58 Act no (36 of 2003), above n 9, para 5.4 & para 5.8.
59 N Kumar, et al., "Barriers in Advancement of Grid Wind-Solar Hybrid in India", *Journal of Emerging Technologies and Innovative Research*, 5(10) (Oct. 2018). Available at https://www.jetir.org/papers/JETIR1810147.pdf. Accessed on 03.01.2025; A Das, et al., "A Comprehensive Review of Wind–Solar Hybrid Energy Policies in India: Barriers and Recommendations", *Renewable Energy Focus* (2020). Available at https://doi.org/10.1016/j.ref.2020.09.004. (Last visited on January 03, 2025).
60 U Shanker & A Basu, "Chronicling Energy Law in India in the Era of Low-Carbon Transition", in G Bellantuono et al. (eds.), *Handbook of Energy Law in the Low-Carbon Transition*, 413 (De Gruyter Handbuch, 2023).
61 The Ministry of Power, Government of India, oversees the implementation of the Electricity Act and plays a supervisory role in the overall development of the electricity sector. However, the Ministry of New and Renewable Energy (MNRE) is responsible for managing the growth and advancement of renewable energy in India. As the central agency for renewable energy, MNRE formulates policies, fosters investment, streamlines project approvals, and encourages private sector involvement. It monitors various renewable energy programs and projects, working in coordination with other ministries, state governments, and private entities to ensure effective implementation. To promote renewable energy, MNRE provides financial incentives such as subsidies, grants, and tax benefits, reducing project costs and accelerating sectoral growth. Additionally, the ministry prioritizes research and development by funding innovative initiatives and collaborating with academic institutions to advance renewable energy technologies. Through its multifaceted role, MNRE acts as a key driver in India's transition to a sustainable energy future, ensuring that policies are effectively implemented.
62 Electricity Act, 2003, Sec. 86(1) (e): this section empowers SERC to specify a minimum percentage of electricity that a distribution licensee must purchase from RE sources.
63 These revisions came into effect from 1 April 2024.
64 Only five states (Karnataka, Andhra Pradesh, Gujarat, Punjab, and Uttar Pradesh) have consistently achieved their solar Renewable Purchase Obligations (RPOs). Eastern states have generally underperformed since 2014, with data often missing (e.g., West Bengal and Odisha) or low integration despite potential (e.g., Jharkhand, Chhattisgarh, and Odisha). Northern states like Rajasthan, despite high solar capacity, failed to meet targets from 2014 to 2019 and have since stopped disclosing their performance. Some northern states (Delhi and

Haryana) may still depend heavily on expensive thermal power purchase agreements, hindering cheaper solar adoption. Eastern states appear particularly ill equipped for the energy transition, lacking market mechanisms to decarbonize their electricity sectors compared to other states. The lack of RPO data suggests weak governance within their State Electricity Regulatory Commissions (SERCs). Several states, including Telangana, Bihar, Chhattisgarh, Odisha, West Bengal, and Tamil Nadu, have not consistently published their data. https://www.downtoearth.org.in/renewable-energy/how-renewable-purchase-obligation-targets-can-advance-energy-transition-in-india-96101, May 2024.

65 The REC mechanism was launched in November 2010, based on the REC Regulations notified by the CERC on January 14, 2010. The REC mechanism has helped promote investments in renewable energy generation and Facilitate compliance with RPO. The Central Electricity Regulatory Commission (CERC) issued the Terms and Conditions for Renewable Energy Certificates for Renewable Energy Generation Regulations, 2022, on May 9, 2022. The Regulations became effective on December 5, 2022.

66 CERC, above n 65.

67 The Electricity Act, 2003, Sec.73 (b).

68 The Electricity Act, 2003, Sec 2(32).

69 The Electricity Act, 2003, Sec 2(33).

70 The Electricity Act, 2003, Sec 2(34).

71 The Electricity Act, 2003, See Chap IX SS 70–73.

72 Except that of a local authority engaged in the business of distribution of electricity before the appointed date.

73 Under the Electricity Act, 2003, the Ministry of Power notified The Green Energy Open Access Rules, 2022, "Electricity (Promoting Renewable Energy Through Green Energy Open Access) Rules, 2022" to accelerate India's renewable energy programs by enabling smaller consumers to access green energy through open access.

74 Sec 2(47): "open access" is defined as the non-discriminatory provision for using transmission lines or distribution systems by any licensee, consumer, or power generator, subject to regulations set by the Appropriate Commission.

75 Detailed process is provided in Electricity Act, 2003, Part VI SS 61–66.

76 Electricity Act, 2003, above n 75.

77 Electricity Act, 2003, above n 75, Secs. 76–81.

78 Electricity Act, 2003, Sec. 76.

79 Electricity Act, 2003, Secs. 82–88.

80 Electricity Act, 2003, Secs. 82–88.

81 Central Electricity Regulatory Commission (Terms and Conditions for Tariff Determination from Renewable Energy Sources) Regulations, 2017. This Regulation defines renewable energy "as grid quality electricity generated from renewable energy sources." The term "renewable energy sources" has been further defined in the Regulation to include "small hydro, wind, solar including its integration with combined cycle, biomass, biofuel cogeneration, urban or municipal waste and other sources as approved by the Ministry of New and Renewable Energy."

82 For instance, the constitution of the Electricity Contract Enforcement Authority, which "shall have the sole authority to adjudicate disputes emerging from power purchase agreements, make establishment of payment security mechanism under power purchase contracts mandatory, levy of penalty on distribution licensees for not adhering to renewable purchase obligations, and empowering the federal government to notify a National Renewable Energy Policy and prescribe a minimum renewable purchase obligation."

83 The Electricity Act, 2003, above n 67, Sections 61(h), 86(1)(e), and 181.

84 Electricity (Rights of Consumers) Rules, 2020, provide a framework for Net Metering, defining key terms like "Prosumer" (a consumer who both generates and consumes electricity) and outlining the process for implementing Net Metering schemes.
85 *The Energy Conservation Act* (Act No. 52 of 2001).
86 The EC Act promotes energy efficiency and conservation by regulating energy consumption in buildings, appliances, industries, and equipment etc. See *The Energy Conservation Act*, Sec. 14 and 15.
87 *The Energy Conservation Act*, Sec 3.
88 S Upadhyay, "Renewable Energy Development in India: The Need for a Robust Legal Framework", in U Tandon (ed.), *Energy Law and Policy*, 155 at 165 (Oxford University Press, 2018).
89 Upadhyay, above n 88, Sec. 14 & 15.
90 *The Energy Conservation Act*, above n 85 at 166.
91 Prime Minister's Council on Climate Change, Government of India, "National Action Plan on Climate Change". Available at https://archivepmo.nic.in/drmanmohansingh/climate_change_english.pdf. Accessed on 3.01.2025.
92 Prime Minister's Council on Climate Change, Government of India, above n 91, para 4.
93 The National Renewable Energy Bill of 2015, Cl. 4. Available at https://indiacorplaw.in/wp-content/uploads/2016/05/draft-rea-2015.pdf. (Last visited on January 05, 2025).
94 However, it did not pass during its session and subsequently lapsed. In the Indian legislative process, a bill pending in the Lok Sabha (House of the People) lapses upon the dissolution of the Lok Sabha, while a bill pending in the Rajya Sabha (Council of States) does not lapse. Since the National Renewable Energy Bill, 2015, was not enacted before the dissolution of the 16th Lok Sabha in 2019, it lapsed. As of December 2024, there is no record of the bill being reintroduced or enacted in subsequent parliamentary sessions.
95 The National Renewable Energy Bill of 2015, above n 93, Cl. 19 (4).
96 The National Renewable Energy Bill of 2015, above n 93, Cl. 36.
97 The National Renewable Energy Bill of 2015, above n 93, Cl. 20.
98 The National Renewable Energy Bill of 2015 likely couldn't be passed due to a combination of factors, including conflicting interests between different stakeholders in the energy sector, lack of political will to significantly disrupt the existing fossil fuel-based system, concerns over the economic implications of a rapid transition to renewables, and potential challenges with land acquisition and grid integration needed for large-scale renewable energy projects; despite the growing recognition of the need for renewable energy, significant legislative changes often face resistance from established industries and may require substantial political push to overcome these hurdles.
99 "Through a separate law, MNRE would get freedom to execute projects and not depend on other ministries and departments for necessary clearances," said a senior official at the MNRE. https://cleantechlaw.com/2015/07/national-renewable-energy-act-to-change/.
100 The National Renewable Energy Bill of 2015, above n 93, Cl. 4 (2), 15.
101 The National Renewable Energy Bill of 2015, above n 93, Cl., 7–10.
102 The National Renewable Energy Bill of 2015, above n 93, Cl. 23.
103 The National Renewable Energy Bill of 2015, above n 93, Cl. 11–14.
104 The National Renewable Energy Credit Support Bill, 2021 (Bill No. 1 of 2021) was introduced by Shri Ritesh Pandey then BSP MP in November 6, 2019 (Bill, being printed in 2021, the year in the title of Bill was changed from 2019 to 2021). Available at https://sansad.in/getFile/BillsTexts/LSBillTexts/Asintroduced/1%20of%202021As%20Int....pdf?source=legislation#:~:text=(1)%20This%20

Act%20may%20be,in%20the%20Official%20Gazette%2C%20appoint. (Last visited on December 28, 2024).

105 The National Renewable Energy Credit Support Bill, above n 104, Cl. 4 (a) states "The appropriate Government and the local authorities, as the Case may be, shall provide such tax-cut or credit on the electricity bill, being not less than ten per cent. and not more than thirty per cent. as it may deem fit, to the individuals, communities, public buildings, residences, companies, commerce, industries and agriculture that use renewable energy for the purpose of personal consumption."

106 The National Renewable Energy Credit Support Bill, above n 104, Cl. 6.

107 The National Renewable Energy Credit Support Bill, above n 104, Cl. 10.

108 The National Renewable Energy Credit Support Bill, above n 104, Cl. 12.

109 The National Renewable Energy Credit Support Bill, above n 104, Cl. 6.

110 The National Renewable Energy Credit Support Bill, above n 104, Cl. 3.

111 The National Renewable Energy Credit Support Bill, above n 104, Cl. 8.

112 M Giannelli, "Trade Secrets: Protecting Clean Energy Technology". Available at https://www.finnegan.com/en/insights/articles/trade-secrets-protecting-clean-energy-technology.html. Accessed on 29.12.2024.

113 R Gachago, "The Effect of Technology on Copyright" (June 2011). Available at https://doi.org/10.13140/RG.2.2.24252.28803. (Last visited on December 29, 2024).

114 VN Vasudeva, "Open Source Software Paradigm and Intellectual Property Rights", *Journal of Intellectual Property Rights*, 17, 511 (Nov. 2012).

115 Green City Times, "How to Secure Sustainable Energy Systems with Cybersecurity". Available at https://www.greencitytimes.com/cybersecurity-in-sustainable-energy-systems/. Accessed on 26.12.2024.

116 RE companies must comply with evolving data protection Regulations, such as the "General Data Protection Regulation (GDPR)" in the EU and the "California Consumer Privacy Act (CCPA) "in the US.

117 Union of Concerned Scientists, "Environmental Impacts of Renewable Energy Technologies" (Mar. 2013). Available at https://www.ucsusa.org/resources/environmental-impacts-renewable-energy-technologies. Accessed on 26.12.2024.

118 International Renewable Energy Agency, (IRENA), "Intellectual Property Rights the Role of Patents in Renewable Energy Technology Innovation" (2013). Available at https://www.irena.org/-/media/Files/IRENA/Agency/Publication/2013/Intellectual_Property_Rights.pdf. (Last visited on January 01, 2025).

119 Open-source RE projects refer to initiatives where the design, technology, and information related to renewable energy systems like solar, wind, or hydro power are freely accessible to the public, allowing anyone to modify, build, and improve upon the technology without proprietary restrictions.

120 The Digital Data Protection Act, 2023, Sec 18.

121 See EC Act, 2003, Electricity Act, 2003, above n 62, Part XI, SS 110–125.

122 Electricity Act, 2003, above n 62, para 28.

123 Electricity Act, 2003, above n 62.

124 *Green Energy Association* v *Himachal Pradesh Electricity Regulatory & ors*, APTEL (Appellate Jurisdiction), Appeal no. 54 of 2015, decided on December 10, 2015. Available at https://aptel.gov.in/judgements/Judg2015/A.No.%2054%20of%202015.pdf.

"In this case appeal has been filed by M/s. Green Energy Association, (a registered association of developers engaged in the generation of electricity through renewable energy sources), under Section 111 of the Electricity Act, 2003 against the order passed by the Himachal Pradesh Electricity Regulatory Commission. The State Commission, on failure to purchase the Solar RECs available in the power exchanges, ordered that the obligated entity is liable to pay compensation in terms of Regulation 9 of the RPPO Regulations, that would be utilised for development of transmission infrastructure for evacuation of power from renewable sources of energy."

125 Green Energy Association, above n 124, paras 7–9.

126 Decided on May 13, 2015 SC. In this case the captive generation companies filed an appeal before the Supreme Court, against High Court of Rajasthan wherein these companies were ordered to purchase minimum energy from renewable sources under 2007 and 2010 Regulations (Raj).

127 Decided on May 13, 2015 SC, above n 126, para 44.

128 *Juniper Hotels Private Limited* v *Delhi Electricity Regulatory Commission*, decided on 3 Nov. 2023.

129 The Delhi Electricity Regulatory Commission (Renewable Purchase Obligation and Renewable Energy Certificate Framework Implementation) Regulations, 2021 ("Regulations") and the Open Access Charges and Related Matters (Fourth Amendment) Order, 2021 ("Order"), inter alia increase the renewable purchase obligations and imposed additional surcharges, *Juniper Hotels Private Limited*, above n 128, para 01.

130 *Juniper Hotels Private Limited*, above n 128, para 05.

131 The Petitioner is an 'obligated entity' as per Regulation 2(16), and hence is required to observe the minimum thresholds prescribed in the Regulations (*Juniper Hotels Private Limited*, above n 128, para 04).

132 *Juniper Hotels Private Limited*, above n 128, para 11.

3 India's Energy Trilemma and the National Policy on Biofuels

A Critical Analysis

Neeraj Kumar Gupta

3.1 Introduction

Sustainability has been a concern of the world since long and its meaning and scope has been well settled now.[1] Strategies to attain sustainability vary on various factors such as climate, topography, and population. India is the world's most populated country and rapid and inclusive economic growth is one of the important aspects of sustainable development for India. Economic development of India is dependent on availability of energy resources.[2] India's energy requirements are largely met with fossil fuels and other non-renewable energy resources. Furthermore, India is the third largest importer of crude oil, which creates excessive pressure on India's foreign exchange reserves. Energy demands of India are yet to peak and there is a growing need to tap all possible avenues, which may provide domestic, affordable, and clean source of energy. India's concerns of energy security, affordability, and sustainability, commonly referred as energy trilemma, can be tackled only when dependence on import is reduced and alternative clean energy sources are deployed to meet the needs of growing energy demands. Further, the issue of energy justice is at the core of the discussion related to an energy-hungry nation as the affordability aspect of energy is directly connected with issues of energy access by the largest and economically vulnerable classes of population.

This chapter is an attempt to analyse India's National Policy on Biofuels, 2018 (hereinafter as NPB, 2018), as amended in the year 2022 with a view to seeing if it is in sync with short-term, medium-term, and long-term goals to deal with the energy trilemma of the country or if it needs fine-tuning. The chapter is divided into five parts. The first part is introductory. The second part deals with, very briefly, the history of biofuels. The third part of the chapter deals with India's energy trilemma issues and concerns in the context of Sustainable Development Goals (SDGs). The fourth part deals with analysis of NPB, 2018. The fifth part of the chapter deals with a critical evaluation of various plans, schemes, and mandates as introduced from time to time under the aegis of NPB, 2018, and arrives at conclusions.

The second part of the chapter highlights that biofuel has been an original source of energy to deal with heating and cooking requirements since the

DOI: 10.4324/9781003733751-5

beginning of human civilization. References can be found in various sacred texts with respect to energy requirements being met with the help of biofuels in one or the other forms. It is pointed out that biofuels were the most obvious choice prior to the discovery of petroleum and other fossil fuels. Even today, it remains an important source of energy, though usage of the same is relegated to those classes who cannot afford to buy mainstream sources of energy.

The third part of the chapter revolves around issues related to the energy trilemma of India in the context of SDGs, as adopted by the United Nations.[3] It is argued that sustainable development, saving planet Earth, protection of environment, and reducing greenhouse gas emission cannot be seen in isolation, especially when a large population of the nation is suffering from poverty and vagrancy. Therefore, developing countries like India need to place due emphasis on economic development, which requires satiating the energy hunger at affordable rates and with secure energy resources. The third part also highlights India's commitment to eliminate polluting energy sources and greening the energy basket while ensuring economic development. Special attention is drawn to various efforts related to energy transition taken by Indian government. It is argued that India has taken big strides in greening the energy mix, yet the reliance on fossil fuel in the recent future will remain dominant due to the large and energy-hungry population of the country.

The fourth part of the chapter deals with an analysis of NPB, 2018, as amended in the year 2022[4] by the Ministry of Petroleum and Natural Gas, Government of India (MoPNG, GoI). It is pointed in this part that NPB, 2018, has drawn inspiration and taken leads from NPB, 2009, policy; NPB, 2018, is a step in the direction of *Aatmanirbhar Bharat*[5] (self-reliant India), Make in India,[6] and Making for India along with waste to wealth[7] and *Swachh Bharat* Missions[8] (Clean India Mission). A major highlight of the policy is that it contains a holistic framework for development, use, consumption, distribution, marketing, and research relating to biofuels of various types such as bioethanol, biodiesel, bio CNG. The last part of the chapter critically reviews the efforts so far taken and the major roadblock which may arise with respect to achieving the long-term goals as envisaged under the policy; the conclusion is drawn on the basis of analysis.

3.2 A History of Biofuels

It is interesting to note that, historically, biofuels have been considered as the first liquid fuel of humanity.[9] The history of biofuel is so old that the same has been cited in mythologies along with the discovery of fire.[10] There are various instances in Indian as well as in world history where vegetable oils,[11] animals fat,[12] animal dung cake,[13] and bio-gas[14] have been utilized as sources of energy in addition to firewood and charcoal, firewood and charcoal also being unprocessed biofuels. Prior to the discovery of petroleum and other fossil fuels, for a long period of time reliance on biofuels was due to scarcity of other sources of energy; even after discovery of fossil fuels, usage of unprocessed biofuels continued. Even today, unprocessed biofuels are the first source of energy for those who cannot afford to pay for costlier sources of energy such as electricity, LPG, or other forms of petroleum.

First, biofuel was charcoal and it has been in use since the very early stages of human civilization. It is also the first solid biofuel. The first liquid biofuel was vegetable oil, which was used around 4000 B.C.[15] Ethanol was used for biofuel after the distillation process and distillation was discovered in the years between AD 721 and 1315.[16] Biogas was used in the early 17th century[17] and whale oil is an example of animal fat being used for energy prior to the discovery of petroleum. Thus, it can be seen that the history of bio-fuel is actually the history of fire and energy itself. However, techniques for use of such sources of energy have gradually changed with the passage of time and advancement of technology has changed the methods of use. Today, the first generation of biofuels, which may also be called processed biofuels, is being used by developed as well as developing countries throughout the world where agricultural and industrial waste, corn and other materials containing sugar, starch, cellulose, and lignocellulose materials are being used to produce bioethanol. Similarly, biodiesel is being produced with the help of oil seeds, which are edible and non-edible. Technology for the second generation of biofuels, which can be produced from other types of vegetation and waste of different types including municipal solid waste, is in its nascent stages.

In India, we can see cow dung cake, firewood, and charcoal made from firewood were and are still in use by the lowest economic class of the population in rural as well as urban areas of the country.[18] If we talk about the Indian history of biofuel, immediately after independence, the discussion on use of biofuel was undertaken under the aegis of the United Nations.[19] It was pointed out that ethanol or alcohol for power purposes cannot be produced from grains, as scarcity of food is the biggest concern of the country. Therefore, there was a need to conduct research on alternative sources, which can be converted into biofuels.[20] Concerns for food availability were well received in legislative enactment of 1948, which provided that power alcohol can be produced from molasses only and any other substance as per the direction of the government.[21] This legislation was enacted to promote and regulate the power alcohol industry in India[22] and it was based on command and control model, where there was absolute prohibition on use of any other source for production of ethanol, except molasses. The enactment did not provide any incentives for research and development, nor encouraged any experiment by industries. The Act provided that blending of petrol can happen in the range of 20–25% in terms of volume[23] but prescription was directory in nature and subject to availability of bioethanol in the market, without providing for creation of market or encouraging the same in the legislation.

There are reports which point towards experiments with ethanol in the 1970s and 1980s.[24] However, there was sudden interest in bio fuels after experimentation with the jatropha plant. There was renewed interest in biofuels, due to jatropha plant being used for biodiesel production.[25] It was mooted as a green alternative to fossil fuels by the industry and it was taken up by government with both hands as such[26] due to compelling reasons of heavy reliance on import of petroleum products and price fluctuation of crude oil in the international market, which led to implementation of the National Biofuel Policy, 2009 (NBP, 2009).

3.3 Sustainable Development, SDGs, and India's Energy Trilemma

Sustainable development requires that development has to be inter-generational as well as intra-generational. The present generation cannot dictate to future generations about the choices that future generations are to make for themselves. Also, present developments cannot be halted to save resources for future generations. Therefore, an appropriate balance between economic development, environment protection, and social development is the core debate relating to sustainable development.[27] The philosophy of sustainability has always been at the heart of Indian knowledge systems[28] and modern jurisprudence with respect to sustainability has been adopted by Indian legislature,[29] higher judiciary,[30] and policy-makers.[31]

India is the world's most populous country as on date and it has always catered to one of the largest populations in the world. The material well-being of such a large population has remained at the core of the law, policy, and governance framework of the country since independence of the country and framing of the Constitution. In the United National Conference on Human Environment, India's position with respect to environment protection was clearly put across by the then prime minister by pointing out that poverty is the biggest polluter and any attempt towards environment protection will remain unfulfilled until measures to deal with poverty are not integrated into the issue of environment protection.[32] This argument resonated with the Brundtland Commission Report, which noted that sustainable development can be ensured when needs of the poor are given priority while determining the policies relating to sustainable development and environment protection.[33] Hence, we find emphasis on Millennium Development Goals and the same have been revised and adopted as SDGs for the period of 2015 to 2030. There are 17 SDGs, which talk about material, social, and environmental well-being and all the SDGs are interconnected. It is difficult to deal with any goal or target in isolation; however, there are a few specific goals which are directly related to energy, which need special mention.

SDG 7 talks about ensuring affordable, reliable, and sustainable and modern energy for all by the year 2030.[34] Target 7.1 clearly provides that emphasis is on ensuring universality of energy access. India has made Herculean efforts in the last decade for energy accessibility to the population, yet energy access is still a concern to be addressed. It is remarkable that India has achieved universal electrification of villages[35]; however, it is not sufficient to claim that energy is accessible to all. Today, life is dependent on access to energy in various ways; integration with the internet, electronic governance, modern technology, education, medical facilities, and every aspect of life is dependent on energy. The aim of the government should be quality, quantity, and affordability of energy along with the energy being sustainable. Energy justice is the need of the hour in the Indian context.[36]

India's energy demand has increased 2.2 fold in less than two decades with CAGR of 4.76%[37]; per capita GDP has grown at the rate of 4.28%.[38] India's population has grown by CAGR of 1.17% in last two decades, whereas[39] energy consumption has grown at more than 10% CAGR per capita. This is indicative of large-scale energy hunger and potential huge demand in recent future. If we look at the consumption of energy and compare the same with expansion of GDP then the consumption has grown merely by CAGR of 2.54% in the last two decades. Thus, it can be inferred that on one hand per capita consumption of energy has grown, which is indicative of the fact that energy supply and energy demand have increased in the economy; on the other hand, GDP is still growing in energy-efficient manner.[40]

It is projected that India's population will continue to rise and it is estimated to be around 1.55 billion by the year 2040.[41] As far as estimates of energy consumption are concerned, the same is projected to be around three times the global average in next two decades.[42] The energy intensity of the country is expected to be much less as the country's GDP is service sector oriented, which requires less energy in comparison to large-scale production-industry-based GDP, which makes the economy energy intensive.[43]

SDG target 7.2 emphasizes on an increased proportion of renewable energy (RE) in the energy mix.[44] It is to be noted that in India, coal still enjoys the largest share in the energy mix. India's share of coal in the energy mix has increased faster than the average demands of energy in the last two decades.[45] The CAGR for coal stands at 5.1% whereas growth in energy demand is at CAGR of 4.76. However, it should be noted that energy demands have doubled in the last two-and-a-half decades. It is also to be noted that India has progressed by leaps and bounds in the last two decades when it comes to growth of RE sources. The renewable sources of energy have increased at a CAGR of 22.38%. Further, nuclear energy sources have increased more than the average energy demands. India has made substantial progress keeping in mind the growing needs of the country. Dependency on coal and other fossil fuels will continue in future as RE sources face various challenges, such as affordability, technology transfer, safety concerns, massive upfront investments and running costs, and regional and global cooperation.

India has targeted to become a net zero emission country by the year 2070[46] and therefore, gradually reliance on coal and other polluting fossil fuels is bound to decrease. It is expected that India's energy mix will be of primarily clean fuels. Further, the issue of energy efficiency is related to the concern of investment and other incentives of various types. The energy intensity of India is much lower in comparison to that of developed countries and the same may be reduced further by ensuring efficiency. The legal framework is already in place, which provides for regulatory[47] and incentive measures to ensure energy efficiency and conservation in the country. Thus, it can be seen that India as a country is still in such a stage of development where demands for energy supply will continue to rise. India's GDP growth will remain energy efficient, yet the demands for energy and energy resources are bound to increase due to the size of the population and economic growth.

As pointed out, the majority of India's energy demands are being met by coal. Largely, domestic coal is being used to meet energy demands. However, import of coal has also seen as upward trend, despite large reserves available in India,[48] due to various factors such as shortage of coal and cost.[49] Further, usage of oil for energy supply has also remained constant, near the average demands of energy, in the last two decades. There are various issues with the coal and oil. Firstly, they are non-renewable resources and therefore will stand exhausted after a while; it is to be noted that India has a very small reserve of oil and most of the demands relating to oil is being met with the help of imports from oil-producing countries. Secondly, the geopolitical issues in the Middle East and Eastern Europe and other factors always become a cause of concern when it comes to import of oil. In recent times, we have seen that geopolitical concerns have created interesting geopolitical scenarios. Thus, concerns of India's energy security remain under scrutiny and largely in threat due to factors beyond control. The vision of *Aatmanirbhar Bharat* will remain unfulfilled until the dependency on imports stands reduced and minimalized.

Thirdly, oil and coal are polluting sources of energy. Usage of coal leads to emission of greenhouse gases in the process of consumption, in addition to pollution in the process of extraction. Oil extraction, refining processes, and transportation are also polluting, in addition to emission of greenhouse gases on being used as a source of energy in relation to production, transportation, and other end-use cases. In comparison to coal, oil is less polluting.[50] It is imperative that reliance on polluting substances for energy must be gradually reduced and phased out, and a shift is made towards greener alternatives keeping in mind that demand for energy in India is bound to grow at the rate of three times the global average, at least in the next two decades. Thus, issues of sustainability, that is ensuring energy comes from cleaner and alternative sources of energy as well as sources which are domestically produced, are one of the biggest concerns for the Indian economy.

Fourthly, aspects related to the affordability highlight concerns related to energy justice. It is to be noted that energy affordability is still a concern for India. A large portion of the population cannot afford to buy energy at higher prices. When it comes to the lower classes of population, expenditure on energy resources is much higher compared to the middle and upper classes.[51] This indicates that affordable energy is the need of the hour for ensuring growth and inclusion of impoverished masses to realize *Viksit Bharat*. Energy affordability can be ensured by various means, including providing subsidies such as *PM Ujjwala* scheme, subsidies on installation of solar panels, and many more other schemes in the country. However, until dependence on import is not reduced, affordable energy will remain a distant dream as large portions of foreign currency reserves are utilized to procure energy,[52] prices of which are always fluctuating due to various geopolitical causes. If energy sources are imported at a higher cost, it becomes highly improbable that affordable energy can be made available by companies providing various types of energy: oil-marketing companies (OMCs), electricity gencoms, transcom, or discoms or any other form of energy-related company. Companies and corporations involved with energy resources, especially public sector electricity distribution companies, are facing huge financial burdens and most of them are loss-making entities.[53] In such a situation, it is not expected that these companies or corporations will offer affordable energy resources to masses without financial assistance from government. As far as oil-refining companies are concerned, though they have shown profits in the recent past, profits are largely dependent on geopolitical factors and it may skew down any time when the geopolitical environment is not favourable.[54] Thus, affordability on the one hand requires that energy resources are made available to masses at cheaper prices, and on the other hand companies and corporations dealing with energy work on an economically sustainable model of business to ensure that financial burdens of affordable energy are not passed on to taxpayers, ultimately in form of bailout packages in budgetary allocations.

3.4 The National Policy on Biofuels, 2018: Analytical Overview

The NPB, 2018,[55] is not the first policy relating to biofuels. However, it can be considered as one of the most progressive policy by far, especially after the 2022 amendments, which bring it closer to being one of the latest and future-oriented policy documents. Prior to NPB, 2018, the NPB, 2009,[56] was adopted. The NPB, 2009, pointed out that energy is a critical input for economic growth, equity, and human well-being and there is need to arrive at an optimum mix of primary resources of energy generation.[57] It recognized that fossil fuels will remain as the dominant source of energy supply; however due to its non-renewable and polluting nature, and dependence on import, their use need to be reduced and should be substituted by green, renewable, and domestically produced sources, which will be cost effective and inexhaustible.[58]

The NPB, 2009, highlights fluctuating oil prices in the world market, and its impact on the Indian economy. It is highlighted that India is dependent on import as nearly 80% of its oil needs are being met by imports.[59] Therefore, if India wishes to ensure energy security, then alternative fuels such as biofuels are required to be discovered, developed, and used. As biofuels are produced from biomasses, they are non-depleting in nature and lead to environment protection.

The NPB, 2009, also points to usage of market mechanisms for promotion of biofuels and related technology, supply of biofuels, storage, and research and development, including research in the second generation of biofuels.[60] The NPB, 2009, envisions mainstreaming of biofuels in the energy and transportation sector by promotion and cultivation of the same as substitute oils by ensuring availability in the market and implementing blending of biofuels in petrol and diesel to the extent of 20%.[61]

The NPB, 2009, adopted the strategy of producing biofuels by use of waste and degraded forest and non-forest land being used for plantation of non-edible oil seeds. It envisaged encouragement in the form of minimum support price, direct and indirect fiscal and financial assistance to producers, and encouragement to research in the field of biofuels. Further, the approach of eliminating market distortions due to subsidies of various types to fossil fuel was adopted for providing level playing field to biofuels.

It was envisaged that the rural community will be involved in the supply chain as the most important intervening institution, which will not only ensure employment at rural parts but will also ensure development of rural areas. NPB, 2009, also talked about carbon financing and joint ventures, foreign direct investment up to 100%, and biofuels being declared as a priority sector for lending by banks including NABARD, SIDBI, and IREDA. The policy also talked about quality standards, international cooperation, concerns and policy with respect to import and export of biofuels and biomass, role of states, awareness of capacity building, and institutional mechanisms.

The NPB, 2018, has taken inspiration from 2009 policy and the same has been prepared keeping in mind developments that have taken place in a decade with respect to biofuel technology and other areas. The NPB, 2018, was amended in the year 2022 to meet the revised targets and commitments of greening of the energy sector by Government of India. The salient features of NPB, 2018, as amended in 2022, are as follows.

3.4.1 Exhaustive Preamble in Line With the Goal of Aatmanirbhar *and* Viksit Bharat

The preamble of NPB, 2018, reiterates that energy is a critical input for economic growth, equity, and human wellbeing. Further, the fact that conventional sources of energy will remain the most dominant sources of energy in future is also highlighted.[62] However, it places higher emphasis on biofuels by giving them the status of "strategic importance,"[63] which points towards a policy shift to aggressive implementation of policy in comparison to previous policy. The policy has been directly linked with Make in India and *Swachh Bharat Abhiyan*, doubling the farmers' income, employment generation, and waste-to-wealth creation.[64]

The interconnectedness of Aatmanirbhar Bharat Abhiyan and Biofuel Policy has been highlighted by the fact that biomass for the biofuel is to be procured from the indigenous sources and the import of the biofuel is permitted only in exceptional circumstances. Further, the EBP has set a target for 20% blending by the year 2025, which indicates that the dependence on imports for oil should reduce. Further, the vision of Viksit Bharat is possible only when all sections of the society are given their due, including enough resources of education, health, nutrition, and clean sources of energy. PM Ujjwala Yojana has been specifically highlighted as one of the schemes in alignment with Viksit Bharat. The policy also acts as a tool for the Viksit Bharat vision.

The preamble of the NPB, 2018, talks about concrete targets of 175 GW of RE by 2022.[65] Further, it emphasizes 100 percent electrification of villages by 2019[66] and supply of electricity 24×7 and reduction of energy emission intensity by 33–35% and non-fossil fuel-based capacity in the electricity mix by 40% by 2030.[67] The preamble talks about a five-pronged approach for achieving the targets with respect to self-reliance in energy sector. These are increasing domestic production, adopting biofuels and RE sources, energy efficiency norms, improvement in refining process, and demand substitution.

All this has to be done keeping in mind the approach of *Samavesh* or inclusivity. The preamble also provides insights into the global context, which is to be kept in mind, and how the developments in science and technology and other variables at global level may be relevant considerations for policy.[68] The preamble provides insights into the micro-level working of the policy, and indicates the history of policy and vision for the future, along with the need to keep in mind the macro and global events that may be directly related to energy security, affordability, and sustainability. It also provides concrete targets which government wishes to achieve in short-term, medium-term, and long-term time horizons.

3.4.2 *Wide, Elastic, and Improved Definitions of Terminology*

The NPB, 2018, has improved some of the definitions used in NPB, 2009. For example, definition of "biofuels"[69] has been amended to explain the meaning of "biofuel" rather than from what it is made of; further, renewable resources[70] have been defined in place of definition of biomass as it was defined in NPB, 2009.[71]

The definition of bioethanol has been amended in 2018 as well as in 2022 keeping in mind the inclusion of new sources from which bioethanol can be produced. There are three main categories of feedstocks for bioethanol. These are materials containing sugar, starch, and cellulose. The definition of NPB, 2009, did not mention rotten potatoes and industrial waste within the ambit of starch-containing biomass; further, the 2022 amendment included agro/food pulp industry waste. Further, in the cellulosic materials category, industrial waste also finds place in the 2022 amendments. Further, definition of bioethanol has been made elastic by use of the phrase "other renewable resources" or "any mix combination of above feedstock" by the 2018 definition.

The definition of advanced biofuels appears in NPB, 2018, for the first time, which has been amended in the year 2022. The fuels produced from lignocellulosic feedstocks such as agricultural and forestry residue, algae, industrial waste, and residue streams, as well as fuels having low CO_2 or high greenhouse gas (GHG) reduction which does not compete with food crops or the land of food crops have been considered as advanced biofuels.[72] Advanced biofuels have been further classified as second-generation (2G) biofuels and third-generation (3G) of biofuels. 2G fuels include 2G ethanol, drop-in fuel, and so forth. 3G fuels include bio-CNG, DME fuels, produced from municipal solid wastes (MSW), and bio-hydrogen. Drop-in fuels[73] refer to any liquid fuel produced from biomass, agri-residue, MSW, plastic waste, industrial waste, and so forth as per Indian standards in this regard for being used as Jet fuel and other vehicles without modification of engines and it can be utilized by existing petroleum distribution system.

3.4.3 Updated Vision and Goals

The NPB, 2018, has the updated vision and goals as envisaged in NPB, 2009. The 2009 policy was primarily concerned with mainstreaming of biofuels. The NPB, 2018, shifts the focus towards utilizing, developing, and promoting domestic feedstock and increased substitution of biofuels in place of fossil fuels, so that concerns related to energy security, climate change, and sustainability can be answered while creating opportunities for employment as well as development of advanced technology for utilization of biofuels.[74] The 2022 amendment further updates on the vision and goals and requires that OMCs mandatorily sell ethanol-blended petrol (up to 20 percent blending of ethanol) from the financial year 2023–24 and 20 percent blending will be the mandate from the year 2025–26. Further, the goal of biodiesel blending is set for 5 percent by the year 2030.[75]

Goals which are envisaged as long-term goals are the reinforcement of supply of ethanol and biodiesel by increasing domestic production, setting up 2G refineries, development of new feedstocks, development of new technologies for conversion to biofuels, and creating suitable environment for biofuels for integration with main fuels.[76] Thus, where vision and goals in NPB, 2009, were recommendatory, the same stands concretized and partially achieved and revised by the 2018 and 2022 amendments; updated goals with higher targets have been incorporated by the 2022 amendments.

3.4.4 Updated and Focused Strategy With an All-Encompassing Approach

The NPB, 2009, had adopted the strategy where focus was on the use of degraded forest and waste land for supply of feedstock for biofuels. Cultivators as well as corporates will be provided with incentives in various forms to undertake farming of non-edible oil seeds and ethanol production form molasses.[77] It was envisaged that minimum support price[78] will be guaranteed to the cultivators and market distortion acting against the biofuels will be done away with by withdrawing direct or indirect subsidies available to conventional fuels.[79] Further, research is one of the strategies for development of biofuels. The NPB, 2018, policy has certainly updated and improved on strategy and approach. It talks about implementation of the Ethanol Blending Programme (EBP) as one of the key strategies. In addition, development of 2G ethanol technology, blending of biodiesel by EBP, exploration into various types of feedstocks for biodiesel, improved engines and machineries for better suitability for usage of biofuels, focus on drop-in fuels from MSW, and so forth, and advanced biofuels have been envisaged as the long-term strategy.[80]

Supply of domestic feedstock is core of NPB, 2018. It focuses on overall development of the supply chain of domestic feedstock for biofuel production, which is to be supported with appropriate financial and fiscal measures. Further, research on new technology, refining, and utilization of new feedstock for conversion purposes have been envisaged as one of the key strategies in NPB, 2018.

3.4.5 *Interventions and Enabling Mechanisms*

The chapter on interventions and enabling mechanisms can be considered as the backbone of NPB, 2018. It provides flesh and blood to the entire policy. This section discusses (a) feedstock availability and its development, (b) blending and bio-refinery programme, (c) financing, (d) financial and fiscal incentives, (e) research & development and demonstration, (f) quality standards (g) distribution and marketing, and (h) pricing of biofuels.

3.4.5.1 *Feedstock Availability, Refining, Blending, Distribution, Marketing, and Allied Concerns*

The emphasis of feedstock availability is based on the philosophy of Aatmanirbhar Bharat and Make in India. The feedstock is to be procured from waste and other resources locally available rather than being imported. Feedstock for bioethanol, biodiesel, drop-in fuels, and advanced biofuels differ from each other, although there is a possibility of overlap. One of the major concerns in biofuel policy has always been the food-versus-fuel debate. When it comes to ethanol production from feedstocks containing sugar, starch, and cellulose-containing materials, they are major source of food for human and animal consumption. The current EBP provides for bioethanol production from non-food sources. The policy has taken an aggressive shift from the existing approach on feedstock supply: damaged food grains, which are unfit for human consumption, and food grains in the surplus phase, as determined by the National Biofuel Coordination Committee (NBCC), may also be utilized for bioethanol production.[81] In addition, algal feedstock, sea weeds, and other industrial waste can work as feedstock with the development and exploration of technology as 2G and 3G biofuels.[82]

The policy also indicates and recognizes competition from the alcohol industry, when it comes to bioethanol usage by industry.[83] Therefore, the policy emphasizes development of 2G technologies for conversion of domestic feedstock usage for bioethanol production.[84] It is expected that 2G technology will use alternative sources of feedstock for ethanol production and there will not be scarcity of resources, which may lead to increased prices of bioethanol. It is envisaged by the policy that research and exploration in the field of 2G technology is required to be supported by financial grants and incentives and provisions for the same have been incorporated in the policy.[85]

As far as feedstock for biodiesel is concerned, the policy has kept the focus on non-edible oil seeds.[86] Along with it, the policy also talks about used cooking oil, animal tallow, and acid oil.[87] Farmers will be encouraged to use their marginal lands and wasteland for production of non-edible oil seeds and other types of feedstocks for biodiesel.[88] The role of local self-governing institutions has been recognized in the supply chain of feedstock.[89] Species such as Jatropa, Neem castor, and so forth have been given due weightage as crops which can be utilized as feedstock for biodiesel production.[90] Farmers will be provided with incentives to undertake cultivation of plants which act as feedstock. Further, minimum support price mechanisms can be explored for encouraging farmers to grow and plant crops as part of intercropping and second crop for feedstock.[91]

Regarding other types of biofuels, such as advanced biofuels and drop in fuels, there is no dearth of feedstock. However, the problem lies in the utilization of such feedstock with the help of technology. The policy recognizes that the technology to use MSW and other vegetation feedstock for conversion into biofuels is still in its nascent stages and the same requires to be explored in detail. Therefore, emphasis on experiments and demonstration of such technology is to be encouraged by providing grants for research and demonstration.[92]

The policy has shifted the focus from non-food to food sources in case of surplus and waste food grains. It is expected that in addition to molasses, which is the primary source of bioethanol, sugarcane juice and other food products will lead to generation of a much higher quantity of biofuels. It is estimated that 700 crore litres of ethanol can be produced each year. Similarly, biodiesel production is expected to rise in recent future. Therefore, the policy envisages periodic decisions with respect to increasing blending mandates, and usage of biofuels will increase in the transportation sector as well other areas of energy consumption.[93]

The OMCs are expected to store, distribute, and market biofuels with the existing or improved infrastructure.[94] Blending is intended to be ensured by offtake assurance by the public sector undertaking (PSU) OMCs. The PSU OMCs are expected to sign MoUs with bioethanol suppliers, which will ensure a market for suppliers and an assured stream of revenue, which will ensure consistent production and supply of biofuels. A similar approach is to be adopted with respect to other biofuels.

3.4.5.2 Incentivization of Biofuel Production and Usage by Market Intervention

The policy adopts multiple strategies to encourage and incentivize feedstock availability, its blending, marketing and consumption of biofuels, and research and development. Oil extraction and processing units for the production of biodiesel is to be declared as a priority sector for lending purposes by banking and non-banking financial institutions.[95] Further, multilateral and bilateral funding is to be encouraged for carbon financing opportunities.[96] Also, 100% foreign direct investment and joint ventures will be allowed if production and use of biofuel is limited to domestic use.[97]

In addition, the government is expected to provide for viability gap funding, subsidies, and grants for biofuels.[98] A fund called National Biofuel Fund is to be created, which will encourage efforts in the direction of 2G technology and other types of incentives as envisaged under the policy.[99] Further, R&D will be promoted with the help of various types of mechanisms, including financial and fiscal incentives. Biofuel mission is to be connected with the employment and rural integration and ensuring local participation by providing various types of support to farmers as and when suitable.[100] Thus, it is expected that biofuel policy will also raise the income of farmers and provide alternative sources of employment to the rural community involved in supply chain of feedstock.

As far as pricing is concerned, the policy envisages the role of government intervention in the market, be it first generation or second generation of biofuels. The object behind such an intervention appears to ensure competitiveness of biofuels in the market and provide encouragement to biofuel producers, distributors, and marketing companies. Thus, subsidies, VGF, and other fiscal measures can be utilized by government for ensuring competitiveness as well as robustness of industry from time to time, depending on various factors and stages of development of the biofuel usage and technology in the country.[101]

3.4.5.3 Research and Quality Standards of Biofuels and Allied Technology

The policy provides for the identification of various thrust areas with respect to biofuels technology and demonstration along with emerging technology and developing new raw material for feedstock.[102] Further, plantation and processing-related research will be promoted. Major areas for research envisaged in the policy include feedstock production, conversion technologies, end use technologies, and utilization of biproduct of biofuels. A focussed group is envisaged in the policy, which is supposed to promote R&D in biofuels and provide insights from experts from academia to industry and act as knowledge connects for NBCC.[103]

As far as the quality of biofuels is concerned, the policy provides that standards along the lines of BIS and other parameters will be developed based on methods, procedures, and protocols meeting the international benchmark for use of biofuels. Further, existing technologies in engine and other machinery will be reviewed in light of development of biofuels. In this regard, timely guidelines will be provided as and when required.

3.4.6 Role of Stakeholders and Institutional Mechanisms

The policy envisages the role of all stakeholders in detail.[104] The policy can be considered as very inclusive in envisaging the roles of various stakeholders at various levels. It elucidates the role of ministries at the central and state levels, farmers, and business or industry professionals. All have something to contribute in the implementation of the policy. The various areas of cooperation and coordination include generation of feedstock, encouragement to farmers and industry, and ensuring suitable supply chain, storage and infrastructure, licencing, and clearances along with incentives and other encouragement by the state and the central governments.

The policy specifically focuses on the role of states in feedstock generation and its supply is directly dependent on an active interest being taken by states. It is expected that states set up biofuel boards and empower them to take various decisions for promotion of biofuels. Further, land usage is an area where states exercise prerogatives and therefore the role of the state on promotion of plantation and cultivation of feedstock becomes crucial. Also, the state has regulatory control over industries and it is expected that administrative hurdles and red tapes will be avoided and single window clearance shall be granted to industries dealing with biofuels.

At central government level, MoPNG has been given charge as the overall coordinating ministry for the development of biofuels. It is also responsible for policy update and implementation, R&D, marketing, distribution, pricing and procurement policy,[105] dispute redressal, international collaboration, and MSW to transportation fuel.[106] The Ministry of Rural Development is expected to link the policy with MGNREGA and rural livelihood and ensure a proper supply chain. The Department of Agriculture and Cooperation is expected to ensure plantation and developments of breeds which can be utilized as feedstocks for biofuels. Similarly, the Ministry of Science and Technology is expected to push R&D. The Road and Transport Ministry and Railways is expected to encourage consumption and usage of biofuels. The Consumer Affairs Ministry is expected to provide for standards and specification of biofuels and the Ministry of Heavy Industry is expected to encourage production and manufacturing of machineries which can utilize biofuels.

The policy envisages a NBCC headed by the minister for PNG. The NBCC comprises representation from various ministries and it is expected to provide synergy and work as a central core for implementation of policy due to wide ranging and interdisciplinary approaches needed for implementation of policy.[107]

3.4.7 Import, Export, and International Cooperation

The primary object of the policy is to reduce dependence on import of oil and thereby become self-reliant. However, under certain special circumstances import of feedstock is to be permitted only to meet the demands of market for biofuels in India with respect to biodiesel on the recommendation of NBCC.[108] Exports of biofuels are strictly prohibited by the policy. As far as international cooperation is concerned, R&D-related cooperation, joint programmes, and field studies may be undertaken to explore the possibilities along with funding opportunities.

3.5 Critical Analysis and Conclusion

Biofuels can be considered as an original source to deal with the energy needs of heating and cooking. Biofuels were the most obvious choice prior to the discovery of petroleum and other fossil fuels and even today it remains an important source of energy for those who cannot afford mainstream sources of energy. Large masses are suffering from poverty and vagrancy. Developing countries like India are necessarily required to place due emphasis on economic development. Economic development requires accessibility and availability of energy resources at affordable prices. India is committed to ensuring economic development, along with greening the energy mix of the energy basket to meet net zero emission target and other environmental commitments as promised at international forums. Efforts related to energy transition taken by the Indian government have ensured that nationally determined contributions (NDCs) are met and emission targets remain within the stipulated goals. India has taken big strides in greening the energy mix, yet its reliance on fossil fuels will remain dominant due to the sheer size of its population. However, the energy intensity of the country is much lower than that of any developed country of the world.

When it comes to NPB, 2018, it has drawn inspiration and taken leads from NPB, 2009. NPB, 2018, is in consonance with *Aatmanirbhar Bharat*[109] (self-reliant India), Make in India,[110] and making for India along with waste to wealth[111] and *Swachh Bharat* missions[112] (Clean India Mission). NPB, 2018, is a forward-looking policy and it contains short-term, medium-term, and long-term goals, which provide the seeds to deal with the energy trilemma of the country. Major highlights of the policy include its holistic nature and framework for development, use, consumption, distribution, marketing, and research relating to biofuels of various types. The policy is an appropriate mixture of command and control and fiscal and financial incentive measures to create a market of and for biofuels. The NPB, 2018, also places due emphasis on the research and development required for developing technologies, which will lead to waste-to-wealth scenarios by converting MSW, agricultural waste, and other industrial waste. The policy envisions a future where the biofuel is to play a central role in energy planning and energy transition. The policy has set and revised various targets, which is indicative of the fact that goals set earlier have been achieved much earlier than estimated and newer targets should be achieved in due course. Current production of biofuels has huge production potential. As per government data, the total ethanol production capacity is of 875 crore litres from molasses and 505 crore litres from grains.[113] Around 23 crore litres of biodiesel is being produced in India.[114]

Development in the field of biofuels is certainly encouraging; however, data relating to import of petroleum products still remains a cause of concern. As per the latest reports available, India is importing 4674 barrels a day[115] or 232.5 million tonnes annually.[116] It is also a cause of concern that dependence on fuel imports is on an all-time high[117] and current production levels of biofuels are not even close to 0.5% of total imports made. Therefore, there is huge potential in terms of production and blending and increase of the use of biofuels.

The MoPNG notified 10% ethanol blending in the year 2019 in all states.[118] Further, a target for 20% blending has been preponed from the year 2030 to 2025 and it is expected that the same will be achieved very soon.[119] The amount of ethanol for achieving 20% blending is around 1016 crore litres of ethanol and the capacity of India has reached 1300 crore litres if ethanol produced from grains is also taken into consideration. Therefore, EBP 20 is certainly within reach. Also, diesel blending is expected to take off and 5% blending of biodiesel is expected by 2030. In the meantime, 5% of ethanol is being targeted to be mixed in the meantime.[120]

Thus, the achievements with respect to the first generation of biofuel usage and consumption in the short term are commendable. It is expected that medium-term and long-terms goals will also be achieved. Consumption of biofuels is certainly beneficial to the economy as reliance on imports is reduced. Further, the waste-to-wealth approach reduces the burden on nature's carrying capacity with respect to assimilation of waste. It appears to be a sustainable solution as well. Other sources of green energy or clean energy, such as hydro projects, require upfront investment and running cost, in addition to land. Further, they may lead to displacement of the local community, which may pose a threat to the natural environment. Solar energy also requires upfront costs and land for solar parks, whereas biofuel production and supply chain can be integrated with the existing infrastructure with some modification; technology for 2G and 3G biofuels, if developed, will be a long-term solution for energy demands and reduce the burden of MSWs and other industrial and agricultural waste in India.

One of the biggest challenges with respect to biofuels is the food-versus-energy debate. The policy envisions use of food grains in surplus period as well as waste grains as feedstock for ethanol production. However, there have been instances in the recent past when the government had to halt the production of ethanol from sugarcane juice due to short supply of sugar and interconnected inflationary reasons.[121] Similarly, when government procures grains by importing to deal with the issues of short supply of grains for inflationary reasons then it becomes difficult to justify usage of grain as feedstock for purposes of biofuels.[122] Thus, NBCC has to be extremely cautious in the decision-making process on the usage of food grains for biofuel purposes; otherwise, the burden on the common man may shift from energy sources to food. It is expected that NBCC will take care of these situations and their decision with respect to the use or non-use of grains in a particular phase will certainly remain under scrutiny.

It is true that India has the domestic feedstock available for conversion into biofuels; however, the technology for use of the same is still in its infancy. The issue of stubble burning and allied concerns of air pollution in the NCR region is one of the examples where the technology is in place but the usage of the same is not economically viable to the farmers: has been pointed out as one of the primary reasons for stubble burning.[123] The financial incentive for the same was discussed by the Supreme Court and an order was made to provide for incentives to farmers by giving INR 100 per quintal of stubble.[124] Thus, creating and incentivizing the supply chain requires efforts for the effective implementation of the policy.

NPB, 2018, as amended in 2022 is a policy document which is in consonance with the visions of *Aatmanirbhar* Bharat, Make in India, and Making for India. The policy has certainly achieved the short-term goals with flying colours and it is expected that mid-term and long-term goals will also be achieved. The chapter has not analysed data with respect impact of policy on employment aspects and increasing farmers' income. However, if goals and targets with respect to feedstock supply chain is achieved, then it is bound to generate sustainable employment opportunities, especially when the same is aligned with the waste-to-wealth mission. The impact of the policy on the reduction in pollution must also be studied from a medium-term perspective; so far, data for the same may not be sufficient.

As far as 2G biofuels and other advanced biofuels are concerned, the initiative of Indian government in terms of international collaboration will certainly yield results. Last year, when India hosted the G20 Summit, India proposed a Global Biofuel Alliance and 37 countries have agreed to be part of such an alliance.[125] This alliance will certainly lead to collaboration in research, capacity building, and finding solutions to the existing problems and technologies for the utilization of green resources for biofuel purposes. In short, the policy has the potential to deal with the energy trilemma if the same is implemented and proper push and encouragement is given to research and development.

Notes

1 The Brundtland Report defined sustainable development as "development that meets the needs of the present without compromising the ability of future generations to meet their own needs. It contains within it two key concepts: the concept of 'needs,' in particular the essential needs of the world's poor, to which

overriding priority should be given; and the idea of limitations imposed by the state of technology and social organization on the environment's ability to meet present and future needs." See Report of the World Commission on Environment and Development, "Our Common Future," p. 41, available at: https://sustainabledevelopment.un.org/content/documents/5987our-common-future.pdf (Last visited on January 09, 2025).

2 Report of the World Commission on Environment and Development, above note 1, at p. 20.

3 United Nations, "Transforming Our World: The 2030 Agenda for Sustainable Development", A/RES/70/1, available at: https://www.un.org/en/development/desa/population/migration/generalassembly/docs/globalcompact/A_RES_70_1_E.pdf (Last visited on January 09, 2025).

4 MoPNG, The National Policy on Biofuels, 2018 (hereinafter as NPB, 2018), Gazette of India, Vide S.O. No. 2492 (E), 4th August 2017, available at: https://mopng.gov.in/files/uploads/NATIONAL_POLICY_ON_BIOFUELS-2018.pdf (Last visited on January 09, 2025).

5 The Aatmanirbhar Bharat Abhiyan was launched officially in the year 2020; however, the emphasis of the government on becoming self-reliant in the arena of energy sufficiency can be seen in various efforts and schemes in the area of self-reliance related to energy. A list of activities on the reduction of oil imports and self-reliance can be accessed at MoPNG, "Steps by Government to Reduce Import Dependency on Crude Oil Posted On: 12 Dec. 2024 2:45PM by PIB Delhi", *PIB*, available at: https://pib.gov.in/PressReleseDetailm.aspx?PRID=2083669&reg=3&lang=1 (Last visited on January 21, 2025).

Some of the examples include Policy under PSC regime for early monetization of hydrocarbon discoveries, 2014; Hydrocarbon Exploration and Licensing Policy (HELP), 2016; Policy to Promote and Incentivize Enhanced Recovery Methods for Oil and Gas, 2018; Policy Framework for exploration and exploitation of Unconventional Hydrocarbons under Existing Production Sharing Contracts (PSCs); Coal Bed Methane (CBM) Contracts and Nomination Fields, 2018.

See the details on Aatmanirbhar Bharat Abhiyan on the website https://transformingindia.mygov.in/aatmanirbharbharat/?sector=OTcz&type=en#scrolltothis (Last visited on January 09, 2025).

6 See the details of Make in India at https://www.makeinindia.com/ (accessed on January 9, 2025), where the new and renewable energy sector has been identified as one of the sectors for Make in India.

7 See the Vision, Mission and details of Waste to Wealth Mission at: https://www.psa.gov.in/waste-to-wealth (Last visited on January 09, 2025).

8 See the details of Swachh Bharat Mission at: https://swachhbharatmission.ddws.gov.in/ (Last visited on January 09, 2025).

9 Prof. Kovarik, "Environmental History/ Biofuels," available at: https://billkovarik.com/bio/cabi/ (Last visited on January 07, 2025).

10 LEC Partners, "A Short History of Biofuels", available at: https://lee-enterprises.com/a-short-history-of-biofuels/ (Last visited on January 07, 2025).

11 LEC Partners, above note 10.

12 LEC Partners, above note 10.

13 LEC Partners, above note 10.

14 LEC Partners, above note 10.

15 Prof. Kovarik, above note 9.

16 Prof. Kovarik, above note 9.

17 Prof. Kovarik, above note 9.

18 The Ujjwala Scheme or the Pradhan Mantri Ujjwala Yojana was launched by Prime Minister Narendra Modi to shift the usage from these polluting sources to clean fuel by women, targeting the elimination of such sources and shifting to cleaner and less polluting sources. The scheme provides for free LPG cylinder and

LPG stove to the women belonging to below-poverty-line households. See details available at: https://mopng.gov.in/en/marketing/pmuy#:~:text=Pradhan%20 Mantri%20Ujjwala%20Yojana%20(PMUY)%20was%20launched%20 by%20the%20Hon,to%20women%20from%20poor%20households (Last visited on January 09, 2025).

19 Prof. Kovarik, above note 9.

20 Prof. Kovarik, above note 9.

21 The Indian Power Alcohol Act, 1948, (Act, No. 22 of 1948), Section 4,

22 The Indian Power Alcohol Act, 1948, above note 21, Preamble.

23 The Indian Power Alcohol Act, 1948, above note 21, Section 6(2).

24 HT Correspondent, "Evolution of Indian Biofuels", *Hindustan Times*, July 2, 2007, available at: https://www.hindustantimes.com/business/evolution-of-indian-biofuels/story-WAq1BJGVzMJq8lpaePXJhI.html (accessed on January 7, 2025).

25 Sunil Kumar, Alok Chaube and Shashi Kumar Jain, "Critical Review of Jatropha Biodiesel Promotion Policies in India", *Energy Policy*, 41 (2012), 775–781, ISSN 0301-4215, https://doi.org/10.1016/j.enpol.2011.11.044, available at: https://www.sciencedirect.com/science/article/pii/S0301421511009219.

26 Kumar, Chaube and Jain, above note 25.

27 T. R. Franks, "Managing Sustainable Development: Definitions, Paradigms, and Dimensions", *Sustainable Development*, 4 (1996), 53.

28 Rudrapatna Shamasastry (tr.), *Kautilya'a Arthashastra* 240 (Government Press, Bangalore, 1915).

29 See generally the National Green Tribunal Act, 2010 (Act No. 19 of 2010).

30 See generally *Indian Council for Enviro-Legal Action and Ors.* v. *Union of India and Ors.* (1996) 3 SCC 212; *Vellore Citizens Welfare Forum* v. *Union of India and Ors.* (1996) 5 SCC 647; *Andhra Pradesh Pollution Control Board* v. *M. V. Nayudu* (1999) 2 SCC 718.

31 See generally MoEF, GoI, National Forest Policy, 1988; MoEF, GoI, Policy Statement for Abatement of Pollution, 1992; Ministry of Water Resource, GoI, National Water Policy, 1987; MoEF, GoI, National Environment Policy, 2006 etc.

32 Indira Gandhi, "Speech at the Stockholm Conference in 1972" (Stockholm, 14 Jun. 1972), available at: http://lasulawsenvironmental.blogspot.com/2012/07/indira-gandhis-speech-at-stockholm.html (Last visited on January 09, 2025).

33 Report of the World Commission on Environment and Development, above note 1.

34 See all the SDG Targets and Indicators available at: https://unstats.un.org/sdgs/indicators/indicators-list/ (Last visited on January 07, 2025).

35 Ministry of Power, GoI, "Year End Review-2024", available at: https://pib.gov.in/PressReleasePage.aspx?PRID=2089243 (Last visited on January 07, 2025).

36 See generally Mansi Verma, "Achieving Redistributive Energy Justice: A Critical Analysis of Energy Policies of India", Usha Tandon (ed.), *Energy Law and Policy* (OUP, 2018).

37 The CAGR has been calculated on the basis of the data available at NITI Aayog Dashboard of India Climate and Energy Dashboard, available at: https://iced.niti.gov.in/energy (accessed on January 7, 2025). CAGR calculation has been done with the help of the calculator available on various websites; the initial year is taken as 2006 and current year 2024.

38 Trading Economics, "India GDP Per Capita," available at: https://tradingeconomics.com/india/gdp-per-capita (accessed on January 7, 2025).

39 Macro Trends, "India Population 1950–2025," available at: https://www.macrotrends.net/global-metrics/countries/ind/india/population (Last visited on January 07, 2025).

40 NITI, Aayog, India Climate and Energy Dashboard, "Per Capita GDP and Per Capita Electricity Consumption", available at: https://iced.niti.gov.in/analytics/per-capita-gdp-vs-consumption (Last visited on January 07, 2025).

41 CEIC, "India Population Projection: Single Year", available at: https://www.ceicdata.com/en/india/population-projection-single-year/population-projection-single-year (Last visited on January 07, 2025).
42 International Energy Agency, "World Energy Outlook 2021: Special Report", p. 12, available at: https://iea.blob.core.windows.net/assets/4ed140c1-c3f3-4fd9-acae-789a4e14a23c/WorldEnergyOutlook2021.pdf (Last visited on January 07, 2025).
43 International Energy Agency, above note 42.
44 SDG Target 7.2.
45 India Climate and Energy Dashboard, "Sector-wise Coal Consumption", available at: https://iced.niti.gov.in/energy/fuel-sources/coal/production (Last visited on January 07, 2025).
46 Ministry of Environment, Forest and Climate Change, "Net Zero Emissions Target", Posted On: Aug. 3, 2023 5:04PM by PIB Delhi, available at: https://pib.gov.in/PressReleaseIframePage.aspx?PRID=1945472 (Last visited on January 09, 2025).
47 See generally the Energy Conservation Act, 2001 (Act No. 52 of 2001).
48 India Climate and Energy Dashboard, above note 45.
49 Ashok Sreenivas, Maria Chirayil and Rohit Patwardhan, "Understanding India's Coal Imports Premium: The Discourse Around Coal Shortages in India Needs Course Correction", *The Hindu*, Mar. 29, 2024, available at: https://www.thehindu.com/opinion/op-ed/understanding-indias-coal-imports/article68003203.ece (Last visited on January 09, 2025).
50 Climate Portal, "Why Does Burning Coal Generate More CO2 Than Oil or Gas?" available at: https://climate.mit.edu/ask-mit/why-does-burning-coal-generate-more-co2-oil-or-gas (Last visited on January 07, 2025).
51 Lydia Powell, Akhilesh Sati and Vinod Kumar Tomar, "Household Consumption Expenditure in India: A Small Share for Energy", *Observer Research Foundation*, 25 Apr. 2024, available at: https://www.orfonline.org/expert-speak/household-consumption-expenditure-in-india-a-small-share-for-energy#:~:text=According%20to%20the%202022%2D23,and%20INR6521%20in%20urban%20households. (Last visited on January 09, 2025).
52 The Hindu, "India Paid $132.4 Billion in Oil Imports for the Financial Year 2023–24", *The Hindu*, 17 Apr. 2024, available at: https://www.thehindu.com/business/Economy/indias-crude-oil-import-bill-falls-but-import-dependency-hits-new-high/article68075642.ece (Last visited on January 07, 2025).
53 India Climate and Energy Dashboard, "Operational and Financial Performance of Discoms", available at: https://iced.niti.gov.in/analytics/discom-accumulated-losses-consumer-sales (Last visited on January 07, 2025).
54 MoPNG, "OMCs'FY24 Combined Profit Rose Over 25-Times Year-on-Year; OMC's Combined Net Profit in FY 24 up 543% vs FY 14; IOCL Reported Historical Best Refinery throughput and Net Profit in FY24", Posted On: 10 May 2024 7:49PM by PIB Delhi, available at: https://pib.gov.in/PressReleasePage.aspx?PRID=2020282 (Last visited on January 09, 2025).
55 The NPB, 2018.
56 Ministry of New and Renewable Energy, GoI, "National Policy on Biofuels, 2009" (hereinafter as the NPB, 2009), available at: https://lpr.adb.org/resource/national-policy-biofuels-2009-india (Last visited on January 09, 2025).
57 Ministry of New and Renewable Energy, GoI, above note 56, Preamble, 1.1.
58 Ministry of New and Renewable Energy, GoI, above note 56, Preamble, 1.1.
59 Ministry of New and Renewable Energy, GoI, above note 56, Preamble, 1.2.
60 Ministry of New and Renewable Energy, GoI, above note 56, Preamble, 1.5.
61 Ministry of New and Renewable Energy, GoI, above note 56, 2.2.
62 The NPB, 2018.
63 The NPB, 2018, Preamble 1.3.

64 The NPB, 2018, Preamble 1.7.
65 Deccan Herald, "All Census Village Electrified by 2018: Govt Tells Lok Sabha", *Deccan Herald*, 17 Dec. 2024, available at: https://www.deccanherald.com/india/all-census-village-electrified-by-2018-govt-tells-lok-sabha-3321614; India has attained the target and as of October 2024, India achieved 200 GW capacity in terms of renewable energy in the energy mix). See Press Information Bureau, GoI, "India's Renewable Energy Capacity Hits 200 GW Milestone: Renewable Energy Now Constitutes 46.3% of Total Capacity", MNRE, Posted On: 14 Oct. 2024 6:26PM, available at: https://pib.gov.in/PressNoteDetails.aspx?NoteId=153279&ModuleId=3&reg=3&lang=1 (Last visited on January 09, 2025).
66 The same was achieved in the year 2018 itself. See Deccan Herald, above note 65.
67 The NPB, 2018, Preamble 1.1.
68 The NPB, 2018, Preamble 1.8.
69 The NPB, 2018, Definitions, Clause 3.1 (i) "Biofuels" are fuels produced from renewable resources and used in place of or in blend with, diesel, petrol or other fossil fuels for transport, stationary, portable and other applications.
70 The NPB, 2018, Definitions, Clause 3.1 (ii) Renewable resources are the biodegradable fraction of products, wastes and residues from agriculture, forestry, tree-based oil other non-edible oils and related industries as well as the biodegradable fraction of industrial and municipal wastes.
71 The NPB, 2009, Clause 3.1 (ii) "biomass" resources are the biodegradable fraction of products, wastes and residues from agriculture, forestry and related industries as well as the biodegradable fraction of industrial and municipal wastes.
72 The NPB, 2018, Clause 3.2.(iii) Advanced biofuels: Fuels which are (1) produced from lignocellulosic feedstocks (i.e. agricultural and forestry residues, for example rice & wheat straw/corn cobs & stover / bagasse, woody biomass), non-food energy crops (i.e. grasses, algae), animal dung or industrial waste and residue streams, or any mix combination of above feedstock. (2) having low CO2 emission or high Green House Gas reduction and do not compete with food crops for land use. Fuels such as Second Generation (2G) Ethanol, biodiesel made from Used Cooking Oil (UCO), non-edible tree borne oils, short gestation non-edible oil rich crops; green diesel from renewable sources and Industrial waste, biofuels produced from synthesis (syn) gas, drop-in fuels from renewable sources and industrial waste, algae based 3G biofuels, halophytes based biofuels, bio-CNG, bio-methanol, Di Methyl Ether (DME) derived from bio-methanol, bio-hydrogen, drop-in-fuels from MSW resource / feedstock material will qualify as – Advanced Biofuels.
73 The NPB, 2018, Clause 3.2. iv. "drop-in fuels": Any liquid fuel produced from Biomass, agri-residues, wastes such as Municipal Solid Wastes (MSW), Plastic wastes, Industrial wastes etc. which meets the Indian standards for MS, HSD and Jet fuel, in pure or blended form, for its subsequent utilization in vehicles without any modifications in the engine systems and can utilize existing petroleum distribution system.
74 The NPB, 2018, Clause 2.1 and 2.2.
75 The NPB, 2018, Clause 2.1 and 2.2.
76 The NPB, 2018, Clause 2.1 and 2.2.
77 See generally The NPB, 2009, Clause 4.
78 The NPB, 2009, Clause 5.
79 The NPB, 2009, Clause 4.
80 See generally The NPB, 2018, Clause 4.
81 The NPB, 2018, Clause 5.2.
82 See generally The NPB, 2018, Clauses 5.20 to 5.28.
83 The NPB, 2018, Clause 5.9.

84 The NPB, 2018, Clause 5.10.
85 The NPB, 2018, Clause 5.10.
86 The NPB, 2018, Clause 5.2.
87 The NPB, 2018, Clause 5.2.
88 The NPB, 2018, Clause 5.5.
89 The NPB, 2018, Clause 5.5.
90 The NPB, 2018, Clause 5.5.
91 The NPB, 2018, Clause 5.6.
92 The NPB, 2018, Clause 5.20.
93 The NPB, 2018, Clause 9.2.
94 The NPB, 2018, Clause 5.28.
95 The NPB, 2018, Clause 5.13.
96 The NPB, 2018, Clause 5.14.
97 The NPB, 2018, Clause 5.15.
98 The NPB, 2018, Clause 5.12.5.
99 The NPB, 2018, Clause 5.16.
100 The NPB, 2018, Clause 5.29.
101 The NPB, 2018, Clause 5.29.
102 The NPB, 2018, Clause 4.5.
103 The NPB, 2018, Clause 5.24.
104 The NPB, 2018, See generally Clause 7.
105 See MoPNG, GoI, "Long Term Ethanol Procurement Policy Under Ethanol Blended Petrol (EBP) Programme: 2019–2024", available at: https://mopng.gov.in/files/uploads/Final_Ethanol_Procurement_Policy.pdf (Last visited on January 09, 2025).
106 The NPB, above note 104.
107 The NPB, above note 104. See generally Clause 9.
108 The NPB, above note 104. See generally Clause 6.
109 See the details on Aatmanirbhar Bharat Abhiyan on the website https://transformingindia.mygov.in/aatmanirbharbharat/?sector=OTcz&type=en#scrolltothis (Last visited on January 09, 2025).
110 See the details of Make in India on the Official Website https://www.makeinindia.com/ (accessed on January 9, 2025), where the new and renewable energy sector has been identified as one of the sectors for Make in India.
111 See the Vision, Mission, and details of Waste to Wealth Mission on the official website at: https://www.psa.gov.in/waste-to-wealth (Last visited on January 09, 2025).
112 See the details of Swachh Bharat Mission on Official Website at: https://swachhbharatmission.ddws.gov.in/ (Last visited on January 09, 2025).
113 Ministry of Consumer Affairs, Food & Public Distribution, Ethanol Production Capacity in the country is 1380 crore litres of which 875 crore litres is molasses based and 505 crore litres is grain based Posted On: 20 DEC 2023 5:18PM by PIB Delhi, available at https://pib.gov.in/PressReleasePage.aspx?PRID=1988727 (Last visited on January 09, 2025).
114 Statista, Production volume of biodiesel in India from 2010 to 2023, with a forecast for 2024, available at https://www.statista.com/statistics/1051902/india-biodiesel-production-volume/ (Last visited on January 09, 2025).
115 CEIC, "India Crude Oil: Imports:1980–2023", available at: https://www.ceicdata.com/en/indicator/india/crude-oil-imports (Last visited on January 05, 2025).
116 The Hindu, "India's Crude Oil Import Bill Falls, But Import Dependency Hits New High," *The Hindu*, 17 Apr. 2024, available at: https://www.thehindu.com/business/Economy/indias-crude-oil-import-bill-falls-but-import-dependency-hits-new-high/article68075642.ece (Last visited on January 09, 2025).
117 The Hindu, above note 116.

118 MoPNG, Ministry of Petroleum and Natural Gas Notification, New Delhi, G.S.R. 107(E), 5 Feb. 2019, available at: https://mopng.gov.in/files/uploads/2019Gazette_Notification.pdf.

119 MoPNG, India's Ethanol Push: A Path to Energy Security Achieving 15% Ethanol Blending in 2024, India Targets 20% by 2025, Posted On: 24 Oct. 2024 1:06PM, available at: https://pib.gov.in/PressNoteDetails.aspx?NoteId=153363&ModuleId=3&reg=3&lang=1 (Last visited on January 09, 2025).

120 MoPNG, "Road Map for Mixing Ethanol in Petrol and Diesel", PIB, Posted On: 04 Dec. 2023 4:25PM by PIB Delhi, available at: https://pib.gov.in/PressReleaseIframePage.aspx?PRID=1982356 (Last visited on January 09, 2025).

121 The Hindu, "Government Directs Sugar Mills to Not Use Sugarcane Juice for Ethanol Production to Keep Prices in Check", *The Hindu*, 7 Dec. 2023, available at: https://www.thehindu.com/news/national/government-directs-sugar-mills-to-not-use-sugarcane-juice-for-ethanol-production-to-keep-prices-in-check/article67614174.ece (Last visited on January 21, 2025).

122 The constant fluctuation is visible in the data of import of rice, wheat, and other cereals available in the last four years. See the details on Ministry of Agriculture & Farmers Welfare, "Import of Food grains and Value-Added Food Products" Posted On: 16 Dec. 2022 6:32PM by PIB Delhi, available at: https://pib.gov.in/PressReleasePage.aspx?PRID=1884237 (Last visited on January 21, 2025).

123 Aparna Jaiswal, Aradhana Kumari, Vinita Parte and S. K. Rai, "Stubble Burning in Agriculture: Current Scenario and Sustainable Solutions", *Harit Dhara*, 6(2) (July–December 2023), available at: https://iiss.icar.gov.in/eMagazine/v6i2/11.pdf (Last visited on January 21, 2025).

124 *M.C. Mehta* v. *Union of India and Ors.*, 2020 SCC OnLine SC29. SCA-4690-2012.

125 MoPNG, "About Global Biofuels Alliance (GBA)", available at: https://mopng.gov.in/en/page/68 (Last visited on January 09, 2025).

4 Legal Mechanisms to Ensure Energy Efficiency in India With Special Reference to Energy Saving and Carbon Credit Certifications Under the Energy Conservation Act, 2001

A Critical Review

Swati Kumari Mawandiya

4.1 Introduction

The concept of sustainable development (SD) documents the negative consequences of globalization and shifts attention to industrialization and population growth. The phrase today is linked to all areas with three main pillars: environmental protection, social development, and economic growth.[1] The goal of SD requires governments to take positive action to guarantee essential needs of present society without compromising those of future generations.[2] While tackling the challenge of food security, drinking water, and education, housing is important, equally important is the need to reduce consumption of fossil fuels and emphasis on usage of renewable resources in order to mitigate urgent concerns related to climate change.

The 2030 Agenda for sustainable development and Paris Agreement on Climate Change were key developments of the year 2015, which pointed towards a sustainable society. Sustainable development goals (SDGs) aim for various transformations such as health, education, equality, sustainable food, land, water, oceans, cities, communities, and so forth. One such transformation that is required is energy decarbonization and sustainable industry.[3] This transformation aims to ensure access to modern energy sources and decarbonization of energy system. Whether the energy profile of a country enables sustainable energy development depends on the analysis of the energy trilemma.

The energy trilemma indicates three variables which are inter-dependant and cannot be considered independent of each other.[4] The three variables are energy security, energy sustainability, and energy affordability.[5] It is the balance of these three goals that will enable energy sustainability, which is indeed the end goal.[6] The increased demand for energy is primarily due to the exponential growth of world population and rising urbanization and economic growth. Energy is required for the purposes of power, heat, and transport, and global energy demand is expected to double by 2050.[7] In order to

DOI: 10.4324/9781003733751-6

efficiently achieve the increased demand, proper allocation of resources will be required.

The government of India is committed towards the goal of sustainable development and energy efficiency and is persistently working towards achieving the goals at global level. The government is also conscious of its rising population and increasing demand for energy. This chapter discusses the goals for energy efficiency and an analysis of the policy and legislative mechanisms to achieve those goals.

This chapter is divided in six parts. The second part analyses the energy dilemma of India by studying factually the data comprising the total final consumption of energy and the GHG emission caused by the various energy sources. It is argued that the two primary sources of energy – coal and crude oil – form the largest part of the energy basket and impact the energy trilemma. Coal, although available, is polluting the environment and has high emission level and thus needs to be phased out. Crude oil too is polluting but its bigger impact is on energy security and energy affordability. It is argued in this part that there is an urgent need to find alternative and innovative ways to secure the energy access and make it cleaner, greener, and more affordable in order to achieve the international goals.

The third part conducts an in-depth analysis of the international commitments made by India and also analyses the steps taken at the national level to achieve these goals. SDG 7 is specifically discussed, as it caters to the goal of energy efficiency and achieving clean and affordable energy. This part argues that there is an interlink between the goals of energy conservation and achieving energy efficiency across sectors. This part also discusses the target of net zero emission by 2070, which focuses on lowering GHG emissions and also the updated nationally determined contributions (NDC). This part argues that there is need of concrete steps to achieve the larger goal of energy efficiency, which will not only boost economy but also protect the environment.

The fourth part discusses the legal mechanism in place in India to achieve the goal of energy efficiency and conservation. This part discusses in detail the enactment of Energy Conservation Act, wherein the powers and functions of the bureau and government have been analysed. This part argues that India has been performing commendably and shows constant progress by developing innovative mechanisms to achieve these stated goals. It is argued that the provisions of the Act provide immense potential for reduction of carbon footprint by adopting energy efficient measures.

It is then followed by the fifth part, which caters to a specific discussion of certain innovative and special initiatives taken by the government in the form of energy saving and carbon credit certificates, which are in the form of tradeable units that aim to incentivize the industry to adopt energy efficient technology for production. Such schemes promote reduction of carbon emission while enabling conservation of energy. It is argued that every unit of energy conserved is equivalent to energy produced and it is the need of the hour for our nation to invest in such measures for a better future. The last part concludes the discussion and ends with an overall impact analysis of the legal mechanism to ensure energy efficiency.

4.2 Analysing Energy Dilemma of India

India is the world's most populated country and its third largest consumer of energy. There is a high demand of energy and India needs to take active steps to conserve energy for efficient usage. A study of last two decades of total final consumption (TFC)[8] in India suggests that the industrial sector is the largest consumer with 41% of the total TFC while residential and transport sectors stand next with 25% and 16%, respectively.[9]

The study depicts the evolution and growth of these sectors with the exponential rise in the industrial sector. An analysis of the study suggests that due to large consumption by a few sectors, CO_2 emissions are also increasing. In order to meet the goals of energy efficiency and decarbonization, India needs to find cleaner, greener, and more affordable alternative.[10]

A study of energy mix and TES[11] is essential to understand the challenges and reasons for energy crisis in India. As per the World Energy Report of 2024, coal is the largest source of energy supply in India, followed by crude oil. While India is the second largest producer and consumer of coal, it has adverse effects on the environment and hence, there is a need for reduction of dependence on coal.[12] The energy sector of India depends largely on coal for generation of electricity and power supply. The consumption of oil too has grown rapidly due to use of vehicles and road transport.[13]

Crude oil is the primary source of energy after coal in India and is used as a raw material for many products. It is used for generation of electricity, to heat buildings, power the vehicles, and many others. The demand for oil is expected to grow exponentially in the next two decades, and it will have a great impact on India's economy.[14] While there is rise in demand and consumption of crude oil, the production has declined considerably. In order to meet domestic needs, crude oil is largely imported. Oil-based fuels are used to power all automobiles, aircraft, and ships, which is integral to maintenance of modern life.

As per the report of the Petroleum Planning and Analysis Cell (hereinafter referred as PPAC), for the financial year 2024–25, India is the third largest consumer of crude oil with the data of 249.3 MT while producing only 32.6 MT.[15] The petroleum products in India are largely produced through processing of crude oil along with other inputs in the refineries. The report also provides that crude oil import in FY 2024–25 H1 registered a 3.2% increase in comparison to the previous financial year, aligning with the rise in consumption of petroleum products.[16]

The dependence of the nation on import of crude oil negatively impacts its foreign reserves; the currency of such a nation also depreciates due to rising inflation. Due to its high dependence on import, the core dimension of energy security is directly threatened due to price and supply shocks.[17] The fluctuation of prices at the international oil market significantly impacts the foreign exchange reserves as well.[18] Additionally, the geopolitical tensions in the Middle East and eastern Europe impact the crude oil import as supply can get affected.[19]

The global energy crisis has intensified post the Russian invasion of Ukraine in 2022, which has led to geopolitical tensions. The war has a significant risk for rise of conflict in the eastern Europe and Middle East, which will directly influence oil prices.[20] The economic impact of geopolitical tensions poses a great challenge to energy trilemma as it will affect both energy security and energy affordability. Access to affordable energy, being a public good, becomes the responsibility of the government. Any obstacle to the supply of energy will hinder industrial and economic growth. Thus, India needs to devise mechanisms to address the growing demand of energy amidst the geopolitical tensions.

The Standing Committee on Petroleum and Natural Gas[21] provided some key recommendations, which can be discussed to analyse the policy on import of crude oil. The first recommendation was to diversify the import of crude oil rather than depending on some specific areas as it becomes a threat to energy security. The second recommendation was to devise mechanisms to reduce the import dependence. Taking note of the fact that the demand for petroleum products is likely to increase, the report stated that India can take steps to either increase the domestic production or promote alternatives such as biofuel and renewable energy. Other relevant steps can be to improve the refinery process and promote energy efficiency by adopting measures for cleaner, greener, and more affordable sources of energy. As for the rising prices of crude oil, the committee recommended that a study should be conducted to audit whether the spot contracts are more affordable than term contracts. While the former seems cheaper, the latter provides energy security by assuring certain volumes for a decided period. The committee suggested that newer and shorter transportation routes can be identified which have lesser geopolitical tensions to reduce the transportation cost that is borne by the importing country.

The committee recommended setting up of an interdisciplinary body of experts, which can devise policy measures to reduce the demand of fossil fuels. Certain mandates and standards can be devised to promote or incentivize use of energy efficient appliances. Due emphasis should be placed on means to conserve energy as every unit of energy saved would be equivalent to energy produced. It is not only an economic but also an environmental requirement to protect the environment and choose sustainable means of energy production.

The target of energy efficiency and energy sustainability is primarily to mitigate climate change and reduce emissions of GHGs. Energy is the biggest contributor to climate change and GHG emission which happen due to burning of fossil fuels.[22] Despite global efforts, the graph of CO_2 emissions remains higher than that required to mitigate the impact of climate change. The IEA Report of 2022 shows that while coal contributes to 72% of total CO_2 emissions due to fuel combustion, oil is the second largest contributor.[23]

The sectoral analysis of such emissions also shows that 53% of such emissions in the year 2022 was due to electricity and heat producers followed by 24% by industry and 13% by transport. An analysis of the study conducted by the International Energy Agency shows the expansive requirement of fossil fuels by various sectors. It also depicts the relevance of energy security and the impact that loss of such security can have on India.[24]

This part analysed various reports to depict the dependence of India on coal and oil for its energy source wherein the import of the latter poses serious challenges and risks that need to be addressed. The next part discusses the international commitments and goals which are set in order to achieve energy sustainability and maintain the axes of the three core dimensions of energy trilemma.

4.3 International Commitment of Energy Sustainability

The previous section discussed the energy dilemma, pointing out the urgent need to ensure access of energy with the rising population along with finding mechanisms to mitigate climate change caused due to emissions by these energy sources. Tackling the need of energy security while balancing energy sustainability thus becomes the major challenge in the global energy scenario.[25] The concept of sustainable development thus provides for a perception of a socioeconomic system which emphasizes on fulfilling the human needs while ensuring growth and improved quality of life.[26]

The analysis of the energy trilemma suggests that there is urgent need to identify energy sources that are safe, accessible, and sustainable. There is need for diversification of energy mix, geographical diversification, and deployment of new technologies or finding energy efficient means which focus on low emissions.[27] Thus, sustainable energy is the pathway to transition to a new energy model that emphasizes energy efficiency and energy conservation, which will help balance the energy trilemma and address the global energy challenges.

There are a number of immediate reasons which make securing sustainable energy models necessary. Environmental protection, economic impact, energy security, clean and affordable energy sources for future generations, and improvement of the quality of life are some of those.[28] Sustainable energy allows for decarbonizing of the energy supply by reducing dependence on fossil fuels.

It was for these reasons that the nations at the global level partnered together to commit to certain SDGs in which SDG 7 talks about ensuring "access to affordable, reliable, sustainable and modern energy for all."[29] There are other targets like net zero emissions to be achieved by 2070. This section will discuss the aim set forth by SDG7, performance of India to achieve the set goals, measures and policies for energy efficiency adopted to achieve the said goal, target for net zero emissions, and updated NDC.

4.3.1 Goal 7 – Affordable and Clean Energy

The member states of United Nations adopted the 2030 Agenda for Sustainable Development, which provide the 17 SGDs that are an urgent call for action by all countries in a global partnership to achieve the common end goals. It is understood that ending poverty and other types of deprivations are aligned with other goals like improvement of health, education, sanitation, and economic growth along with tackling the concern of climate change and environmental protection.

Goal 7 aims largely at ensuring clean and affordable energy sources, which form an important key to development in the fields of agriculture, healthcare, education, transportation, business, and so forth.[30] While our everyday life depends on reliable and affordable energy, it is the consumption of energy which is the dominant contributor to climate change. The previous part of this chapter discussed the dependence on fossil fuels and the emission of GHGs caused due to it.

Goal 7 sets five targets to be achieved by 2030[31] (i) Target 7.1 discusses universal accessibility to modern forms of energy which can be reliable and also affordable. (ii) Target 7.2 discusses diversifying the energy mix and substantially increasing the share of renewable energy. (iii) Target 7.3 discusses increasing the global rate of improvement by two times in energy efficiency. (iv) Target 7.A aims to foster international cooperation to facilitate access to clean energy and cleaner fossil fuel technology. (v) Target 7.B aims at expanding the infrastructure and upgradation of technology for supply of sustainable energy services to developing and least developed countries.

Energy is a major contributor to problems of human health resulting from pollution caused due to fuel combustion. The environmental impact of the energy system depends on the primary energy source, the conversion technology, and other allied factors. As per the UNEP report on SDG7, a billion people still live without electricity, which significantly limits their developmental opportunities. Billions of people depend on heavily polluting sources such as charcoal and wood for the purpose of cooking and heating.[32] It is submitted that improved energy efficiency could lead up to 40% reduction of emission. Goal 7 targets mandate the government to opt for energy-efficient means in their countries and uptake clean energy technologies.

While all SDGs are inter-related and inter-dependant, SDG7 happens to be the catalyst for achieving a number of other SDGs. The achieving of cleaner and affordable means will allow the people living in impoverished states to choose better modes of living and access to cleaner services. It will improve the quality of life, health, and wellbeing through cleaner energy sources reducing air pollution. It can also indirectly provide means for inclusive and equitable education by providing for electrification and save time of families arranging unhealthy and cheaper modes of living. It will undoubtedly boost industrial and economic growth by reducing economic pressure due to import of raw materials such as crude oil. Other goals such as sustainable cities, responsible consumption, and climate action will directly be impacted by making a move towards to energy efficiency.

This brief analysis reflects the importance of SDG7 and the impact it can have in achieving the larger goal of sustainable development. However, the recent SDG progress report shows that there is a long way to go to achieve these targets. The latest report states:

> In the area of climate and biodiversity, despite some reductions in greenhouse gas emissions in developed countries, greenhouse concentrations hit record highs in 2022, with real-time data in 2023 indicating a continued rise. Carbon dioxide levels have surged to 150% above pre-industrial levels. Public funding for oil, coal, and gas production and consumption more than doubled from 2021 to 2022 and tripled since 2015, hindering progress towards a net-zero transition. Ocean acidification is on the rise and will persist if CO_2 emissions continue to climb. Species extinction risk is worsening, with a 12% decline in the aggregate Red List Index since 1993.[33]

Nations set targets based on their NDCs, energy policies, and corporate strategies. The Government of India too has made massive efforts to achieve the targets set in goal 7. As stated, while all goals are inter-related; the central theme of this chapter is to analyse the action agenda for energy efficiency. Hence, an analysis of the target 7.3.1 is conducted as it deals with ensuring energy efficiency in the sectors such as agriculture, industry, business, transport, and also energy efficient appliances and equipment. This part discusses the agenda action of government of India – its ambition, actions taken, outcomes, impact, and the next decade's action agenda.

In order to achieve the stated target, the Government of India aims largely at implementation of the National Mission for Enhanced Energy Efficiency (NMEEE) by the year 2030, which aims at enhancing energy efficiency.[34] It was noted in a report that India has achieved emission reduction of 28% and might exceed its NDC commitments by 2030.[35] The report provides for ways in which the government of India aims to achieve enhancement of energy efficiency:

a The emphasis of NMEEE is on achieving energy efficiency in all sectors of economy and provide for innovative business models in the energy sector.
b The Roadmap of Sustainable and Holistic Approach to National Energy Efficiency (ROSHANEE), a broader version of NMEEE, was released in 2019 and includes all potential areas of energy efficiency.
c The Bureau of Energy Efficiency (BEE) is the agency which implements ROSHANEE. Enhanced energy efficiency will lead to lower energy consumption across sectors and allow for mitigation of GHG emissions.

d The energy efficiency targets for the industry are set under the Perform, Achieve and Trade (PAT) scheme. The scheme provides Energy Savings Certificates (ESCerts) if the specific industry achieves efficiencies higher than targets, which are tradeable on power exchange. PAT can be extended to more sectors to incentivize other sectors to choose energy-efficient sources of production.
e The Street Lighting National Program (SLNP) aims at replacing all present streetlights with LED streetlights with an aim to conserve energy and also reduce GHG emission. It is expected that India would have 100% LED streetlights by 2030.
f The Standards & Labelling (S&L) programme, which was launched in 2006, aims to create awareness by informing the end consumers about the products by labelling their energy performance ("Star Labels"). These labels allow the consumers to know the energy and cost they would save by purchasing an energy-efficient product. Mechanisms are to be increased to provide such labelling to more appliances and equipment to incentivize both manufacture and usage of energy efficient products.
g In order to push for energy efficiency in the building sector, the government launched the Energy Conservation Building Code 2017 (ECBC) and Eco-Niwas Samhita 2021 for commercial and residential buildings respectively. Stringent mandates are needed to ensure compliance with ECBC.

These actions will not only allow enhanced energy efficiency but also lead to reduction of consumption of energy across sectors while mitigating CO_2 emissions. These steps will lead to creation of new and skilled jobs and also upskilling of people employed in these sectors through certification programmes or online courses. Adopting these measures can lead to improved competitiveness in the production and reduction in energy costs. It will help achieve energy sustainability through reduction of energy consumption and GHG emissions. Lastly, it will help in achieving other SDGs, as discussed.

The report also provides a checklist of guiding principles to achieve the set ambitions.[36] It requires the government to step up and accelerate the implementation of the SDG7 targets aligning with national development plans, NDCs, and target of net zero emission strategies. In order to successfully achieve these targets, the government must choose measures that are technically sound and also feasible to be adopted. The government must commit to these measures and constantly update and advance these targets.

4.3.2 Target of Net Zero Emission by 2070

Historically, India's cumulative emissions have always been on the lower side even when it has larger population as compared to many other nations. In this light, the global responsibility of India towards global warming has always been minimal. Although the GHG emissions are bound to increase with rising population and economy, India follows the low carbon development strategy. In CoP 26 of UNFCCC held in November 2021, India announced the target to achieve net zero emissions by the year 2070.[37]

This target aligns with the principles of equity and climate justice and is based on certain steps that form the low carbon pathways. The targets aim at certain transitions such as low carbon development of electricity, energy efficient transport system, sustainable urbanization using energy efficient material, low industrial emission, and engineering solutions to remove carbon dioxide and also enhance the forest cover to encompass the ecological considerations.[38]

4.3.3 Updated India's NDC

In the year 2015, as a party to UNFCCC and its Paris Agreement, India had submitted its first NDC, which was to reduce the emission intensity of its GDP by 33–35% by 2030 based on its 2005 level. However, India was able to achieve this target well before time and by December 2023, it had already achieved the goal of 33% reduction. Recently, in August 2022, India has updated the NDC to reduce the emission intensity of its GDP to 45% by 2030. A latest PIB report mentions:

> Further, as of 31 May 2024, the share of non-fossil sources in the installed electricity generation capacity has reached 45.4 per cent up from 32 per cent in April 2014. India is also on track to make an additional carbon sink of 2.5 to 3.0 billion tonnes through tree and forest cover by 2030, with a carbon sink of 1.97 billion tonnes of CO_2 equivalent having already been created from 2005 to 2019.[39]

The aforementioned analysis shows that the efforts taken to achieve the goals of energy efficiency in all sectors to achieve the SDG7 are in sync with the target of net zero emissions and the updated target to reduce the emission intensity. India has made immense effort and must be appreciated for its commitment towards energy efficiency and environmental protection. It is understood that the rising energy needs of the nation can only be met if we switch to cleaner and more affordable energy alternatives and by attempting to conserve energy by adopting measures for energy efficiency. The next part discusses the legislative enactment on energy conservation and the measures contained therein.

4.4 Legal Mechanism of India for Energy Efficiency and Energy Conservation

The Constitution of India does not specifically mention terms like "energy" or "energy conservation" but deals with some forms of energy sources. The Centre and State lists make mention of a few sources such as atomic energy,[40] mines and minerals,[41] gas,[42] and so forth. The statute on energy conservation was enacted in 2001[43] through exercise of the residuary power provided under Art. 248 read with Entry 97 in the first list.[44] The Energy Conservation Act (hereinafter, Act) was made with the object to promote both energy efficiency and energy conservation as it was the need of the hour.

An inter-ministerial group on energy was constituted in the year 1983 for the first time to consider the challenges of energy conservation in different sectors. In 1987, the advisory board raised a demand for a specific legislation dealing with energy concerns and the matter was referred to the ILI to prepare a Bill for the same. ILI submitted the draft in 1988. The need for such law was reiterated in the ninth five-year plan (1997–2002) and thereby in 2000, the Bill was then introduced in the Parliament.

The Act consists of 10 chapters with 62 sections and 1 schedule. The preliminary chapter contains certain definitions which mostly are in form of reference of main sections.[45] The demand for electricity and fossil fuels has been on the rise leading to increase in energy consumption and impacting the GHG emissions as well. However, India shows great potential of reducing its energy consumption in various sectors by adopting energy efficient measures. Such measures have been provided through statutory mandates making it mandatory and universally applicable. These steps have not only reduced economic pressures but also substantially benefited the environment by reducing emissions. In order to achieve the objective, the Act has established the BEE to provide for regulations and standards for consumption of energy. These regulations would become enforceable by the central government.

The second chapter contains provisions with regard to establishment of the bureau – its membership, management, meetings, and so forth. The BEE succeeds the previously existing body, Energy Management Centre.[46] The Act empowers the central government to establish the BEE, which will be a body corporate with perpetual succession.[47] The head office of BEE will be in Delhi while it can establish offices in other states as well.[48]

According to the Act, the affairs of the bureau – superintendence, direction and management – will be vested in the Governing Council (GC) consisting of minimum 31 and maximum 37 members, who will be appointed by the central government.[49] The Central Minister of Power would be the ex-officio Chairperson of the GC and other members would be appointed by the central government, like Secretaries of various ministries like Power, Coal, Atomic Energy, Forest and Climate Change, Consumer Affairs, Housing and Urban Affairs, Road Transport and Highways, Steel, Civil Aviation, Ports, and Shipping and Waterways. Other ex-officio members would be from other departments like Railway, Electricity Authority, Central Power Research Institute, Supply, Commerce and Industry, Indian Renewable Energy Development Agency Limited, and few others.[50]

The Director-General of the Bureau shall be the Chief Executive Authority of the Bureau, who shall exercise and discharge such powers and duties of the Bureau as may be determined by regulations. The appointment of Director-General is based on ability and standing. It should be a person with adequate knowledge and experience in the area of energy and matters connected therewith – production, supply, energy management, standardization, measures for conserving and efficient use of energy.[51] The analysis of these provisions reflect excessive bureaucratization, and the parent Act does not mention the qualifications, which were added in the rules.

The GC exercises all powers to perform all acts that can be done by the Bureau[52] and would convene meeting in accordance with the norms provided in the regulations.[53] Advisory committees can be constituted for efficient discharge of functions[54] and the Bureau may also constitute certain technical committee of experts to decide the standards and norms regarding energy consumption or norms with regard to appliance or equipment.[55] The function of the Bureau is to essentially and effectively coordinate with the DC, designated agencies in order to identify, and utilize the existing set of resources and infrastructure.[56] It performs the functions as assigned under the legislation which are in the nature of being recommendatory and action based and with certain residuary powers.

The standards set by BEE are to apply to appliances, equipment, vehicles, buildings, and all establishments such as commercial, residential, and industrial. These norms are the key to achieve mitigation of climate change and reduce the GHG emissions by following a model that conserves energy. The energy-efficient measures thus allow for conservation of energy, which is equivalent to production of energy. This would in turn be a positive step towards energy security, which is a serious concern for a nation like India, which largely depends on import for its energy needs. The BEE also specifies qualification of energy auditors and procedures for certification whereby the audit of use of energy by industries would be conducted. The mission of BEE thus is to develop policies and strategies to promote self-regulation and market-based principles to achieve reduction of energy intensity of the Indian economy within the legal framework of the Act.[57]

The fourth chapter of the Act provides a list of powers and functions to be performed by the BEE. A select few have been hereby discussed[58]:

- a Recommendation of norms for standards of energy consumption for appliances and equipment
- b Recommendation of designated consumer (DC) based on intensity and quantity of energy used
- c Recommendation of particulars to be displayed on labels of such appliances and equipment and also the manner of its display
- d Prescribing guidelines for building codes for energy conservation
- e Arrange for training of personnel and specialists in the techniques as required to achieve the objective and also facilitate a strong consultative process in the area of energy conservation
- f Provide for measures for creation of awareness and dissemination of information for steps towards energy efficiency and energy conservation and also prepare educational curriculum for educational institutions

g Promote innovative financing of the projects working on energy-efficient models and processes or give financial assistance to institutions working on such fields
h Development of certification and procedure and facilities for testing of appliances for energy consumption
i Establishment of an Energy Conservation Fund for both Central and State governments
j Imposition of penalties on contravention of provisions of the Act
k Adjudication procedure for penalties and appeal to the High Court

The standards as set by the BEE and enforced by the government are done in the name of the Bureau, which creates a system of faith in the consumers who have inbuilt confidence in its name. Through a recent amendment, any deceptive use of name of BEE has been prohibited and the provision states that any usage of a name even nearly resembling the Bureau would require prior permission.[59] While the powers of the BEE are mainly recommendatory, the central government has the power to direct enforcement of the standards by establishing mandates, in consultation with the Bureau.[60]

As provided in the fifth chapter of the Act, the central government has the power to specify the standards for energy consumption for any equipment or DCs or even prohibit manufacture or import of any such appliance or equipment which does not conform with such standards. The central government can direct display of particulars on the label of such appliances to inform consumers of the compliance with energy efficiency standards. The central government, being the implementing authority, has to enforce the recommendations proposed by the Bureau. A few amendments to this part allowed for innovative ways for energy conservation (discussed in the next part).

The sixth chapter encompasses provisions regarding powers of the state government. The state government, although limited, has similar powers of direction in consultation with the BEE. It can also amend the codes for energy conservation and sustainable building to accommodate with the climatic conditions at regional and local level and specify it through rules and bye laws.[61] The state government can also designate a specified agency, which would coordinate and enforce the provisions of the Act in the State. Such an agency has to perform the function of preparing a budget for the next financial year as per the prescribed form along with estimated receipts and expenditure and forward to the state government to include in the annual budget of the state. The Agency will have the powers of inspection of the DC to ensure compliance of the energy consumption standards.[62]

The Act provides for establishment for funds at both levels – Centre[63] and State[64] – to be utilized towards realization of objective of the Act. At the State level, such funds are utilized to meet expenses by the designated agency to discharge the functions while at the central level, it is also utilized for salary, fee, and allowances of the members of the BEE. The grant of loan made to the BEE of the state government is out of the central fund.[65]

The eighth chapter provides for penalties for non-compliance with the standards and regulations prescribed under the Act.[66] While a person has to pay maximum 10 lakhs for such non-compliance, the Act provides for factors in which additional penalties will have to be paid. This makes the implementation of the law strict and in sync with other environment related legislations. The adjudicating officer will be appointed by the State from among the members of the state commission, which shall conduct the enquiry in accordance with the procedure prescribed by the central government.[67] The Act excludes the jurisdiction of the civil court[68] and also enlists the procedure and the factors[69] that the adjudicating officer would consider while conducting the adjudication. The ninth chapter provides for the process of appeal to the Appellate Tribunal and the Supreme Court.[70]

The last chapter of the Act contains 17 provisions under the heading of "Miscellaneous," which provides for the rule-making powers,[71] power to issue directions, power to exempt,[72] power to obtain information,[73] and so forth. It states that the central government has power to issue directions to the BEE with which the bureau would be bound. However, a proviso mentions that an opportunity to express views will be given to the Bureau before such directions.[74] In situations of grave emergency, defaults by the Bureau, or in public interest, if the circumstances so require, the central government shall supersede the Bureau after due publication for a maximum period of six months.[75]

This section discusses the present legal mechanism for energy efficiency and energy conservation. While the design and scheme of the Act seems comprehensive, it requires effective implementation to ensure that the object of the Act is achieved. The next section deals with the innovative provisions of the legislation by subsequent amendments. These special initiatives pose a great potential to achieve both national and international goals.

4.5 Special Initiatives for Energy Efficiency Under the Act

In the years 2010 and 2022, certain important amendments were made in chapters 4 and 5 of the Act, which allows the BEE and the central government to provide for innovative ways of energy conservation. In order to achieve the goals of enhancement of energy efficiency of the industrial sector, the Ministry of Power has undertaken several initiatives, now incorporated in the legislation. It aims at reduction of emphasis on usage of fossil fuels and coal thereby reducing emissions and also achieving energy security. This is in sync with India's climate goals decided in the Paris agreement.

PAT is one such flagship initiative, which is a market-based scheme under NMEEE. This scheme is a regulatory instrument designed with the objective to reduce specific energy consumption of certain energy intensive industries. It also aims at reducing the energy use per unit of production for the DCs in the energy intensive sectors. The PAT scheme is a multi-cycle process and each cycle lasts for three years. Each cycle, on a rolling basis, identifies a new set of DCs and target sectors. This section discusses the market-based mechanisms associated with PAT which aims specifically at achieving the goal of energy efficiency. The Act has been amended to incorporate these norms.

4.5.1 Energy Saving Certificates

Energy saving certificate (ESC) is an instrument issued by an authority or agency which guarantees that a stated amount of energy saving has been achieved under the PAT scheme. Such an ESC becomes a tradeable commodity giving property right over additional units of intangible bundle of benefits (societal and environmental) created by such savings.[76] ESC, being tradeable, provides a platform to trade its attributes earned through energy savings. Thus, ESC has the capacity to mobilize the market forces and capital incentivizing investment in energy conserving and efficient means.

The 2010 amendment empowered the central government to issue ESCs to industries which consumes energy lesser than the maximum allowable limit. Such certificates are tradeable and can also be sold to consumers whose consumption cross the maximum allowable limit.[77] The Act provides that the central government may issue such ESC to the DC whose consumption is lesser than the prescribed norms and standards.[78] Also the DC whose consumption is more than the prescribed standards can purchase such ESCs.[79]

This scheme is to incentivize cost effectiveness in industries which are energy intensive through certification of excess energy that is saved. It thereby attempts reduction of specific energy consumption in such industries.[80] The ESC is issued post verification of the baseline data, done through the energy audit, and when it is found that the industry has saved energy beyond its targets. After issuance, the DC can register as an eligible entity for trading at the Power Exchange Platform.

The PIB report states:

> As per the audit reports received by BEE, total energy saving was more than 14 million tonnes of oil equivalent (MTOE) which has also avoided 66 million tonnes CO_2 emission. This initiative has resulted in energy saving of Rs. 31,445 crores and industries have reported an investment of over Rs. 43,721 crores. In order to incentivise the exemplary performance, Ministry of Power issued Energy Saving Certificates (ESCerts) to those units who exceeded their targets.[81]

ESC has proved to be a key instrument for improving the overall energy efficiency. However, there are certain challenges such as lack of ex ante estimation of prices that can help computation of the savings potential. A prior estimate can help in knowing the price signals which would increase the likeliness and incentivize investment in energy-efficient mechanisms by such industries,[82] a minimum pricing by the government, and involvement in maintaining competitive prices of ESC. In order to achieve the maximum success through such schemes, there is a need to deepen the ambit of the PAT scheme by lowering the threshold level to identify more DC.

The PAT scheme needs to be widened in phases to be inclusive of more sectors than it initially began with. The price stability also needs to be maintained to reduce the uncertainties faced by the industry players.[83] Apart from these, the industries can be incentivized to invest in energy efficient technologies by giving tax concessions such as excise duties, customs, VAT, and so forth; by providing financial support through extension of Partial Risk Guarantee Fund; and through dedicated fund for research and development.[84] The scheme of ESC is applaudable and has been immensely beneficial in achieving the target of reduction of carbon foot print and reduced dependence on import of fuel.

4.5.2 *Carbon Credit Certificates*

As discussed in the previous section, in CoP-26, India made certain commitments to reduce its emission intensity leading to reduction of carbon footprint. An amendment of the Act was done in the year 2022 to achieve the CoP-26 goals for which a concept of carbon credit certificate (CCC) was introduced with which carbon credit earned by usage of non-fossil sources can be traded. Such a scheme can lead to faster decarbonization of the industry and help achieve the goals set forth at the international level. Such a certificate is issued under the carbon credit trading scheme (CCTS), which empowers the government to issue certificates of carbon credit.[85]

Carbon credit would imply a tradeable permit to certain registered entities to produce a specified amount of CO_2 or any other GHG. Similar to ESC, such a certificate can also be purchased by any other person on a voluntary basis. Through the CCTS scheme, the government also provides for a compliance mechanism, whereby the obligated entities must comply with the GHG emission intensity targets as notified by the central government.

The compliance mechanism under the Indian Carbon Market operates on the principle of regulating Greenhouse Gas Emission Intensity through a system of targets and tradable carbon credits. Each obligated entity is assigned a baseline emission intensity and a corresponding reduction target for a given compliance period. At the end of this period, the entity's actual emission intensity is measured against the prescribed target. Where an entity achieves an emission intensity lower than the target (i.e., over-compliance), it generates surplus reductions, for which it is issued CCCs.

These certificates represent quantified emission reductions beyond the mandated level and can either be traded in the carbon market or be banked for use in future compliance cycles. Conversely, if an entity fails to meet its target and records an emission intensity higher than the prescribed limit (i.e., under-compliance), it is required to offset the excess emissions by purchasing CCCs from the market. This creates a system of market-based flexibility, wherein entities with surplus reductions can sell credits to those facing deficits, thereby enabling cost-effective compliance across the sector. After trading, entities that still fall short of compliance are subject to penalties, while those with surplus credits may retain or bank them for future use. Overall, the mechanism incentivizes emission reductions by rewarding efficient entities and imposing economic costs on those that exceed their permissible emission levels, thereby aligning environmental objectives with market-based instruments.[86]

The obligated entity is notified of the mandatory emission intensity by the government for each year and the entities which overachieve their target are issued CCC which become tradeable unit. This CCC, if surplus, can be banked by the entities at the end of the compliance year or be traded with other participants. The CCC scheme under CCTS shows great potential to reduce the carbon emission benefitting not only the nation but also the entire world by reducing the carbon footprint.

4.5.3 Labelling of Appliance and Equipment

The BEE has the power to make regulations consisting of particulars that need to be displayed and manner of display of label of appliances or equipment.[87] As per the regulations, the central government may then direct such display on the appliances and in the manner prescribed.[88] The central government also has the power to specify the class of such equipment or appliance which can be a vehicle, a vessel, an industrial unit, or a building.[89] In exercise of these powers, the central government has notified certain "Star Labels" for certain equipment and appliances (briefly discussed here).

The Star Rating by BEE is a labelling process displayed on the appliances, which help consumers make informed choice of products that are energy efficient and also determine the cost-saving potential equipment. The appliances will be labelled in a star rating ranging from 1 to 5, with 5 being the most energy efficient. Such a label is now mandatory for appliances such as refrigerators, washing machine, air conditioners, and ceiling fans. The central government has notified regulations providing for mandates for specific appliances. A higher number of stars depict that the product would consume less energy and be more productive.[90] The S&L scheme, launched in 2006, covers around 34 equipment out of which 11 are mandatory as of now. The scheme aims to reduce the end-use energy consumption while not impacting or reducing the service levels of the appliance. This scheme definitely has the potential for cost-saving and conserving energy.

The star labelling system classifies appliances such as refrigerators based on their annual energy consumption (AEC), calculated using total volume (Vtot), and a prescribed formula. Each star band (1 to 5 stars) corresponds to a specific range of AEC, with lower energy consumption earning higher star ratings. For example, a 5-star refrigerator consumes less than (0.080 × Vtot + 124) kWh/year, indicating maximum efficiency, while 1-star models consume the most energy within the defined range. This labelling helps consumers compare energy efficiency and encourages manufacturers to design appliances with lower electricity consumption.[91]

The label must contain the logo of BEE, name of the manufacturer of the appliance, type, series code and model number, total volume and electricity consumption, star level, and label period. The notification also specifies the design, colour, size, and so forth of the label.[92] This scheme is innovative and definitely assists in making the correct choice and move closer to achieving the goal of energy efficiency. This section thus clearly shows the persistent effort made by India to not only fulfil its stated commitments but also go beyond it.

4.6 Critical Review and Conclusion

This chapter aimed at a critical analysis of the existing legal mechanism in India to achieve the goals of energy sustainability, efficiency, and conservation. The overall analysis reflects that India is committed to the goal of mitigation of carbon emission and reach its target of net zero by 2070. The Act helps achieve this goal by setting benchmark limits and enabling market mechanisms to boost trade on saving of energy and reducing carbon emission. The innovative schemes of ESC and CCS are a boon and touch the nerve of the industry as they incentivize investment on technology for energy efficiency.

The second section of the chapter conducted an analysis of the data, which shows the rising demand for energy with the population growth. The percentage of coal is the highest in the energy mix followed by crude oil. It has been found through statistical analysis that there is need to find cleaner, greener, and more affordable alternatives as both these energy sources are not sustainable for India in the long run. The carbon emission caused by reliance on these energy sources can be reduced only when the various sectors such as industry, transport, agriculture, and housing commit to this goal and choose cleaner means of production.

The import of crude oil is not only impacting the Indian economy but also impacting energy security as the geopolitical tensions are a mark of uncertainty and the heavy reliance of India will directly impact production. The second section thus concludes that this a serious and an urgent concern, which can be resolved only by strict mandates and a mechanism to ensure its enforcement. Energy concerns today leave no room for non-compliance as it impacts the national economy and global environment.

The third section of the chapter focussed on understanding factually the international commitments made by India and the steps taken towards them. It was found through the analysis of policy reports that India is vehemently working towards achieving the international target of SDG 7 to ensure efficiency across sectors. Additionally, it was found that India has already achieved its target of reduction of carbon emission and updated its share of NDC. This clearly shows the strong commitment of India and its work towards energy sustainability. This part also discussed the target of net zero emission by 2070. The steps taken with regard to policy and recent legislative amendments show that India is working towards this target and the day is not far that India would achieve its target.

The fourth and fifth sections of the chapter conduct an analysis of the legislation and recent amendments in order to understand the steps taken to secure the targets through norms and standards. The Act establishes the Bureau and empowers it with the power to recommend such standards of energy consumption which the various sectors need to abide by. The Act enlists the powers of the Bureau and the government along with penalties for non-compliance with the standards set by the BEE.

The fifth section specifically discusses three special initiatives brought about by recent amendments, which aim at incentivizing the industry by creating tradeable units of energy saved. The government issues certificates of energy saving when it saves more energy than its target and certificates of carbon credit when it reduces the GHG emission by choosing cleaner technology. Such certificates are market-based initiatives and tradeable at the power exchange. These schemes are extremely innovative and would definitely lead to achieving the targets, as discussed in the third section. Apart from this, the Act provides for the mandate of displaying labels on appliance and equipment, which aims at helping consumers in making an informed choice of energy efficient products.

The overall analysis suggests that India is moving in the right direction in achieving its target of net zero emission. The legislative amendments suggest that it is not only incentivizing industry to choose and invest on cleaner technology but also the consumers to make better choices to save energy at all levels. This effort is truly applaudable and it needs to be appreciated. However, there are certain limitations of the PAT scheme and also the initiative such as ESC which should be taken note of.

The PAT scheme sets targets for industries, which are not stringent and are achieved quite easily. The PAT scheme, as discussed, is a market-based mechanism and aims to incentivize the energy-intensive sectors by allotting tradeable certificates. However, with such easy targets, most sectors achieve them without much effort. This, in turn, leads to over-supply of ESCs, creating lesser demand and also lower prices at which these can be bought. There is no scarcity of ESC and it does not fulfil the purpose of incentivizing enough as there is low investment in energy-efficient technology.

The PAT compliance mechanism too needs to be stringent as there is a considerable delay in PAT cycles. While in the first PAT cycle, the ESC trading was completed in a year, all later cycles are yet to complete the process. There is delay in both issuing of ESC and trade post issuance. The implementation and governance system needs to be improved. This can be done by reducing the period of cycle and ensuring compliance. The legal mechanism needs to incorporate heavy penalties for non-compliance and under-achievers.

The legal mechanism must provide a minimum floor price so that the industries get the price signals along with higher targets so that there is greater push for investments in energy efficient technology. Also, the Act provides for constituting of conservation funds. The implementation mechanism needs to ensure that the states are provided with the necessary financial support to enable use of funds for energy efficient projects. The lack of funding is also due to lack of awareness and expertise in the banking sector about the profitability of such projects. Hence, concluding the critical analysis, it is submitted that the scheme is designed with a positive goal and can only achieve its purpose if these inter-related anomalies are resolved.

Notes

1 Luisa Marti and Rosa Puertas, "Sustainable Energy Development Analysis: Energy Trilemma", *Sustainable Technology and Entrepreneurship* 1, no. 1 (2022): 100007.
2 World Commission on Environment and Development, "Our Common Future", p. 41, available at: https://sustainabledevelopment.un.org/content/documents/5987our-common-future.pdf (Last visited on January 09, 2025).
3 Jeffrey D. Sachs, Guido Schmidt-Traub, Mariana Mazzucato, Dirk Messner, Nebojsa Nakicenovic and Johan Rockström, "Six Transformations to Achieve the Sustainable Development Goals", *Nature Sustainability* 2, no. 9 (2019): 805–814.
4 Sachs, Schmidt-Traub, Mazzucato, Messner, Nakicenovic and Rockström, "Six Transformations to Achieve the Sustainable Development Goals".
5 Puspanjali Behera, Litu Sethi and Narayan Sethi, "Balancing India's Energy Trilemma: Assessing the Role of Renewable Energy and Green Technology Innovation for Sustainable Development", *Energy* 308 (2024): 132842.
6 See United Nations, "Transforming Our World: The 2030 Agenda for Sustainable Development", A/RES/70/1, SDG 7, available at: https://www.un.org/en/development/desa/population/migration/generalassembly/docs/globalcompact/A_RES_70_1_E.pdf (Last visited on January 14, 2025).
7 Leonid M. Grigoryev and Dzhanneta D. Medzhidova, "Global Energy Trilemma", *Russian Journal of Economics* 6, no. 4 (2020): 437–462.
8 TFC is the energy consumed by end users such as individuals and businesses to heat and cool buildings; to run lights, devices, and appliances; and to power vehicles, machines, and factories.
9 See International Energy Agency, available at: https://www.iea.org/countries/india/efficiency-demand (Last visited on January 14, 2025).
10 International Energy Agency, above note 9.
11 Total energy supply (TES) includes all the energy produced in or imported to a country, minus that which is exported or stored.
12 Ministry of Coal, "Reducing Coal Imports, Boosting Exports", *Press Information Bureau*, 29 July 2024, available at: https://pib.gov.in/PressReleasePage.aspx?PRID=2038365 (Last visited on January 14, 2025).
13 Energy Institute, "Statistical Review of World Energy, 2024", available at: https://www.energyinst.org/statistical-review (Last visited on January 14, 2025).
14 MoPNG, "India's Energy Demand Will Continue to Provide Fuel for Future Economic Growth and Is Bound to Grow Exponentially in Coming Years", *Press Information Bureau*, 10 October 2023, available at: https://pib.gov.in/PressReleaseIframePage.aspx?PRID=1966227 (Last visited on January 14, 2025).
15 Petroleum Planning and Analysis Cell, "Ready Rockener Report", *PPAC*, available at: https://ppac.gov.in/#imp-report-sec (Last visited on January 14, 2025).
16 Planning and Cell, "Ready Rockener Report".
17 Planning and Cell, "Ready Rockener Report".

18 Planning and Cell, "Ready Rockener Report".
19 Sufang Li, Dalun Tu, Yan Zeng, Chenggang Gong and Di Yuan, "Does Geopolitical Risk Matter in Crude Oil and Stock Markets? Evidence from Disaggregated Data", *Energy Economics* 113 (2022): 106191.
20 Larry Elliot, "Middle East Conflict Risks Sharp Rise in Oil Prices, Says IMF", *The Guardian*, 16 April 2024, available at: https://www.theguardian.com/business/2024/apr/16/middle-east-conflict-risks-a-sharp-rise-in-oil-prices-says-imf (Last visited on January 14, 2025).
21 PRS, "Review of Policy on Import of Crude Oil: Standing Committee Report", available at: https://prsindia.org/policy/report-summaries/review-of-policy-on-import-of-crude-oil (Last visited on January 14, 2025).
22 Fossil fuels such as coal, oil; and natural gas for power generation or to fuel vehicles and machines contribute to such GHG emissions.
23 International Energy Agency, "Total CO2 Emissions from Energy", available at: https://www.iea.org/countries/india/emissions (Last visited on January 14, 2025).
24 International Energy Agency, "Total CO2 Emissions from Energy".
25 World Energy Council, "World Energy Scenarios: Composing Energy Futures to 2050", available at: https://www.worldenergy.org/assets/downloads/World-Energy-Scenarios_Composing-energy-futures-to-2050_Executive-summary.pdf (Last visited on January 14, 2025).
26 United Nations, "Transforming Our World", above note 6.
27 Fang Wang, Jean Damascene Harindintwali, Zhizhang Yuan, Min Wang, Faming Wang, Sheng Li, Zhigang Yin, et al., "Technologies and Perspectives for Achieving Carbon Neutrality", *The Innovation* 2, no. 4 (2021).
28 Krishna Kumar Jaiswal, Chandrama Roy Chowdhury, Deepti Yadav, Ravikant Verma, Swapnamoy Dutta, Km Smriti Jaiswal and Karthik Selva Kumar Karuppasamy, "Renewable and Sustainable Clean Energy Development and Impact on Social, Economic, and Environmental Health", *Energy Nexus* 7 (2022): 100118.
29 United Nations, "Transforming Our World", above note 6.
30 United Nations, "Transforming Our World", above note 6.
31 United Nations, "Transforming Our World", above note 6.
32 UNEP, "17 Goals to Transform Our World: SDG7 issue Brief", available at: Home – United Nations Sustainable Development (accessed on 14 January 2025).
33 UNGA, ECOSOC, "Progress Towards the Sustainable Development Goals Report 2024", 2 May 2024 [A/79/79E/2024/54], available at: https://unstats.un.org/sdgs/files/report/2024/SG-SDG-Progress-Report-2024-advanced-unedited-version.pdf (Last visited on January 14, 2025).
34 Bureau of Energy Efficiency, "National Mission for Enhanced Energy Efficiency", available at: https://beeindia.gov.in/en/programmes/nmeee (accessed on 14 January 2025).
35 MNRE, GoI, "SDG7 Energy Compact Report", available at: https://www.un.org/sites/un2.un.org/files/2021/09/energy_compacts-_india-_22nd_september_2021.pdf (Last visited on January 14, 2025).
36 MNRE, GoI, "SDG7 Energy Compact Report".
37 Ministry of Power, GoI, "Government Rewarded Energy Efficiency Industries through Energy Saving Certificates", *Press Information Bureau*, Posted On: 19 Aug. 2021 5:32PM by PIB Delhi, available at: https://pib.gov.in/PressReleasePage.aspx?PRID=1747407 (Last visited on January 14, 2025).
38 Ministry of Power, GoI, "Government Rewarded Energy Efficiency."
39 Ministry of Finance, GoI, "Despite Being One of the Fastest-Growing Economies in the World, India's Annual Per Capita Carbon Emission Is Only About One-Third of the Global Average India Is the Only G20 Nation in Line with 2-Degree Centigrade Warming: IFC Report India Achieves Its NDC Target of Reducing Emission Intensity 11 Year in Advance Before the Target Year 2030

India Successfully Decouples Its Economic Growth from Greenhouse Gas Emissions Mission Life: Mass Movement to Address Climate Change and Foster Sustainable Living India Leading Several International Initiatives Towards Climate Change Mitigation and Building Resilience", Posted On: 22 July 2024 2:21PM by PIB Delhi, available at: https://pib.gov.in/PressReleasePage.aspx?PRID=2034915 (Last visited on January 14, 2025).

40 The Constitution of India, Seventh Schedule, List I, Entry 6.
41 The Constitution of India, Seventh Schedule, List I, Entry 54, Constitution of India.
42 The Constitution of India, Seventh Schedule, List II, Entry 23.
43 The Energy Conservation Act, 2001 (52 of 2001).
44 C. M. Jariwala, "The Indian Energy Conservation Law: A Critical Overview", Usha Tandon (ed.), *Energy Law and Policy* (OUP, 2018), 242–269.
45 The Energy Conservation Act, 2001, Section 2.
46 The Energy Conservation Act, 2001, Section 12.
47 The Energy Conservation Act, 2001, Section 3.
48 The Energy Conservation Act, 2001, Section 3.
49 The Energy Conservation Act, 2001, Section 4.
50 The Energy Conservation Act, 2001, Section 4(2).
51 The Energy Conservation Act, 2001, Section 9.
52 The Energy Conservation Act, 2001, Section 4.
53 The Energy Conservation Act, 2001, Section 5.
54 The Energy Conservation Act, 2001, Section 8.
55 The Energy Conservation Act, 2001, Section 8.
56 The Energy Conservation Act, 2001, Section 13(1).
57 Ministry of Power, GoI, "Government of India Developed Robust Mechanism for Online Trading of Energy Saving Certificates Government Rewarded Energy Efficiency Industries Through Energy Saving Certificates", Posted On: 19 Aug. 2021 5:32PM by PIB Delhi", available at: https://pib.gov.in/PressReleasePage.aspx?PRID=1747407 (accessed on 14 January 2025).
58 Jariwala, "The Indian Energy Conservation Law", Section 13(2).
59 Jariwala, "The Indian Energy Conservation Law", Section 13A.
60 See generally, chapter V of the Energy Conservation Act, 2001.
61 The Energy Conservation Act, Section 15.
62 The Energy Conservation Act, Section 17.
63 The Energy Conservation Act, Section 20.
64 The Energy Conservation Act, Section 16.
65 The Energy Conservation Act, Section 19.
66 The Energy Conservation Act, Section 26.
67 The Energy Conservation Act, Section 27.
68 The Energy Conservation Act, Section 29.
69 The Energy Conservation Act, Section 28.
70 The Energy Conservation Act, Sections 31, 31A, 44, 45.
71 The Energy Conservation Act, Sections 56 & 57.
72 The Energy Conservation Act, Section 53.
73 The Energy Conservation Act, Section 52.
74 The Energy Conservation Act, Section 46.
75 The Energy Conservation Act, Section 47.
76 Tanushree Bhattacharya and Richa Kapoor, "Energy Saving Instrument–ESCerts in India", *Renewable and Sustainable Energy Reviews* 16, no. 2 (2012): 1311–1316.
77 Edward Vine and Jan Hamrin, "Energy Savings Certificates: A Market-Based Tool for Reducing Greenhouse Gas Emissions", *Energy Policy* 36, no. 1 (2008): 467–476.
78 Jariwala, "The Indian Energy Conservation Law", at Section 14A.

79 Jariwala, "The Indian Energy Conservation Law", at Section 14A.
80 SEC is the energy used per unit of production in energy intensive large industries.
81 Ministry of Power, GoI, "Government Rewarded Energy Efficiency Industries through Energy Saving Certificates", *Press Information Bureau*, available at: https://pib.gov.in/PressReleasePage.aspx?PRID=1747407 (Last visited on January 14, 2025).
82 Vine and Hamrin, "Energy Savings Certificates". *Supra* note 77
83 Vine and Hamrin, "Energy Savings Certificates".
84 Vine and Hamrin, "Energy Savings Certificates".
85 Jariwala, "The Indian Energy Conservation Law", Section 14(w) and Section 14AA.
86 MoEFCC, GoI, "Detailed Procedure for Compliance Mechanism under the Indian Carbon Market", available at: https://beeindia.gov.in/sites/default/files/Detailed%20Procedure%20for%20Compliance%20Procedure%20under%20CCTS.pdf (Last visited on January 14, 2025).
87 Jariwala, "The Indian Energy Conservation Law", Section 58(i).
88 Jariwala, "The Indian Energy Conservation Law", Section 14(d).
89 Jariwala, "The Indian Energy Conservation Law", Section 14(b).
90 Bureau of Energy Efficiency, "Star Label", available at: https://beeindia.gov.in/en/star-label (Last visited on January 14, 2025).
91 Ministry of Power, GoI, Notification S.O. 4554(E) dated 26 September 2022, available at: https://beeindia.gov.in/sites/default/files/REF_Notification_0.pdf (Last visited on January 14, 2025).
92 Ministry of Power, GoI, Notification S.O. 4554(E) dated 26 September 2022.

Part III

Tribal Knowledge

5 Indigenous Ecological Knowledge and the Law – Uncovering Its Modernity

Philippe Cullet

5.1 Introduction

The model of development followed for the past two centuries is in large part the cause of the triple planetary crisis of climate change, biodiversity loss, and pollution.[1] Development was, and remains to a large extent, centred on science and technology as the drivers of economic growth, which is underlying human progress.

The increasingly negative environmental impacts of development were progressively recognised but this did not lead to questioning the model of development itself. Rather, the concept of sustainable development introduced in the 1980s,[2] around which environmental policy has been conceived for several decades worldwide, has been framed around the idea that a correction course was necessary but within the same overall framing.

Several decades later, the world is a much worse position and facing a fast worsening environmental polycrisis. Yet there has been no real questioning of the role of economic growth in making this worse. Further, the belief remains that it is more new technology that can address the consequences of the crisis caused.[3] In effect, there has been no real questioning of the fact that decades of "sustainable development" have not brought about sustainability.

At this juncture, it is increasingly apparent that it is this model of endless growth that must be questioned.[4] This requires new conceptual and practical tools going beyond what has already been experimented with for years. It is in the context of the search for new thinking to revive environmental policy-making that the wisdom, practices, and rules of indigenous societies have been given increasing attention. It has nearly become common place to argue that indigenous communities have sustainably managed their environments, sometimes for millennia, and that this should constitute a template for addressing the environmental polycrisis.[5]

The specific focus on indigenous ecological knowledge is linked to the fact that indigenous communities have in a number of cases maintained closer links with their environment than more mainstream communities. The understanding is thus that they have a much better understanding of nature and the place of human societies within nature than other human communities.

DOI: 10.4324/9781003733751-8

This chapter starts by examining the limitations of sustainable development in terms of its capacity to offer effective solutions to the environmental polycrisis. It then moves on to examine two main ways in which indigenous ecological knowledge has been framed in national and international policy-making. The first is through so-called "traditional knowledge," a notion developed both in the context of environmental conservation and in the context of intellectual property. This mainstream engagement with indigenous ecological knowledge is valuable for having contributed to give more visibility to the contribution that indigenous societies can make to "development," but traditional knowledge conceives indigenous ecological knowledge largely as a tool from the past that can be relevant in the present but something that remains opposed to the modern innovations that are supposed to address the environmental polycrisis. The second framing of indigenous ecological knowledge is in the context of the development of "rights of nature," an evolving ecocentric framing, which is in large part linked to the ethos, norms, and practices of indigenous communities. This chapter argues that the modernity of rights of nature reflects the modernity of indigenous ecological knowledge, which has solutions to offer where the mainstream (sustainable) development has failed. In other words, indigenous ecological knowledge needs to be considered and integrated because it is modern and not because it is traditional. It is its modernity that makes it relevant conceptually and practically to address the environmental polycrisis.

5.2 From Sustainable Development to Traditional Ecological Knowledge

Sustainable development was introduced specifically to address the shortcomings of the model of development focused mostly on economic growth that had prevailed for the previous decades. Development was structured around the idea that there were "developed" and "developing" countries, the latter being supposed to catch up with the former through economic growth.[6] In this context, one of the main points of focus of development policy was poverty alleviation, something that remains true in relation to sustainable development today.

In the first place, the rationale behind sustainable development was that there could be an "integration" of development and environmental protection.[7] This was supposed to indicate that there could be a way forward whereby the promise of modernity linked to development could be achieved without destroying the environment. This was premised on an understanding of development driven by innovations that would lead to the overall progress of human societies.[8] The idea of modernity that the notion of development has contributed to propagate is one that is based on economic growth as the driver of innovation. As a result, the "new" is better than the "old" and the new is largely understood as reflecting sustainability.

The modernity of sustainable development is largely linked to its reliance on science and technology as providing the basis on which sustainability will be built. On the one hand, science and technology are seen as the central tool to minimise environmental damage caused by the process of development. On the other hand, new "environmental sound" technologies are understood as tools to set off a process of development that is less inimical to the environment.

The reliance on natural sciences as the arbiter of necessary protection measures has proved to be both a boon and a bane.[9] On the one hand, scientific data understood as apolitical has been the central pillar justifying environmental protection measures since the 1970s, providing the basis on which policymakers have acted. As long as policy-makers were willing to take heed of the signals given by natural sciences, this allowed for the relatively quick development of environmental protection measures in the last decades of the twentieth century. Yet this reliance on natural sciences has also proved to be a limitation to the expansion and effectiveness of environmental law. On the one hand, natural sciences are not the only effective arbiter of what policy measures are needed to address environmental degradation, and the social and human dimensions of environmental protection need to be given much more emphasis.[10] On the other hand, the diminishing willingness of policy-makers to act in the face of scientific evidence indicates that new bases to foster environmental protection are needed.

With regard to the belief that new technology would help solve the environmental harm caused by economic development, this has also proved to be something that cannot be accepted without further contextualisation. This is illustrated in the context of the climate crisis. At this juncture, there is a growing understanding that the world needs to move towards a net-zero greenhouse gas emission economy.[11] Yet, instead of leading to a rethinking of economic growth as the engine of well-being, this is leading to the introduction of new flawed solutions, such as carbon storage facilities in which billions of dollars have already been invested.[12] Such projects reflect an understanding that there is no need to address the problem causing the harm but that it is enough to "hide" the harm caused so that, in this case, further carbon emissions can be allowed into the atmosphere.[13]

This links back to the inability of sustainable development to shake off the focus on economic growth, which was the marker of "development" earlier. In fact, economic growth has taken centre stage on repeated occasions, such as after the late 2000s and after the pandemic-related economic crises. In other words, the very first rationale for the introduction of "sustainable development," which was to move away from the emphasis on economic growth, has not succeeded.[14] Governments around the world have repeatedly gone back to giving the economy priority over environmental considerations,[15] despite the fact that it is economic growth that has been at the root of the crises the world experiences. This was, for instance, reflected in the attempt to introduce the concept of green economy in the context of the Rio+20 summit (highlighting an attempt to give economic growth more importance).[16]

Overall, reliance on sustainable development for nearly four decades has failed to lead to the eradication of (extreme) poverty or ensure positive environmental outcomes. It is today unclear that even the narrow goal of eradicating extreme poverty reflected in SDG 1 will be achieved by 2030.[17] With regard to environmental outcomes, indicators for the triple planetary crisis are worsening rather than improving, as reflected, for instance, in the breaching of an increasing number of planetary boundaries.[18]

One of the reasons for the failure of the policy paradigm framed around "sustainable" development is that it never moved away from the ideology of "development."[19] This is identifiable, for instance, by considering the fact that two of the three crises highlighted as the main issues that need to be addressed today are the climate change and biodiversity loss crises. These happen to be the subject of two of the most important treaty regimes adopted during the moment of highest hope for sustainable development as a framing notion for environmental law at the Rio 1992 summit.[20] These two regimes have also been given most attention by policy-makers over the past few decades. The overall lack of success in tackling the climate crisis and the biodiversity crisis is thus in part an indictment of the policy framework adopted to tackle these crises.

In this context, it is unsurprising that the modernity implied in mainstream environmental policy is increasingly challenged. On the one hand, there is a growing challenge to the dogma that economic growth based on technological innovation is the only way forward and out of the environmental polycrisis. Various alternative framings have been debated, such as degrowth, *buen vivir*, and ecological democracy.[21] These different debates have in common to highlight the fundamental lack of coherence of existing environmental policies, which fail to give the environment a clear priority in policy-making and implementation. On the other hand, there has been an increasing turn towards a new understanding of the environment that starts from communities who live in close proximity to their environment, who depend on it for their well-being and livelihoods, and for whom the environment has direct cultural or religious links.[22] Indigenous ecological knowledge finds itself at the confluence of these debates and offers one of the potential new anchors to a different modernity to address the challenges of the environmental polycrisis.

5.3 Traditional Knowledge – A Largely Non-Modern Framing of Indigenous Knowledge

Traditional knowledge has become a mainstream policy concept over the past three decades. It refers to knowledge which is traditional because it is not modern. This basic distinction is linked in part to a dichotomy between static and evolving knowledge.[23] The gist of developments since the last decade of the 20th century is that the knowledge of indigenous people has been increasingly deemed to be a subject of mainstream interest, one of the triggers being the rapid development of biotechnology. At the same time, this interest has not translated in either effective positive protection of the knowledge of indigenous people or its integration in mainstream systems of knowledge. Rather, policy developments over the past few decades reflect a continuing relegation of traditional knowledge to something that is seen as "backward" rather than "modern,"[24] and "a valuable resource for technological knowledge" but not novel in itself.[25] This is reflected in the legal framings around traditional knowledge.

5.3.1 Conservation and Protection of Traditional Knowledge – Policy Developments

The discourse around traditional knowledge has had two distinct, though related, framings at the international level. In both cases, the prevailing understanding of traditional knowledge has been of a system that builds upon the past to provide answers up to the present but not of a system that is forward-looking and able to offer what would count as innovative solutions to the crises of the day.

The first framing of traditional knowledge emerged from an article of the Biodiversity Convention dealing with the conservation of biodiversity. Article 8j became the basis of an extensive series of developments leading to the establishment of a permanent subsidiary body in 2024.[26] No binding legal instrument has emerged from these efforts but a series of guidelines and tools have been adopted, such as a Code of Ethical Conduct, to ensure respect for the cultural and intellectual heritage of indigenous and local communities, and guidelines related to prior and informed consent.[27] These instruments provide the foundation for strengthening the position of indigenous peoples and local communities but are framed largely as negative protection.

Some efforts were made to go beyond this and consider, for instance, the possibility of recognising traditional knowledge as "collective property."[28] This could have provided the basis for a more balanced framing around different types of property rights. Yet, by conceptually distinguishing individual and collective property, it was never likely that these different types of property rights would be given the same level of recognition. Further, efforts to develop *sui generis* protection did not go beyond the framing of "elements" for such protection.[29] Further developments have taken place in the past decade, such as in the context of the recognition of peasant rights. The UN General Assembly peasant rights' resolution specifically provide that they have the right to control, protect, and develop their traditional and local knowledge, which includes methods of production or technology.[30]

The second policy framing of traditional knowledge emerged in the context of intellectual property, in part in reaction to concerns about biopiracy, an issue which became particularly controversial with the rapid development of biotechnology and the use of diverse forms of traditional knowledge.[31] In

India, this became particularly controversial following the case of a patent granted on the healing properties of turmeric in the United States on knowledge that was codified in India.[32] This led to a surge of activity to ensure that existing knowledge could not be appropriated by others. This is what led, for instance, to the setting up of the Traditional Knowledge Digital Library, which has been used to prevent more than 300 patent applications on existing knowledge.[33]

Policy-makers reacted to concerns about biopiracy by starting a policy process around genetic resources and associated knowledge in the broader context of the intellectual property rights system. This led to more than two decades of work in the context of the World Intellectual Property Organization (WIPO) culminating in the adoption of a treaty in 2024.[34] The adoption of a legal instrument related to traditional knowledge is a landmark development but it is limited with respect to recognising the contribution of traditional knowledge to innovation. In essence, the treaty is built around the idea that traditional knowledge is a tool to be used for "modern" innovations. In practice, the treaty is essentially focused on improving transparency.[35] The objectives of these measures are not to foster the positive protection of traditional knowledge but rather to enhance the legitimacy of the patents system. This is reflected in the objectives of the treaty, which emphasise the need to enhance "the efficacy, transparency and quality of the patent system" and to ensure that the novelty claimed is not in fact knowledge already in existence.[36]

There are some other legal instruments seeking to protect traditional knowledge. These include the Protocol of the African Regional Intellectual Property Organization on traditional knowledge, which should be more progressive, since it emerged from within countries that are expected to value their own traditional knowledge.[37] Its objective reflects a closer engagement with traditional knowledge holders and their interests, but this only goes as far as providing for protection against infringement.[38] The difference with the WIPO treaty lies in the extensive provisions to make this negative protection regime something that effectively addresses the problems associated with a system where traditional knowledge was essentially free for all to use.

More broadly, the development of "access and benefit sharing" (ABS) in the context of the Biodiversity Convention and the International Treaty on Plant Genetic Resources for Food and Agriculture attests to the understanding of policy-makers that the unrecognised use of genetic resources and the knowledge related to their use had become unacceptable.[39] As a result, the main policy development since the 1990s has been the development of safeguards that are meant to increase the transparency of inter-state transactions and ensure that some benefits flow back to the original holders of the knowledge. At the same time, the one thing that ABS does not do is offer traditional knowledge holders rights that are similar to the intellectual property rights granted, for instance, to patent holders for their "modern" innovations.[40] On the side of "access," the system is constructed around a single point of contact where the value of the knowledge must be assessed a priori. This can be opposed to a royalty system where a rent can be extracted at regular intervals and where the decision on the value of the knowledge licensed is determined at a later point. On the side of benefit sharing, what is shared in practice is in the overwhelming majority of cases a form of monetary compensation. There exist some exceptions, like the provision of the Biodiversity Act, 2002, which offers the "grant of joint ownership of intellectual property rights" as one of the options for benefit sharing.[41] Yet such provisions have largely remained theoretical options that are not used in the practice of ABS. As a result, ABS remains conceptually lopsided since it does not put the holders of "traditional" and "modern" knowledge on a level playing field.

5.3.2 Traditional Knowledge as Un-Modern

The basic shortcoming of policy developments highlighted earlier is that they build squarely on the understanding of "traditional" being opposed to "modern," the latter being what provides solutions to today's problems. The general understanding of traditional knowledge is thus of knowledge which is relevant to the modern world to the extent that it provided solutions in the past. In other words, it largely reflects an understanding that the past should not be ignored in building the future, while maintaining a division between the old and the new, the "traditional" and the "scientific."

One of the reasons for the limited progress regarding the positive protection of traditional knowledge is that such protection is often seen as going against the tenets of scientific research and technological innovation. In legal terms, this is enshrined in the patents system whose premise is that knowledge should be free for use, as an incentive to spur research and innovation. Intellectual property rights are seen as an exception that is justified by the need to reward inventors for the time and expense engaged in providing the "innovation." This exceptional status is enshrined in the limited time during which patent holders benefit from the property rights system set up through the patent system.[42]

A system of innovation that is premised on free flows of knowledge is structured against the recognition of other forms of appropriation, such as by traditional knowledge holders. Here lies the irony of the current system. On the one hand, knowledge sharing has been at the root of innovation in many contexts throughout human history. On the other hand, the mainstream legal framework only rewards certain specific innovations, which can generally be identified as being based on "modern" science and technology. This is what leads to the exclusion of traditional knowledge from the scheme of protection. In this sense, the legitimacy of the current system depends on this clear distinction between "modern" and "traditional" knowledge, which provides the basis for excluding the latter from the more evolved forms of protection. As a result, traditional knowledge has mostly been protected only through "negative" protection framings. In addition, attempts to use mainstream intellectual property protection for the benefit of traditional knowledge holders constitute at best a stopgap measure. This is because intellectual property protection is premised on the commercial exploitation of knowledge, sidelining other reasons for protection.

Traditional knowledge often holds the key to addressing some of the biggest challenges that humanity faces today, which mainstream framings have failed to effectively tackle. In response, debates could have moved towards a recognition of the value of the systems of knowledge covered by the term "traditional knowledge" as forward looking and put on a par with "modern" or "scientific" knowledge. Yet the recent WIPO treaty indicates that there has been little progress towards recasting "traditional" in a more positive form. Yet, policymakers still understand "traditional" as being less than something known as "modern", and science and technology remain the yardstick by which progress is measured.

This is in a sense expected because traditional knowledge is often associated with poverty. Poverty itself is understood as the lack of development. Put differently, development is what provides the means to reduce poverty and this is achieved through "modern" science and technology. In this context, the knowledge of poor people, known as traditional knowledge, can logically only be understood as something of a lesser quantity. It is thus unsurprising that mainstream legal instruments have on the whole failed to provide a more positive framing of the knowledge, techniques, and technologies of people living in close contact with their environment, deriving their sustenance and livelihoods from the same and having an understanding of their surrounding nature building on decades or centuries of proximity.

Overall, traditional knowledge in its mainstream understanding remains the knowledge of poor or marginalised people. Since the development of biotechnology, the use of such knowledge has come under the purview of life patenting, in which context the need to formalise the distinction between "modern" and "old" took a new urgency. In the process, while "traditional knowledge" became a more important subject of debate, the knowledge of indigenous peoples and local communities remains devalued in the mainstream framing, which offers a system of hierarchical rewards to different types of innovations centred around the intellectual property rights framework.

5.4 Rights of Nature – Finding Modernity in Indigenous Ecological Knowledge

The framing of indigenous knowledge as traditional knowledge has led to its increasing visibility but has contributed in only limited ways to enhancing the rights that its holders can exercise. In addition, such knowledge remains largely understood as knowledge that is relevant to help modern science and technology provide the kinds of innovations that will allow the world to progress, or in mainstream jargon, to develop.

The outrage over biopiracy that marked the 1990s progressively subsided thanks to the policy initiatives around traditional knowledge and to the development of legal instruments guiding the process of ABS. Yet, while attempts to stymie biopiracy have been undertaken, "modern" innovations and mainstream environmental policy have not only failed to stem the growing polycrisis but also allowed it to significantly worsen over the past few decades.

5.4.1 *Indigenous Ecological Knowledge, Norms, and Governance to Ensure "Harmony With Nature"*

The increasingly widespread disenchantment with sustainable development has led to the search for alternatives to address the worsening environmental crises. This has led to an increasing interest for the knowledge, practices, and norms of local communities and indigenous peoples. Indigenous ecological knowledge has been particularly in focus because indigenous people are understood to have protected and managed their lands, environment, and natural resources without leading to ecological collapse, often for centuries. In fact, it has been increasingly recognised that indigenous ecological knowledge is at the roots of the sustainability that these peoples have managed over centuries.

The crucial point is not only that indigenous ecological knowledge has proved to be a source of sustainability but also that it is an evolving body of knowledge that has adapted to an ever-changing environment. There is thus in-built modernity in indigenous ecological knowledge. This is, however, a different modernity from the one structured around the knowledge produced by natural sciences and validated by experts. It is also knowledge that is not necessarily dissociated from the human context, within which it evolves and is used. In other words, there is no necessary strict division between the technical knowledge and the social or political dimensions of its application in practice. In a context where environmental policy has been built for more than half a century on the premise of apolitical scientific data driving environmental policy,[43] this provides a distinct framing.

The reason for moving to a different framing based on the individual and collective knowledge of indigenous societies is illustrated by the case of the climate crisis. The limited measures adopted until now in the context of the UN Framework Convention on Climate Change and derived instruments have been based on the scientific consensus emerging from the reports of the Intergovernmental Panel on Climate Change. Despite the increasingly certain science, the margin of uncertainty remaining with regard to many long-term impacts has been used by policy-makers as an indirect reason to delay action, as reconfirmed by the decision of the United States to withdraw for a second time from the Paris Agreement.[44] The lack of willingness to take the necessary action is not due to the absence of necessary regulatory tools. In fact, mainstream environmental law has the answer to scientific uncertainty since the precautionary principle offers the necessary tool to take regulatory measures in such cases.[45] In the case of the climate crisis, existing scientific evidence is more than sufficient to trigger the application of the precautionary principle, one of the central principles in the climate change regime.[46] Yet even the insufficient objective of the Paris Agreement is not being met through nationally determined contributions.[47]

Reliance on natural sciences as a basis for environmental protection measures has led to sidelining social and cultural aspects of environmental protection.[48] One of the reasons for the distinction made has been in terms of an opposition between the scientific and the non-scientific. This is how indigenous ecological knowledge has often been relegated to a lesser kind of knowledge because it is not understood as having scientific validity. Another related reason has been the distinction between rational measures based on scientific data and other framings.

The characterisation of indigenous ecological knowledge as something of a lesser variety has been increasingly challenged, including at the level of international policy-making. The Harmony with Nature framework in the context of the UN constitutes one way in which the knowledge, experience, norms, and practices of indigenous peoples has been given much more visibility.[49] This is based on the understanding that what was for long dismissed as "old" and "unscientific" may in fact hold the keys to the elusive sustainability that sustainable development has not delivered over the past few decades.

Indigenous ecological knowledge is largely based on an understanding of nature based on proximity. It is thus opposed in some ways to what has become a globalised understanding of the environment. The fact that environmental protection measures are largely similar in most parts of world is a case in point. While this provides for easier coordination of such measures in terms of their transboundary nature, this tends to obliterate the great variety of conditions and responses given at the local level to environmental problems. In this way, indigenous ecological knowledge is a reminder that environmental policy should give much more weight to experiences from the ground.

Overall, the central contribution of indigenous ecological knowledge is to provide the basis for a new environmental policy that gives nature priority over human uses. At the same time, it is not a conservationist ecocentric framing that considers human separately from nature. Rather, it is based on the understanding of humans being part of nature, dependent on it and contributing to its upkeep.

5.4.2 *Rights of Nature as an Embodiment of the Modernity of Indigenous Ecological Knowledge*

Over the past couple of decades, the disenchantment with sustainable development has coalesced in legal terms in various initiatives generally framed under the umbrella of rights of nature. This new framing is often understood as giving environmental policy an ecocentric basis in contrast with the anthropocentric perspective that has dominated environmental policy and law. In doing so, rights of nature give priority to nature over developmental considerations, reversing the de facto priority given to economic growth under sustainable development,[50] and the technological scientism that conceives nature as something that can be modified at will through the application of science and technology.[51]

Rights of nature emphasise the utmost importance of conservation. At the same time, in a number of cases, the inspiration for their development has been indigenous ecological knowledge and the ethics of conservation that come with it. The link between rights of nature and indigenous peoples is crucial because it reflects an understanding of nature being "formed and vitalized through human interaction."[52]

In the understanding of rights of nature being linked with people's understanding and management of nature, one of the novel aspects is to provide a basis for rethinking environmental policy away from expert-led state-based law towards law emerging from communities' own governance frameworks.[53] This environmental governance embedded in indigenous ecological knowledge is sometimes also known as biocultural rights, a framing guided by the idea the rights recognised promote the interests of both nature and indigenous peoples.[54] This reflects to an extent the kind of autonomy given to autonomous district councils in areas falling under the Sixth Schedule of the Constitution of India. An example of this would be the land law in Meghalaya framed in such a way that it cannot be transferred to non-indigenous, save under specific exceptions.[55]

The modernity of rights of nature resides in the fact that they provide the basis for considering indigenous ecological knowledge and indigenous environmental management as a form of regulation that is no less than "mainstream" environmental regulation. This is interesting not only in relation to the framing of norms and the priority given to nature but also in relation to the implementation and enforcement of norms. One of the criticisms of environmental law has been that it fails not so much in its conceptualisation but in its enforcement.[56] The latter has been well documented but this only provides one part of the answer. The insufficient enforcement of environmental norms is not due to only insufficient capacity, for instance, of the pollution control boards or unwillingness of some actors to enforce standards. It is also due to the framing of the norms as something imposed on polluters rather than something which is framed around the needs and interests of all actors.

The idea of rights of nature or biocultural rights overall builds on existing categories already in place to address environmental harm. The central contribution is to provide a new lens through which to enter the debates, which is framed squarely around nature as the central element within which human societies live. This is also forward looking because the integration model has failed. Further, the belief in economic growth based on technological innovation has not only failed to stem the triple planetary crisis but has contributed to its rapid worsening.

In other words, the modernity of indigenous ecological knowledge and its legal avatars, such as rights of nature or similar framings putting nature at the apex of policy-making, does not depend on being more advanced scientifically or technologically. Rather, it depends on an understanding that human societies must not only live in harmony with nature but more profoundly live within the confines of the resilience of the ecosystems they depend on.

5.5 Conclusion

Indigenous ecological knowledge has come a long way from being considered as something potentially exotic and relevant but overall "old" and only marginally relevant to the bigger challenges that the world is facing today. The sidelining of indigenous ecological knowledge was linked to the fact that development has been exploitation of natural resources and nature. In this perspective, nature is there primarily to be used to foster development.

The belief that "modern" technology is the only real tool we have to address the current environmental crises remains strong, in part because technological development has transformed the world over the past centuries in ways that have made life much better for a limited proportion of the world's population. One of the reasons why the world now needs to look beyond this framing is that the planet cannot sustain more "development" in the way it has unfolded over the past two centuries.

In this context, looking towards indigenous ecological knowledge and its legal framings, such as rights of nature, is not looking backward but looking forward towards a different kind of engagement with nature and the ways in which it can sustain human societies. The lessons from indigenous ecological knowledge in general is that it has always been responsive to evolving natural and social conditions. This is in fact what explains the resilience of indigenous societies over centuries.

Today's challenges in the context of the accelerating planetary crises may be beyond what indigenous ecological knowledge can address on its own. Relying on its wisdom does not imply rejecting the benefits that mainstream modernity can bring. In fact, there are only few indigenous societies that live in complete autarchy. Others have already embraced certain aspects of mainstream development as part of their own process of adaptation to environmental and social change.

Notes

1 For example, Ministerial Declaration of the United Nations Environment Assembly at Its Sixth Session, UN Doc. UNEP/EA.6/HLS.1 (2024).
2 World Commission on Environment and Development, *Our Common Future* (Oxford University Press, 1987).
3 Sustainable Development Goals and Targets, in United Nations General Assembly Resolution 70/1, Transforming Our World: The 2030 Agenda for Sustainable Development, UN Doc. A/RES/70/1 (2015) repeatedly highlight the importance of technology.

4 For example, Judith Bueno de Mesquita, 'Reinterpreting Human Rights in the Climate Crisis: Moving Beyond Economic Growth and (Un)Sustainable Development to a Future with Degrowth', 42/1 *Netherlands Quarterly of Human Rights* 90–115 (2024).
5 Daniel Shilling, 'Introduction: The Soul of Sustainability', *in* Melissa K. Nelson & Daniel Shilling (eds), *Traditional Ecological Knowledge – Learning from Indigenous Practices for Environmental Sustainability* 3–14 (Cambridge University Press, 2018).
6 Compare Deepak Nayyar, *Catch Up – Developing Countries in the World Economy* (Oxford University Press, 2013).
7 Rio Declaration on Environment and Development, 14 June 1992, UN Doc A/CONF.151/26/Rev. 1 (Vol. I), Annex II (1992).
8 Rio Declaration on Environment and Development, princ. 9.
9 For example, Sam Johnston, 'The Role of Science', *in* Lavanya Rajamani & Jacqueline Peel (eds), *Oxford Handbook of International Environmental Law* 249–266 (Oxford University Press, 2021).
10 For example, Sheila Jasanoff, 'A World of Experts: Science and Global Environmental Constitutionalism', 40 *Boston College Environmental Affairs Law Review* 439 (2013).
11 For example, James Meadowcroft and Daniel Rosenbloom, 'Governing the Net-Zero Transition: Strategy, Policy, and Politics', 120/47 PNAS e220772712 (2023) <https://doi.org/10.1073/pnas.2207727120>.
12 For example, Statista, Carbon Capture and Storage Investment Worldwide from 2018 to 2023 (2025) <https://www.statista.com/statistics/1489995/carbon-capture-and-storage-global-investments-per-year/>.
13 For example, Earthjustice, 'Carbon Capture: The Fossil Fuel Industry's False Climate Solution' (19 September 2023) <https://earthjustice.org/article/carbon-capture-the-fossil-fuel-industrys-false-climate-solution>.
14 For example, Jorge E. Viñuales, 'Sustainable Development', *in* Lavanya Rajamani and Jacqueline Peel (eds), *Oxford Handbook of International Environmental Law* 285 (OUP 2021).
15 For example, Nina Eisenmenger et al., 'The Sustainable Development Goals Prioritize Economic Growth Over Sustainable Resource Use: A Critical Reflection on the SDGs from a Socio-Ecological Perspective', 15 *Sustainability Science* 1101–1110 (2020).
16 UN General Assembly Resolution 66/288, The Future we Want, UN Doc A/RES/66/288 (2012).
17 Sustainable Development Goals and Targets, in United Nations General Assembly Resolution 70/1, Transforming Our World: The 2030 Agenda for Sustainable Development, UN Doc. A/RES/70/1 (2015), Goal 1.
18 Stockholm Resilience Centre, 'Planetary Boundaries' (2023) <https://www.stockholmresilience.org/research/planetary-boundaries.html>.
19 For example, Louis J. Kotzé and Sam Adelman, 'Environmental Law and the Unsustainability of Sustainable Development: A Tale of Disenchantment and of Hope', 34 *Law and Critique* 227 (2023).
20 Convention on Biological Diversity, Rio de Janeiro, 5 June 1992, 1760 UNTS 79; UN Framework Convention on Climate Change, New York, 9 May 1992, 1771 UNTS 107.
21 Ashish Kothari, 'Radical Well-being Alternatives to Development', *in* Philippe Cullet and Sujith Koonan (eds), *Research Handbook on Law, Environment and the Global South* 64 (Edward Elgar, 2019).
22 Philipp Wesche, 'Rights of Nature in Practice: A Case Study on the Impacts of the Colombian Atrato River Decision', 33 *Journal of Environmental Law* 531–556 (2021).

23 For example, Pranav N. Desai, 'Traditional Knowledge and Intellectual Property Protection: Past and Future', 34/3 *Science and Public Policy* 185–197, 188 (2007).
24 Dennis Martinez, 'The Value of Indigenous Ways of Knowing to Western Science and Environmental Sustainability', *The Journal of Sustainability Education* (2010). <https://www.proquest.com/openview/0f530882066a2d62933ff0788d017214/1?pq-origsite=gscholar&cbl=6704831>.
25 Kamrul Hossain & Rosa Maria Ballardini, 'Protecting Indigenous Traditional Knowledge Through a Holistic Principle-Based Approach', 39/1 *Nordic Journal of Human Rights* 51–72, 56 (2021).
26 Draft decision, Institutional Arrangements for the Full and Effective Participation of Indigenous Peoples and Local Communities in the Work Undertaken under the Convention on Biological Diversity, Conference of the Parties to the Convention on Biological Diversity, Sixteenth meeting, UN Doc. CBD/COP/16/L.6 (30 October 2024).
27 The Tkarihwaié:ri Code of Ethical Conduct to Ensure Respect for the Cultural and Intellectual Heritage of Indigenous and Local Communities, Decision X/42, Conference of the Parties to the Convention on Biological Diversity, Tenth Meeting, UN Doc. CBD/COP/DEC/X/42 (2010); Mo'otz Kuxtal Voluntary Guidelines, Conference of Parties to the Convention on Biological Diversity, Decision XIII/18, UN Doc. CBD/COP/DEC/XIII/18 (2016).
28 Ad Hoc Open-Ended Intersessional Working Group on Article 8(j) and Related Provisions of the Convention on Biological Diversity, Possible Elements of *Sui Generis* Systems for the Protection of Traditional Knowledge, Innovations and Practices of Indigenous and Local Communities, UN Doc. UNEP/CBD/WG8J/8/6/Add.1 (2013).
29 Ad Hoc Open-Ended Intersessional Working Group on Article 8(j) and Related Provisions of the Convention on Biological Diversity, Possible Elements of *Sui Generis*.
30 UN Declaration on the Rights of Peasants and Other People Working in Rural Areas, 17 December 2018, UN Doc. A/RES/73/165, art. 26.
31 For example, Vandana Shiva, 'Bioprospecting as Sophisticated Biopiracy', *in* Beth Burrows (ed), *The Catch – Perspectives in Benefit Sharing* 15–36 (Edmonds, WA: Edmonds Institute, 2005); Daniel F. Robinson, 'Biopiracy and the Innovations of Indigenous Peoples and Local Communities', *in* Peter Drahos & Susy Frankel (eds), *Indigenous Peoples' Innovation – Intellectual Property Pathways to Development* 77–93 (ANU E Press, 2012).
32 US Patent No. 5,401,504, Use of Turmeric in Wound Healing (issued 28 March 1995).
33 For example Allison Fish, 'Conceptualizing Nature as Information – Rendering South Asian Genetic Resources and Associated Knowledge Searchable and Legible', *in* Jose Bellido & Brad Sherman (eds), *Intellectual Property and the Design of Nature* 330–350, 333 (OUP, 2023).
34 WIPO Treaty on Intellectual Property, Genetic Resources and Associated Traditional Knowledge, Geneva, 24 May 2024.
35 WIPO Treaty on Intellectual Property, Genetic Resources and Associated Traditional Knowledge, art 3 (disclosure requirement).
36 WIPO Treaty on Intellectual Property, Genetic Resources and Associated Traditional Knowledge, art 1.
37 Protocol on the Protection of Traditional Knowledge and Expressions of Folklore Within the Framework of the African Regional Intellectual Property Organization, Swakopmund, 9 August 2010.
38 Protocol on the Protection of Traditional Knowledge and Expressions of Folklore Within the Framework of the African Regional Intellectual Property Organization, art. 1.

39 Convention on Biological Diversity (note 20), art 15; International Treaty on Plant Genetic Resources for Food and Agriculture, Rome, 3 November 2001, 2400 UNTS 303.
40 For example, Philippe Cullet, 'Environmental Justice in the Use and Exploitation of Genetic Resources', *in* Jonas Ebbesson & Phoebe Okowa (eds), *Environmental Law and Justice in Context* 371–389 (Cambridge: Cambridge University Press, 2009).
41 The Biological Diversity Act, 2002, s 21.
42 Agreement on Trade-Related Aspects of Intellectual Property Rights (TRIPS), Marrakesh, 15 April 1994, 1869 UNTS 299, art 33.
43 For example, Sam Johnston, 'The Role of Science', *in* Lavanya Rajamani & Jacqueline Peel (eds), *Oxford Handbook of International Environmental Law* 249 (Oxford University Press, 2021).
44 Executive Order, *Putting America First in International Environmental Agreements*, 20 January 2025.
45 For example, Jacqueline Peel, 'Precaution', *in* Lavanya Rajamani & Jacqueline Peel (eds), *Oxford Handbook of International Environmental Law* 302–318 (Oxford University Press, 2021).
46 UN Framework Convention on Climate Change, New York, 9 May 1992, 1771 UNTS 107, art 3.
47 UN General Assembly Resolution 79/206, Protection of Global Climate for Present and Future Generations, UN Doc. A/RES/79/206 (2024), para 8.
48 Report of the Secretary-General, Harmony with Nature, UN Doc. A/75/266 (2020), para 16.
49 For example, UN General Assembly Resolution 79/210, Harmony with Nature, 19 December 2024, UN Doc A/RES/79/210.
50 For example, Mumta Ito & Massimiliano Montini, 'Nature's Right and Earth Jurisprudence – A New Ecologically Based Paradigm for Environmental Law', *in* Elia Apostolopoulou & Jose A. Cortes-Vazquez (eds), *The Right to Nature – Social Movements, Environmental Justice and Neoliberal Natures* 221–233 (Routledge, 2018).
51 Giulia Sajeva, *Rights for Ecosystem Services Local Communities and the Rights of Nature* 46 (Routledge, 2024).
52 Andreas Gutmann, 'Pachamama as a Legal Person? Rights of Nature and Indigenous Thought in Ecuador', *in* Daniel P. Corrigan & Markku Oksanen (eds), *Rights of Nature – A Re-examination* 36–50, 42 (Routledge, 2021).
53 For example, Margaret Davies, *EcoLaw – Legality, Life, and the Normativity of Nature* (Routledge, 2022).
54 Giulia Sajeva, 'Environmentally Conditioned Human Rights – A Good Idea?', *in* Daniel P. Corrigan & Markku Oksanen (eds), *Rights of Nature – A Re-Examination* 85–100, 88 (Routledge, 2021).
55 Meghalaya Transfer of Land (Regulation) Act, 1971.
56 For example, Ipshita Chaturvedi, 'Why the Ganga Should not Claim a Right of the River', 44/6–7 *Water International* 719–735, 728 (2019).

6 Exploring the Intersection of Technology and Indigenous People for Sustainable Development

The Way Forward

Erimma Gloria Orie

6.1 Introduction

Across African communities and several other parts of the world it is observed that many rural communities comprising indigenous communities and indigenous peoples (IPs) are endowed with urbane understanding of their environment and the adaptation system. They also have mechanisms for survival as well as the conservation and sustainability of natural resources.[1] Through trial and error aided by practical experience these rural communities have acquired invaluable knowledge and traits on survival strategies and preservation of the ecosystems. This knowledge of how to use natural resources in a sustainable way, capacity to preserve the ecosystem, promote and ensure "survivability" from one generation to another is their local way of sustaining their environment and a reasonable contribution to the modern concept of sustainable development (SD). To this extent therefore, IPs can meaningfully contribute to the formation of SD policies and schemes.[2] The IPs are endowed with indigenous knowledge (IK), which provides resourceful ways of solving contemporary glitches. Indeed, this invaluable resourcefulness has been acclaimed over the years by the international community in the framework of Agenda 21,[3] the United Nations Framework Convention on Climate Change (UNFCCC),[4] and even the World Bank Report of 1994[5] and several other regulations. Where such knowledge is thus combined with modern technology, it will greatly enhance the effectiveness of indigenous solutions as well as facilitate scaling up of these solutions.[6] In other words, such integration assists in overcoming resource constraints and providing access to information as well as the benefits of research and development (R&D) carried out by others.[7]

Thus, it is generally accepted that "development initiatives that defer to and incorporate local perceptions and ways are more likely to be germane to people's needs and to engender sustainable interventions."[8] Indeed, a fundamental component of any country's knowledge system is its indigenous knowledge. Notwithstanding, there exists a reasonable level of reticence towards IK. This is in spite of the benefits derivable from the acclaimed contributions of IK towards the attainment of SD. For instance, several countries

DOI: 10.4324/9781003733751-9

including Nigeria appear to relegate IK as far as its contributions to SD is concerned.[9] The IK is hardly incorporated in government laws, policies, and implementation processes, which would have enhanced the attainment of SD. In agriculture, for example, the government will prefer the use of modern approaches that alienate the IP like irrigation practices, crop preservation, and forest conservation, instead of incorporating the indigenous alternatives successfully practiced over centuries and which have the potential to secure the cooperation and endorsement of the IP for government programmes. Such relegation encourages loss of IK and non-inclusiveness of IP in governments' sustainable development projects and ultimately constitutes a barrier to attainment of SD objectives. The challenge is how to reverse the trend and work in synergy with the IP and integrate such valuable Ik into the modern technology for inclusive approach to SD objectives.

For effective SD outcomes which guarantee inclusiveness of the IP, there is need to assimilate indigenous knowledge systems with modern technological approaches. The benefits accruable from such integration constitutes what I have termed the "intersection between technology and IK." Herein lies the task before the participants at this conference. I expect robust interrogation of issues such as empowerment of indigenous communities, legal and policy frame work, technology gaps, customary and ethical considerations, Ownership and Control, capacity development and R&D, amongst others. I will now set the stage by making some conceptual clarifications. This will be followed by discussing integration of indigenous knowledge, cultural expressions and legal framework, benefits of integrating indigenous knowledge and technology, and also highlighting of the issues and strategies that will enhance IK integration for effective SD.

6.3 Conceptual Clarifications

The concepts to be clarified are indigenous peoples, indigenous knowledge, technology, and sustainable development.

i **Indigenous peoples:** Although there is no formal definition of IPs in international law, the IPs form unique social and cultural groups who share an ancestral connection to landscapes and natural resources inextricably linked to identity, culture, livelihoods and health.[10] They have over millennia developed, maintained, and developed systems of knowledge through direct experience interacting with biophysical and ecological processes, landscapes, ecosystems, and species.[11] Therefore, indigenous people's respect for nature ensures environmental sustainability that must be preserved for future generations. In deed, the worldviews of indigenous peoples have also challenged the prevalent discourse on sustainable development, calling for recognition and respect of their traditional knowledge and collective rights to use and control the lands and natural resources that they depend on and strive to protect.[12]

ii **Indigenous knowledge:** Indigenous knowledge systems are a totality of systems, practices culture, beliefs, values, and so on exhibited in indigenous people's lives. Indigenous knowledge, though a global phenomenon and significantly heterogeneous in its existence, is not limited to rural dwellers but exists in all communities that have built a body of knowledge over generations.[13] When compared to modern technologies, sundry traditional solutions have been adjudged "effective, inexpensive, locally available and culturally appropriate; and in many cases are based on preserving and building on the patterns and processes of nature."[14] Hence, IK is a very complex set of knowledge, skills, and practices[15] that has been passed on from generation to generation.

Indigenous knowledge represents the accumulated experience, wisdom, and know-how unique to cultures, societies, and for communities of people, living in an intimate relationship of balance and harmony with their local environments. These cultures have roots that extend into history beyond the advent of colonialism. They stand apart as distinctive bodies of knowledge, which have evolved over many generations within their particular ecosystem, and define the social and natural relationships with their environments. They are based within their own philosophy and cognitive system, and serve as the basis for community-level decision-making in areas pertaining to governance, food security, human and animal health, childhood development and education, natural resource management, and other vital socio-economic activities. It is embedded in community practices, rituals, and relationships.[16]

iii **Modern technology:** Technology is the application of science to address the problems of daily life, from hunting tools and agricultural advances, to manual and electronic ways of computing, to today's tablets and smartphones.[17] Modern technology refers to modern products, services, and infrastructure that are designed and built using scientific knowledge and engineering.[18] The purpose of modernization in technology or technological advancements is to make every area of our daily lives like business productive and convenient.[19] Its benefits include improved means of communication, improved educational structure and system, and seamless task execution. Meanwhile a number of Western corporations are beginning to consider indigenous knowledge in designing products specifically for developing countries. For instance, Hindustan Shampoo that was developed by Unilever and introduced into India is sold in small packets to reduce cost and was developed to work best with cold water due to the lack of heated water. There is a need to see more cases like this. Nevertheless, some of the drawbacks of modern technology manifest in dealing with deforestation, climate change, and environmental degradation.

iv **Sustainable development:** There are various definitions but the working definition for this discussion, which is also more generally acceptable in various climes, is the Brundtland definition.[20] The definition states that SD means "meeting the needs of the present generation without compromising the ability of future generations to meet their own needs."

It is thus not far-fetched to see the interconnectedness of the variables conceptualised. On the one hand the IPs are the repository of IK, some of which are imbued with some SD mechanisms, like in the areas of flood-harvesting, mangrove preservation, and forest conservation. On the other hand, modern technology has the potential to improve on IK resulting in more efficient and effective applications to SD programmes. Furthermore, modern technologies play a significant role in documenting knowledge and in empowering people and communities with tools and information that can help them protect their lands, their culture and know-how. The tools and devices (like smartphones, drones, geographic information systems,) enable IPs to survey, map, measure, track their territories and land use and for the exploitation of natural resources.[21] It therefore goes without saying that modern technology can improve IK and when both are infused produce benefits that enhance SD outcomes.

6.4 Integration of Indigenous Knowledge and Cultural Expressions/ Legal Frameworks

Indigenous people are the wardens of IK. The knowledge is vital for an appreciation of the operations of the natural environment and for guiding the use of ecological services and natural resources towards sustainable relations between human societies and the environment.

The survival and generation of traditional knowledge essentially depends on the indigenous peoples' rights to lands, territories, resources, and self-determined development. Furthermore, their knowledge and sustainable practices, the lands, and waters they continue to manage account for over 80% of the Earth's biodiversity.[22] Notwithstanding this achievement, the indigenous peoples' lands are under threat from extractive industries, large-scale infrastructure development, monocrop agriculture, and cattle raising. Indeed, according to the World Wildlife Fund's (WWF) Living Planet Report 2024, between 1970 and 2020, there is a 73% decline in the average size of monitored wildlife populations, while in Latin America and the Caribbean during that same period it reduced by 95%.[23] Nigeria and indeed Africa as a region is not free from this chaos. In Nigeria, the Land Use Act[24] gives the Governor of each state in Nigeria ownership of all lands in that state and the original hitherto owners can only sell the land with the Governor's consent. This is absurd to IP.

The IP in recent years have engaged in advocacy programmes as a way of resolving the issues. This approach has brought the IP into global limelight and also created a situation where their rights and that of the autonomous governed territories are recognised. The International Labour Organization's Indigenous and Tribal Peoples Convention1989 (No. 169), the United Nations Declaration on the Rights of Indigenous Peoples (UNDRIP),[25] and the Rio Environmental Conventions all recognized the rights of IP to their lands. Furthermore, the Paris Agreement[26] in Article 7 provides that "governments of member countries acknowledge that action on climate adaptation . . . should be based on and guided by the best available science and, as appropriate, traditional knowledge, knowledge of indigenous peoples and local knowledge systems, with a view to integrating adaptation into relevant socioeconomic and environmental policies and actions, where appropriate." Also, Article 8(j) on Traditional Knowledge, Innovations and Practices in the Convention on Biological Diversity has been fundamental in advancing recognition, protection, and transmission of such knowledge. The Conference of the Parties at its Third Meeting in Buenos Aires in 1996, in its Decision III/14, "Recogniz[ed] that traditional knowledge should be given the same respect as any other form of knowledge in the implementation of the Convention." The work of the Working Group on Article 8(j) and Related Provisions led to the approval of three indicators to monitor the standing and developments in traditional knowledge: (i) status and trends of linguistic diversity and numbers of speakers of indigenous languages; (ii) status and trends in land-use change and land tenure in the traditional territories of indigenous and local communities; and (iii) status and trends in the practice of traditional occupations.

The Working Group was also influential in the espousal of two guidelines – among others – the Mo'otz Kuxtal Voluntary Guidelines (2016) and the Rutzolijirisaxik Voluntary Guidelines (2018). The Mo'otz Kuxtal Voluntary Guidelines is for the development of mechanisms, legislation, or other appropriate initiatives to ensure the free, prior, and informed consent of indigenous peoples and local communities for accessing their knowledge, innovations, and practices, for fair and equitable sharing of benefits arising from the use of their knowledge, innovations, and practices relevant for the conservation and sustainable use of biological diversity, and for reporting and preventing unlawful appropriation of traditional knowledge. The objective of the Rutzolijirisaxik Voluntary Guidelines for the Repatriation of TK Relevant for the Conservation and Sustainable Use of Biological Diversity is to facilitate the recovery of traditional knowledge relevant for the conservation and sustainable use of biodiversity. Aichi biodiversity target 18 also provides that by 2020, the traditional knowledge, innovations, and practices of indigenous[27] and local communities relevant for the conservation and sustainable use of biodiversity, and their customary use of biological resources, are respected, subject to national legislation and relevant international obligations, and fully integrated and reflected in the implementation of the Convention with the full and effective participation of indigenous and local communities, at all relevant levels.

Similarly, since 1992, indigenous peoples have engaged directly in UN processes on sustainable development, including in the Commission on Sustainable Development (1993–2013) and its successor, the High-Level Political Forum on Sustainable Development. At the UN Conference on Sustainable

Development (Rio+20) in 2012, the outcome document, "The Future We Want," recognized the importance of the participation of indigenous peoples in the achievement of sustainable development. It is important to also note that the rights of IP are equally recognized not only in United Nations Declaration on the Rights of Indigenous Peoples (UNDRIP),[28] but also in many other international human rights instruments and multilateral environmental agreements.[29]

However, the adoption of the Nagoya Protocol on Access and Benefit-sharing[30] was a milestone in the recognition of the intrinsic link between genetic resources and indigenous knowledge and associated rights. The protocol provides that traditional knowledge and associated genetic resources can only be accessed with the prior informed consent (PIC) of indigenous peoples, and if such access is authorized, fair and equitable sharing of benefits must be ensured. Furthermore, The United Nations High Commissioner for Human Rights (OHCHR) in May 2012 released the "Guidelines for the Protection of Indigenous Peoples Involuntary Isolation and Initial Contact of the Amazon Region, Gran Chaco and Eastern Paraguay."[31] The Guidelines constitute the first United Nations document focused specifically on indigenous peoples living in voluntary isolation and initial contact.[32] On the other hand, the World Intellectual Property Organization (WIPO)[33] is the world's number one source for *global intellectual* property (patents, industrial designs, copyright, trademarks, etc.) information, resources. It serves the world's innovators and creators, ensuring that their ideas travel safely to the market and improve lives everywhere. WIPO member states on 24 May 2024 adopted the first WIPO Treaty[34] to address the interface between intellectual property, genetic resources, and Associated traditional Knowledge, thus creating for the first time, a bridge between environmental law and IP law. The treaty is quite young and so too early for assessment.

The poser is on how do we handle this knowledge, nationally, regionally, and globally. Certainly not all IP and IK can be recognized in all these legislations and sifting them requires using equitable, uniform, and acceptable standards that may not be very comprehensive and acceptable. Is it feasible to have such a uniform standard for accessing the IK that qualify to be incorporated in the legal framework? In addition it will be interesting for this audience to assess how far the extant legal frameworks have been accomplished or are being implemented.

6.5 Benefits of Integrating Indigenous Knowledge and Technology for SD

Integrating indigenous knowledge with technology has become a global trend with tangible benefits, which invariably constitute the intersection between the two. A number of advantages are derivable in mainstreaming IK with technology for SD. These include sustainable resource management, disaster risk reduction, and enhancing implementation of environmental policy framework.

i **Sustainable resource management:** This is one of the benefits of integrating indigenous knowledge with technology. For example, the International Savanna Fire Management Initiative (ISFMI)[35] in Australia is a partnership that harnesses traditional practices with modern science. Coalescing their traditional knowledge with modern science and technology, IPs (i.e. Aboriginal Peoples in Australia) using the traditional fire management, known as "cultural burning" or "cool burning," keep fuel loads down and reduce destructive wildfires resulting in a decline in greenhouse gas emissions, which in turn provides carbon market opportunities. This is similar to what some indigenous people are practicing in Nigeria across the country, where selective bush burning using chemicals is practiced as against the indigenous practice of burning an entire area. Also, the indigenous people are able to determine in advance, the year that will yield bumper harvest by observing the nature of the weather early in the year. In many West African countries like Burkina Faso, Mali and Nigeria, IPs have evolved sophisticated agroforestry systems over times in which crops cohabit with trees, help reduce soil erosion and enhance soil fertility. Meanwhile, crop diversification reduces the risk of total crop failure due to pests, diseases or extreme weather events. These strategies ensure the availability of food and protect livelihoods in the face of climate extremes, like heavy rains and droughts.[36]

In Africa as well as in some parts of Asia, women leverage their knowledge in safeguarding household food security, in seasons of drought and famine. For instance in India, the Adivasi women's knowledge of forest management enables them to know exactly which type of product to collect depending on the season and the time of the day in order not to overexploit the forest.[37] In Tanzania the Masa people, on account of acute observations and readings, can identify clouds that bring heavy rains or that warn of droughts. Such knowledge is similar to what obtains in the Pacific region where relying on their ancient craft of indigenous navigation they traversed the ocean guided by stars, winds, waves, and the behaviour of birds, fish, and whales.

The Arctic region is also not left out. Indigenous forest management practices are being integrated with satellite monitoring to protect and manage forest resources more effectively. Indigenous communities use their traditional knowledge to monitor ecological changes, while satellite technology provides real-time data on deforestation and illegal logging activities. A good example is the PolArctic project which combines local ecological knowledge with artificial intelligence (AI) and satellite data to detect optimal areas for mariculture, helping to sustain the local ecosystem and economy. This collaboration has improved conservation efforts and bolstered local resilience to climate change, leading to a $1 million grant from the National Science Foundation (NSF) in 2023.[38]

Furthermore, in the spirit of infusing technology into IK, the UNESCO developed the Local and Indigenous Knowledge Systems LINKS Programme,[39] tagged "The Canoe Is the People: Indigenous Navigation in the Pacific," to contribute to the safeguarding and development of IK of non-instrument navigation, canoe building, and open-ocean voyaging in the Pacific.[40] These are examples of the benefits of sustainable resource management, which goes to affirm that integrating IK with scientific advancement has the potential to create a holistic approach to resource management.

ii **Disaster risk reduction:** With regards to disaster risk reduction, the IK has been very useful for mitigation purposes. The Island of Simeuleu near West Aceh was able to escape the 2004 Indonesian tsunami leveraging this knowledge. As soon as the earth stopped shaking the people cried "Smong Smong," which is their local word for a tidal wave.[41] The islanders, mostly from the Nias people, began heading to the mountains relying on an old song about smong passed down by parents and grandparents after the Island was hit by a tsunami in 1907.[42]

iii **Enhancing implementation of environmental policy frameworks:** Integrating IK into environmental policy frameworks, regulations, strategies, and measures will enhance acceptability and inclusiveness of the IP by making them stakeholders to and bring to bear their knowledge and experiences for implementation of the policy frameworks. This underscores the inclusiveness aspect of SD.

What can be distilled from the foregoing is that there are enormous benefits derivable from integration of IK with modern technology. The IK through knowledge partnerships between IPs and scientists can promote new knowledge to address SD issues. A good cooperation enables the sharing of information, enhances dialogue, facilitates network building, and leads to greater resilience. However, these are not without challenges, some of which are discussed subsequently.

6.6 Issues and Strategies for Integration of Indigenous Knowledge and Technology Towards SD

There are a number of issues that must be addressed in order to optimally benefit from the integration of IK and technology. These include legal framework, limitation of IK, technological gaps, and empowerment of indigenous communities. Others are ownership and control, customary and ethical consideration, and loss of traditional knowledge.

i **Legal framework:** The legal framework is very important as the basis for recognition of IK and its application in the law. Certainly not all IP and IK can be recognised in all the national and international legislations on the subject matter. There is therefore the need to create and adopt a uniform standard of legal framework for the required integration. This conference could serve as a melting pot for the initiation and articulation of the desired legal framework.

ii **Limitation of IK:** It is important to note that IK is not without limitations. One of the major disadvantages is the fact that this form of knowledge is restricted and not diffused. This means that such knowledge being rooted in a given community where it evolved rarely spreads among other communities, with the result that communities can hardly learn from the experiences of each other, thereby limiting the benefits derivable from such knowledge, the potential of the innovations and the opportunities of further developing or improving these advances. The strategies for addressing this limitation could be to create fora for documentation of the relevant knowledge, use of R&D, and creation of a data bank of such knowledge ready for dissemination at national and international levels.

iii **Technological gaps:** Whereas IK is seemingly archaic, unrefined, and slow in evolving, modern technology is advancing at an astronomical pace with time. This has created a technological gap between the former and the later. This gap makes it difficult to carry the indigenous people along in the quest for the attainment of inclusiveness in the SD programmes. For example, in some rural Nigerian communities, agriculture extension workers find it difficult to convince indigenous people to adopt modern farming techniques including the use of improved seedlings. In essence this does not augur well for the integration of IK and modern technology unless the necessary interventions are made. Any attempt to integrate IK and modern technology will entail the upgrading of the IK and practices to the standards of development already attained by modern technology. This can be achieved through R&D, science education, and various capacity development initiatives targeted at the locals towards attainment of SD objectives.

iv **Empowerment of** indigenous **communities:** This refers to the process of providing IP with the required tools, resources, and agency to control their own lives, enjoy self-governance, make decisions about their communities, and participate in society, including securing their land rights, economic development, and preservation of culture, while respecting their unique traditions and knowledge systems. This issue of empowerment brings along some challenges like lack or limited access to essential resources/services like education, healthcare, and infrastructure; historical dispossession of natural resources like land loss and cultural disruption; and also the problem of internal disagreements due to intra-community conflicts and power imbalances. These issues must be confronted with strategies that will require consultation and inclusiveness, reasonable funding of "home-grown/domesticated" indigenous communities development projects and capacity development. It will also require partnerships with governments, development partners, and private sector investments to empower the indigenous communities for effective participation in SD programmes.

v **Ownership and control of intellectual property rights:** A major issue in integrating IK and technology is the complex problem of ownership and control, where questions arise about who has the right to access, utilize, and profit from traditional knowledge, often held collectively by indigenous communities, potentially leading to exploitation if not properly addressed through ethical practices and clear intellectual property rights frameworks. The WIPO's new regulation and national intellectual laws of each country should be adapted to resolve such disputes.

vi **Customary and ethical consideration:** There are tensions surrounding the engagement of IP in scientific research. These tensions cover areas of ethical engagement, data sharing, acknowledgement of contribution, and benefits for indigenous peoples in providing support to science. Scientific research, historically and currently, relies on IK in many aspects of fieldwork and research. This contribution and collaboration has not always been accredited. The research process should not be one of "academic mining," but of respectful and active engagement of IP. By ethically engaging with IPs, researchers can enhance scientific research by working with IPs in a symbiotic relationship that is beneficial to all. This will eliminate top–down environmental policies which often clash with realities on the ground.[43] There is therefore a need to intentionally fashion out a national blueprint to protect the interest of the indigenous people, scientific researchers, and the government towards effective SD.

vii **Loss of indigenous knowledge:** IK is rapidly disappearing, and indigenous languages are endangered. It is said that the language of a people is tied to their knowledge and culture. So where there is a disconnect between the IK/culture and the language of the people, the integration of technology may likely result in confusion, which will not be in sync with the targets of SD. For example, teaching children in the language and culture of other people could erode the intrinsic value and meaning of the IK leading to loss of some substances of the IK. Undoubtedly, IK and cultural expressions must be maintained and sustained in the quest to address climate change, sustainable agriculture, and other environmental challenges. The strategy to preserve IK is to teach IK using local language. It may also be necessary to sustain the practice of oral tradition by which knowledge is transmitted from one generation to another.

6.7 Conclusion

In conclusion, this study sought to explore the linkages between Indigenous Knowledge (IK) and modern technology and to investigate the enablers and inhibitors involved in integrating IK with modern technological knowledge systems, as well as the processes required to achieve effective outcomes for the attainment of Sustainable Development (SD). It has been asserted that by capturing relevant IK within various national and international legislations, and by integrating it with modern scientific processes and technology, such knowledge and its associated techniques can be safeguarded, improved upon, and ultimately applied for effective sustainable development outcomes.

However, the integration of Indigenous Peoples' (IPs') unique worldviews, knowledge systems, and practices for SD remains fraught with uncertainties, including cultural misgivings, technological gaps, intellectual property rights issues, and the gradual waning of this essential indigenous knowledge. As a way forward, this paper has proffered strategies aimed at preventing the erosion of indigenous knowledge, facilitating its integration with modern technology, and enhancing the empowerment of global communities towards the attainment of effective sustainable development. This can best be achieved through the use of IPs' institutions, practices, and an inclusive approach. Indeed, it cannot be overemphasized that integrating IK with modern technology is the way forward.

Finally, this paper hopes to inspire innovative perspectives on these pressing global issues and to stimulate critical thinking beyond conventional approaches. It also seeks to generate thought-provoking discussions on the peculiar circumstances of Indigenous Peoples and their IK vis-à-vis modern technology and the broader objective of enhancing sustainable development. Overall, notwithstanding the issues and concerns raised in this study, it is believed that continued scholarly engagement in this area has the capacity to generate the strategies and processes required to address the challenge of effectively integrating modern technology and indigenous knowledge for sustainable development.

Notes

1 S. Subba Rao, 'Indigenous Knowledge Organization: An Indian Scenario.' *International Journal of Information Management*, 2006, 26 (3): 224–233; UNESCO Report, *World Conference on Science*. Paper presented at the Science Agenda-Framework for Action Budapest, 1999; D. M. Warren & B. Rajasekaran, 'Putting Local Knowledge to Good Use.' *International Agricultural Development*, 1993, 13 (4): 8–10.

2 Sanchita Bansal, Tapan Sarke, Anshita Yadav, Isha Garg, Mansi Gupta & Harsha Sarvaiya, 'Indigenous Communities and Sustainable Development: A Review and Research Agenda.' *Global Business and Organizational Excellence*, 2023, 43 (4): 65–87, https://onlinelibrary.wiley.com/doi/full/10.1002/joe.22237.

3 Agenda 21 was adopted in 1992 as a non-binding plan from the United Nations (UN) that outlines a program for sustainable development https://sustainabledevelopment.un.org/agenda21 (Last visited on February 02, 2025).

4 The United Nations Framework Convention on Climate Change (UNFCCC), https://unfccc.int.

5 World Bank Report, *Indigenous Knowledge for Development: A Framework for Action: Knowledge and Learning Centre*, Africa Region, 1994, https://documents.worldbank.org/en/publication/documents-reports/documentdetail/388381468741607213/indigenous-knowledge-for-development-a-framework-for-action (Last visited on February 02, 2025).

6 *UNDP: Indigenous Knowledge Is Crucial in the Fight Against Climate Change – Here's Why*, 2024, https://climatepromise.undp.org/news-and-stories/indigenous-knowledge-crucial-fight-against-climate-change-heres-why (Last visited on February 02, 2025).

7 L. Santamaría, M. Jesús Nieto & Andrés Barge-Gil, 'Beyond Formal R&D: Taking Advantage of Other Sources of Innovation in Low and Medium Technology

Industries.' *Research Policy*, 2009, 38 (3): 507–517; Isla Sibanda, *The Intersection of Traditional Knowledge and Modern Tech: Indigenous Approaches to Sustainability*, 2024, https://www.computer.org/publications/tech-news/trends/indigenous-sustainability (Last visited on February 03, 2025).

8 B. Allibay, *Critical Examination of Community Participation in Development Projects*, 2018, https://repository.mdx.ac.uk/download (Last visited on February 02, 2025).

9 Indigenous systems for understanding, learning, teaching, and experimenting were pushed to the periphery. Ethnologists studied indigenous belief systems and traditions for their folklore or antiquarian value, not because they had something to teach to the colonizers. 15; M. Suneetha, 'Traditional Knowledge in Policy and Practice: Approaches to Development and Human Well-Being.' In S. Subramanian & Balakrishna Pisupati (eds), *Traditional Knowledge in Policy and Practice: Approaches to Development and Human Well-Being* (New York: United Nations University Press, 2010), 19.

10 E. Nyadzi, O. C. Ajayi & F. Ludwig, 'Indigenous Knowledge and Climate Change Adaptation in Africa: A Systematic Review.' *CABI Reviews*, 2021, 16 (29): 118.

11 L. Jennings, *Governance of Indigenous Data in Open Earth Systems . . .*, 2025, https://www.nature.com/.../perspectives (Last visited on February 07, 2025).

12 Eugenia Recio & Dina Hestad (Dr.). *Indigenous Peoples: Defending an Environment for All*, 2022, https://www.iisd.org/articles/deep-dive/indigenous.

13 Subba Rao, 'Indigenous Knowledge Organization: An Indian Scenario.' *International Journal of Information Management*, 2006, 26 (3): 224–233.

14 L. Grenier, 'Working with Indigenous Knowledge: A Guide for Researchers.' *International Development Research Centre*. Ottawa, ON, Canada, 1998, 7.

15 A. T. Johnson & M. F. Mbah, '(Un)Subjugating Indigenous Knowledge for Sustainable Development: Considerations for Community-Based Research in African Higher Education.' *Journal of Comparative and International Higher Education*, 2021, 13 (1): 10, https://doi.org/10.32674/jcihe.v13iSummer.3707.

16 Suneetha M. Subramanian & Balakrishna Pisupati (eds), *Traditional Knowledge in Policy and Practice: Approaches to Development and Human Well-Being* (UNU Press, 2010), 19 (Last visited on June 8, 2026).

17 Cathy Matresse & Lumen Learning, *Introduction to Sociology*, https://courses.lumenlearning.com/atd-mvcc-intro-to-sociology/chapter/reading-technology-today/.

18 FounderJar, https://www.founderjar.com/modern-technology (Last visited on February 06, 2025).

19 Martin Luenendonk, *Modern Technology: Definition, Examples, and Applications*, 2024, https://www.founderjar.com/modern-technology/ (Last visited on February 05, 2025).

20 The Brundtland Commission is also known as the World Commission on Environment and Development (WCED).

21 Henrietta Marrie AM, *Emerging Trends in the Generation, Transmission and Protection of Traditional Knowledge*, https://www.un.org/development/desa/indigenouspeoples/wp-content/uploads/sites/19/2019/04/TK-Emerging-trends-in-the-generation-transmission-and-protection-of-TK-final-paper.pdf (Last visited on February 06, 2025).

22 Global Environment Facility, "Indigenous Communities and Biodiversity" available at: https://www.thegef.org/sites/default/files/publications/indigenous-community-biodiversity_0.pdf (Last visited on June 8, 2026).

23 World Wildlife Fund, *Catastrophic 73% Decline in the Average Size of Global . . .*, 2024, https://www.worldwildlife.org/press-releases/catastro...; Henrietta Marrie AM, *Emerging Trends in the Generation, Transmission and Protection of Traditional Knowledge* (accessed 6 February 2025), 1–15, https://www.un.org/development/desa/indigenouspeoples/wp-content/uploads/sites/19/2019/04/TK-Emerging-trends-in-the-generation-transmission-and-protection-of-TK-final-paper.pdf.

24 The Land Use Act 1978. Chapter 202, Laws of the Federation 2004.
25 This UN Resolution is a compendium of different rights accorded the indigenous people, https://www.un.org/2018/11/UNDRIP_E_web (Last visited on February 06, 2025).
26 Paris Agreement, https://www.un.org/climatechange/paris-agreement (Last visited on February 06, 2025).
27 P. Fajardo, D. Beauchesne, A. Carbajal-López, R. M. Daigle, L. D. Fierro-Arcos, J. Goldsmit, . . . R. A Christofoletti, 'Aichi Target 18 beyond 2020: Mainstreaming Traditional Biodiversity Knowledge in the conservation and sustainable use of marine and coastal ecosystems.' *PeerJ*, 2021, 9: e9616.
28 This declaration was adopted on 13 September, 2007, during the initial session of the Human Rights Council. The declaration was drafted and debated for about two and a half decade before it was adopted.
29 Like the *Article 27 of the United Nations Covenant on Civil and Political Rights.*
30 It was adopted in 2010 in Nagoya, Japan.
31 UN Guidelines on Isolated Indigenous Peoples (2012), https://iucn.org/content/un-guidelines-isolated-indigenous-peoples (Last visited on February 07, 2025).
32 International Day of the World's Indigenous Peoples, https://social.desa.un.org/sites/default/files/Concept%20Note%20Intl%20Day%202024.pdf.
33 World Intellectual Property Organization (WIPO), https://www.wipo.int/web/sdgs; https://www.wipo.int/portal/en/index.html (Last visited on February 07, 2025).
34 The treaty, which was born after 25 years of deliberations, will come into force three months after 15 eligible parties (countries or intergovernmental organizations, as stipulated in the Treaty) have either ratified or acceded to the Treaty.
35 http://isfmi.org (Last visited on February 07, 2025).
36 *UNDP: Indigenous Knowledge Is Crucial in the Fight Against Climate Change – Here's Why*, 2024, https://climatepromise.undp.org/news-and-stories/indigenous-knowledge-crucial-fight-against-climate-change-heres-why (Last visited on February 07, 2025).
37 http://medind.nic.in/ibl/t11/i3/iblt11i3p226.pdf (Last visited on August 04, 2025).
38 I. Sibanda, *The Intersection of Traditional Knowledge and Modern Tech: Indigenous Approaches to Sustainability*, 2024, https://www.computer.org/publications/tech-news/trends/indigenous-sustainability (Last visited on February 07, 2025).
39 One of the main goals of the LINKS Programme, which is to bring indigenous knowledge into the formal education system.
40 https://en.unesco.org/links-transmission/canoe (Last visited on February 05, 2025).
41 https://foreignpolicy.com/2018/10/15/indonesias-indigenous-languages-hold-the-secrets-of-surviving-disaster/ (Last visited on February 07, 2025).
42 The song admonished that when earthquake is accompanied by a low tide, that they should not go near the coast to collect the fish on the shore, because there will be a [tsunami]. Rather they should run to the mountains for safety.
43 Top-down environmental policies often clash with realities on the ground; for example, in Canaima National Park, Venezuela, fire suppression policies led to increased occurrences of destructive fires and marginalization of the Indigenous Pemon land management techniques using patch burning in a flammable landscape (Bilbao et al. 2010); K. M. de Freitas, G. A. Winter, T. A. Kennedy, A. Johnny & J. Mistry, 'Working with: The Ethics and Practices of engaging Indigenous Peoples in Scientific Research', *Past Global Changes Magazine*, 2022, 30 (1): 50–51, https://pastglobalchanges.org/publications/pages-magazines/pages-magazine/129036 (Last visited on February 07, 2025); M. del Mar Delgado-Serrano, J. Mistry, B. Matzdorf & G. Leclerc, 'Community-based management of environmental challenges in Latin America and the Caribbean.' *Ecology and Society*, 2017, 22 (1).

7 Bridging Traditional Wisdom and Modern Sustainability

An Empirical Study of PVTGs of Jharkhand and Their Understanding of Sustainability

Prakash Tripathi

7.1 Introduction

In this surreal time, when the world is facing a range of interrelated quandaries and predicaments such as war, climate change, economic disparities, health challenges, and migration, the very notion of development has opened a new dialogue in the academic discourses. Scholars have realized that the anthropogenic pursuits, entrenched in the socio-political and economic needs, have resulted in the biggest upheavals of Anthropocene and reconsidered the Modern Capitalist approach of economic dominance. In the Anthropocene, humankind became enabled to control energy and reproduce new technology by combining pre-existing technologies. The meaning of development is restricted to materialistic growth, whether at the national level or at the individual level. The rapid industrialization and urbanization had become significant reasons for depleting resources and dwindling nature. To meet these changes and adapt them, new technological advancements are required, which further deteriorates the environment. The world is trapped in this vicious cycle of non-renewable-energy-driven, GDP-oriented unsustainable growth.[1] This capitalist model of development has developed a new approach of growth and development which is restricted to statistical measurement of the development, ignoring a holistic model. To address this Anthropocene crisis, in 1987, The United Nations World Commission for Environment and Development (UN-WCED) came out with a historical report under the chairmanship of Gro Harlem Brundtland, then Norwegian prime minister. The report, popularly known as the "Our Common Future" or Brundtland Report, took a holistic approach to define "Sustainable Development" by stating, "Sustainable development is development that meets the needs of the present without compromising the ability of future generations to meet their own needs."[2]

This holistic definition on the one hand focused on the essential needs and aspirations of the people, especially of poor and marginalized people; on the other hand it also urged the participating countries to address the inequality which is a causing factor of the ecological and other crises.[3] On the

DOI: 10.4324/9781003733751-10

Figure 7.1 Sustainable Development Goals
Source: Global Goal Centre

recommendations of the Brundtland report, the first organized effort to address the ecological and associated socio-economic challenges began in 2000 with Millenium Development Goals (MDGs) with specified targets and goals to be met by 2015. This promising effort achieved success in various parts but fell considerably short of achieving the MDGs in the Low and Middle Income Countries. The momentum generated by the MDGs transitioned into Sustainable Development Goals in 2015 and set a target of 17 goals to be achieved in 2030, and 195 nations agreed to achieve these environmentally, economically, and culturally sustainable targets. These 17 goals, with 169 targets, emphasize on 5Ps (People, Planet, Peace, Prosperity, and Partnership), which brings balance between humans and nature.[4] The report also emphasized on the indicators and features which highlight the balance between social, economic, and environmental aspects.[5]

The SDGs have been made for all but what place the tribal or indigenous communities have? The tribal and indigenous communities across the globe, including those in India, have a deep relation with nature and rely majorly on the natural resources due to inability to meet the pace of technological advancement and rapid economic growth. The SDGs have direct mention of these custodians of natural resources in the Goal 2 – End Hunger, Achieve Food Security, and Improved Nutrition – and in the Goal 4 – Ensure Inclusive and Equitable Quality Education. Target 2.3 clearly aims to "double the income of the small-scale food producers including indigenous people by securing access to land and other productive resources and knowledge system." Target 4.5 asks for providing quality education to the indigenous people along with other vulnerable sections of the society. Apart from this, Goal 2 (Good Health and Well-being), Goal 6 (Clean Water and Sanitation),

Goal 13 (Climate Action), and Goal 15 (Life on Land) are also found to be directly linked with the tribal and indigenous communities (Figure 7.1). The author believes that all the 17 goals directly or indirectly affect the lives of the tribal communities across the world, particularly in India.

7.2 Particularly Vulnerable Tribes in India

India is a land of vast and vivid cultural diversities. Nearly 705 indigenous or tribal communities are residing in different states and union territories of India who are characterized by their unique tangible and intangible cultural practices. Out of these 705 communities, 75 tribal groups are classified as Particularly Vulnerable Tribal Groups (PVTGs) of India. They reside in 18 states and the Union Territory of Andaman and Nicobar Islands. These indigenous groups are repositories and custodians of rich, varied, and diverse cultural traditions and practices. The major areas inhabited by these PVTGs are located in the remotest part of India, like forests, mountains, highlands, islands, and so forth. These habitations seclude them from the mainstream population of India. Further, these isolated areas are poor in infrastructural development, which leaves them to survive with the traditional and primitive technologies. They rely, mostly, on the natural resources for their sustenance and livelihood. Their social and cultural practices are unique and different from others. They encompass a wide diversity of culture; these unique cultural communities are believed to be the descent of autochthonous, who resided in a particular territory from a historical period. They claimed that the other residents of that territory were descendants of later migrants who afterward denied first right on the land and natural resources. The PVTGs are socially, culturally, linguistically, and economically vulnerable because of their smaller population, lack of documentation, and outside influences. These groups have a common language, uniform social organization, organic solidarity, unique socio-political structure, a contiguous territory, uniform culture, egalitarian social structure, and a tradition of common descent. Their religious practices and rituals are closely associated with the natural and social milieu and these are diminishing gradually. Their threatened traditional culture, performances, crafts, and languages need to be preserved and protected and wherever possible revived to promote and sustain their ecology, biodiversity, knowledge system for sustainable development and growth (to evolve policy that links sustainable development with their traditional knowledge in consonance with social, cultural, economic, and religious aspirations of communities).

7.3 Research Methodology

The present study is an empirical study in which data was collected through extensive fieldwork. The fieldwork was carried out in 2021, as a part of a project in joint collaboration with Ministry of Tribal Affairs and Ministry of

Culture, which was focused on the people's understanding about development and their socio-economic practices. Qualitative research provides an opportunity to understand the real picture on the ground and enables researchers to examine the indigenous knowledge which has been protecting natural resources since time immemorial. Indigenous communities are not literate enough to fill the questionnaire and survey forms, so this research relies on the extensive participant observation, in-depth interview, focus group discussion, and narratives of villagers and local scholars. Jharkhand state is the abode of 32 tribal communities, out of which 9 tribal communities – Asur, Birjia, Birhor, Korwa, Parharia, Hill Kharia, Sauria Paharia, Maal Paharia, and Savaar – are categorized as PVTGs, due to their secluded, small, and scattered habitat, use of traditional nature-dependent technology, tumbling population, and diminishing culture and language.

The study was carried out in three districts, namely Gumla, Latehar, and Lohardaga, in which these tribal communities were residing. Nearly 200 individuals, from Asur and Birjia communities, were interviewed to get information. Door-to-door interview and village-wise discussions were carried out to gather information on demographic variables and traditional ecological knowledge. Information was collected to have first-hand information on various cultural practices which promote sustainability and protect natural resources. Discussion and narrative analyses were carried out to gather information on the cultural practices and belief systems of these communities which promote sustainable practices.

7.4 Historical References and Contemporary Status of Studied Tribal Communities

Asur and Birjia are two moieties of the same ancestral group (Bir Asur, Birjia Asur, and Agaria Asur are supposed to have originated from the same group who settled in different territories). They are believed to be autochthonous to the Indian subcontinent, who ruled over Northern India before Aryan immigration. The Vedic references suggest that they are the sons of Kahsyapa and originated from Varuna, who became the principal deity of the Asura. It is believed that those who accepted the body-politics (Varna system) became Sura (Gods) and those who refrain themselves from this were known as Asura (demons/rakshasa/daitya).[6]

These tribes belong to the Proto-Australoid race (Munda Group) who inhabited Chhota Nagpur Plateau before other tribes of Munda group settled here. The maximum concentrations of these neighbouring tribes are found in the *Pat* (means Flat) regions of Latehar, Lohardaga, and Gumla districts of Jharkhand. The pat region is formed of crystalline rocks having laterite on the top, from where these traditional iron smelters used to extract iron ore for their livelihood and sustenance. The region is covered with the trees such as *Sal* (*Shorea Robusta*) and other timber from which they extract timber, fuel, and other non-timber forest products for daily uses.[7] They have been protecting these forests and natural resources since time immemorial and extracting their livelihood in a culturally woven sustainable way. Their knowledge about iron-ores and its extraction is very unique and culturally intertwined. The iron extracted by them was stainless and it is believed that they used to supply iron to the royal court in ancient and medieval periods. However, due to industrialization of iron-smelting their traditional iron-smelting practices had vanished. They could not compete with the industrial iron products, which are cheap and easily available. An Asur family has to spend the whole day collecting iron ore and the whole night on extracting iron from that ore. The manual extraction of iron ore was done in the night, which involved dances and songs. This entire endeavour resulted in few kilograms of "virgin iron."[8] The market economy had severely impacted their traditional livelihood and forced them to become daily wagers or migrant labours. These Kings of the Forest became slaves of the industries. The bauxite mining companies started taking bauxite areas on lease at very cheap prices and forced them to shift to agriculture or to become the labourers of those mines.

7.5 Socio-Economic Condition and Sustainable Development

Market volatility, land alienation, and loss of resources had played a pivotal role in shifting the mode of economy,[9] which ultimately killed the traditional knowledge associated with this craft.[10] There are several evidences of furnaces present in the Chhota Nagpur region. The traditional practices were not only viable but also promoted equality among the tribal communities which was sustainable in nature. Economic inequality generated by the industrial market had pushed them to the verge of poverty and restricted their average income to less than Rs. 150 per day.

The classical development theorists believe that industrial and market expansion are core to development[11] and ignore other aspects of developments. Per capita income had become a major indicator of the calculating development for any nation or community. Another variable of calculating "development" is asset. Assets and per capita income are directly proportional, which means when one increases the another will automatically increase. It has been observed that those individuals or communities which have more assets have more per capita income, and it allows them to acquire more assets. Whenever assets decline, per capita income also declines and results in social and economic inequality, which ultimately impose a challenge to the sustainable development. The shifting of occupation (from iron-smelting to daily wagers) due to encroachment of natural (forest and land) and economic resources (bauxite ores) questioned the states' commitment to sustainable development.

Figure 7.2 presents the different occupations of the interviewed individuals ($n = 200$).

The pie chart shows the present status of occupation, which has changed drastically in the past three decades, especially after the introduction of industrial mining and change in agricultural practices. Elders of the studied area told that hunting, gathering, and shifting cultivation, along with iron-smelting, was their traditional occupation that had been practiced since time

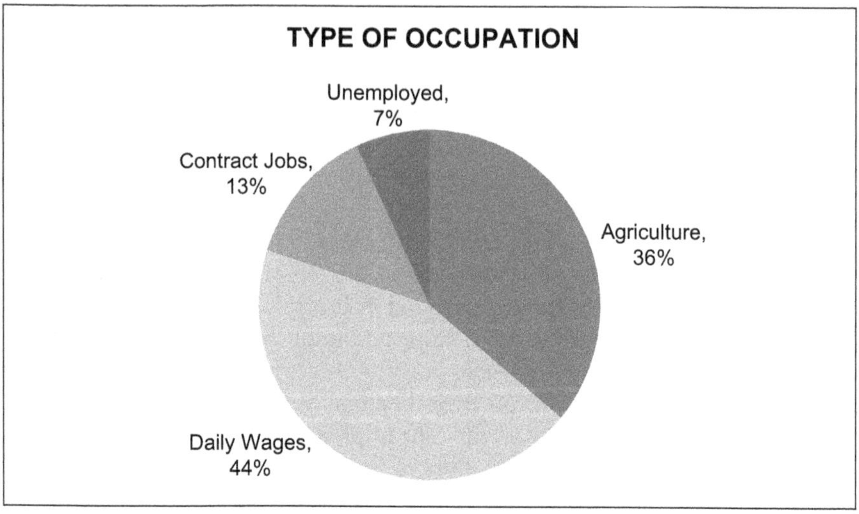

Figure 7.2 Types of occupations among the studied population
Source: Fieldwork

immemorial and their ancestors adhere to the practice of gathering resources solely to meet the need of family or for client. They consciously avoid the pursuit of profit making or monetary gains. This practice was not only sustainable but also in accordance with the natural balance. The occupational data shows that approximately 65% people are unsure about their economic security. Hence, social inequality had become a major challenge here.

Women's conditions are more vulnerable than men's. They are majorly involved in domestic chores and timber and fuel collections, which is not counted as "occupation" as it did not provide any monetary payments. However, they are the custodians and protectors of the traditional wisdoms. Women who are involved in agricultural activities are aware of different varieties of seeds and other edible plants, which are full of nutrition. The region receives very less rainfall. These indigenous women have ample knowledge about how to cultivate more crops in scarce resources. They have amical bonds with nature and their cultural beliefs and practices have evolved with the surrounding nature, which they protect and conserve.[12] *Peora-Pathar* (a type of shifting cultivation in which lands are cleaned after slash and burning) was created before the first rain and seeds were shown by digging with use of traditional agricultural tools. In this process of cultivation, land did not lose its moisture and crops such as maize, beans, and ragi were grown. This type of traditional knowledge can help in sustainable practices and conservation strategies.[13] The conversion of land and forest resources in to bauxite mines has brought cascading effects on these indigenous communities. The lost lands and forest resulted in loss of resources, traditional occupation, crop biodiversity, and most importantly their traditional knowledge. Their meal plates are bereft of nutritious foods and they rely on the government-supplied grains. The SDGs emphasizing on food security, poverty, and nutrition have a direct relation with their present socio-economic conditions.

After the induction of bauxite mining, villagers, by choice or force, started giving their lands to the private companies who ceased their rights and exploited their labours. Their agricultural lands had become a mining site where entire biodiversity vanished and the mining site became useless for agricultural activities due to loss of the soils. Further, the remaining mining sites are filled with water in the name of water harvesting and fishing cultivation to promote the economy of the region. Tribals do not have any rights over these water reservoirs and these have become a fertile ground for malarial parasites, which is a very common disease in the region.

Education is another important social development goal, which emphasizes equitable education among the weaker sections of the society. Target 4.5 of the Sustainable Development Goals talks about providing quality education to the indigenous people. Literacy and school enrolment are considered as important indicators of assessing education. In the studied area, it was found that schooling is very new phenomenon and the studied tribals do not pay much attention to send their children in the school. Government has opened a "residential tribal school" in the region but most of the students in that school are non-tribals. It was found that out of 1290 students present in five different schools, only 73 were tribals and others were non-tribals. Government had introduced several schemes and programmes to ensure maximum enrolment but the tribals are still reluctant to school education and rely more on socio-cultural learnings. Figure 7.3 shows school enrolment of tribal kids in various classes in one of the tribal residential schools.

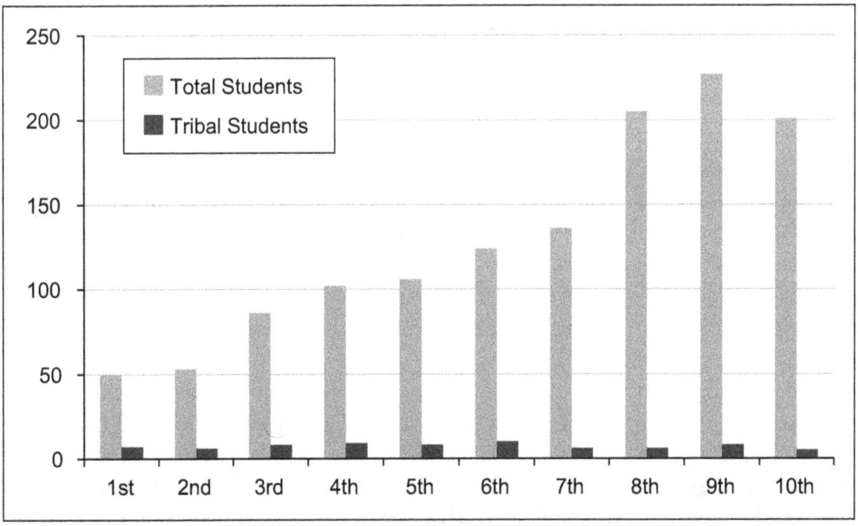

Figure 7.3 School participation of tribal students in different classes

Source: Fieldwork

7.6 Cultural Beliefs and Practices

The indigenous communities of Jharkhand, including the Asur and Birjia tribes, have evolved cultural practices intrinsically linked to their natural environment, fostering a harmonious coexistence with biodiversity. Central to their worldview are their religious beliefs and practices, which revere nature as sacred. These tribes worship a pantheon of deities, including Sing-Bonga (the Sun God), Pola Pahad, Paar Bonga, Dharti Mata, and Sarna Devi, which reflect their deep ecological ethos. Natural objects, trees, animals, and groves hold profound spiritual significance, leading to their preservation as totems or sacred entities. This reverence extends to Sarnas – sacred groves that function as biodiversity hotspots and sanctuaries for flora and fauna. These groves, untouched by industrial or human intervention due to religious sanctity, harbour rich ecosystems and serve as reservoirs of genetic diversity.

The Asur and Birjia's traditions underscore a vital form of biodiversity conservation rooted in spiritual and cultural principles. Unlike modern conservation strategies driven by regulatory frameworks, their practices are sustained by a sense of moral and religious duty. The prohibition against felling trees or hunting sacred animals ensures the survival of critical species and maintains ecological balance. Furthermore, these practices offer a counter-narrative to the ecological degradation triggered by industrialization and capitalist development. The relentless exploitation of resources for industrial purposes has disrupted atmospheric compositions, exacerbated climate change, and led to resource depletion.[14] These changes disproportionately impact vulnerable communities like the Asur and Birjia, exacerbating social deprivation and inequalities. In the face of mounting ecological crises, the cultural practices of these tribes present invaluable lessons. Their traditional ecological knowledge offers sustainable alternatives to the exploitative patterns of modern development. For instance, their totemic worship discourages overexploitation of resources, while the Sarnas act as natural carbon sinks, mitigating climate change effects. By integrating such practices with contemporary conservation policies, we can address biodiversity loss while respecting indigenous rights and knowledge systems. Moreover, the Asur and Birjia's cultural practices highlight the critical need for an inclusive approach to sustainable development, where marginalized communities are not merely seen as beneficiaries but as custodians of ecological wisdom.

The intersection of industrial development and environmental exploitation starkly contrasts with the indigenous ethos of coexistence. Industrial activities have not only stripped the land of its natural wealth but have also marginalized indigenous populations, displacing them from their ancestral lands and severing their cultural ties to nature. The resulting social and ecological imbalance underscores the urgency of revisiting development paradigms. The cultural practices of the Asur and Birjia tribes, rooted in biodiversity conservation, highlight an alternative pathway – one that aligns economic growth with environmental sustainability and social equity. The cultural traditions of the Asur and Birjia tribes, deeply embedded in their spiritual connection to nature, offer a paradigm of ecological stewardship. Their practices underscore the importance of preserving sacred natural sites and respecting traditional knowledge as a means to safeguard biodiversity. As industrialization continues to reshape ecosystems and exacerbate inequalities, integrating such indigenous wisdom into broader sustainability frameworks holds the potential to address ecological crises and foster a more equitable and harmonious relationship between humanity and nature.

7.7 Meaning of Development and Government Initiatives

Development, in the contemporary capitalist model, is understood by the availability and accessibility of infrastructure in any region. Government also measures development through infrastructural variables such as road, electricity, water, housing, and schools. These quantifiable variables help to provide statistical data to substantiate the developmental initiatives. However, these are subject to various factors. Many of the government initiatives either do not reach the targeted communities or failed because of poor implementation and sporadic financial supports. In the studied area, road, water, and electricity are present but their conditions are not promising. Education and dropouts are another major challenge in the area. To ensure quality education, government has opened primary and secondary schools in region but student's dropout is a major concern there. In the three schools visited, out of 176 students only 12 tribal students were found to be enrolled in primary and secondary classes. Accessibility of pure water was found to be another challenge in the studied area. Solar-based water tanks had been installed in the village but when they got damaged, due to any reason, they remain unrepaired for a uncertain time.

Indigenous people's perspective on development is more holistic in nature and revolve around the harmonious relationship between nature and man. They have been subscribing to this understanding from time immemorial. But external influx of information through media and governmental policies had changes their way of thinking and now their believe on development revolves around the infrastructure. Elders of the studied areas still believe that development is about growth of all without affecting the nature. They refuse to accept the present concept of development, which has increased their income but also taken away their forests and lands.

7.8 Sustainable Development and Traditional Knowledge

The understanding of development among tribal communities often centres around infrastructural and materialistic progress, which they perceive as the government's responsibility. They value governmental benefits and subsidies and express a desire for more such initiatives. For instance, programs like the *Dakiya Programme* (which provides grains and other essential edible items) or the *Matritva Yojana* (Maternity Programme), along with various health schemes, are seen as pivotal to their development. When these benefits are received – whether in monetary form or otherwise – their conception of development tends to be limited to these tangible, programmatic interventions.

However, when asked about sustainable development, their perspective takes on a deeper and more nuanced meaning. For them, sustainable development is intrinsically linked to the preservation of their land, forests, water resources, kinship structures, and, most importantly, their culture and identity. They strongly believe that their resilience to change and capacity to cope with external shocks are inherently tied to the availability and preservation of their natural resources. In this sense, sustainability is less about economic growth and more about maintaining the ability to recover and ensuring that natural resources continue to support their way of life. They firmly believe that nature has the capability to fulfil human needs and they have learnt what natural resources are required to mitigate various challenges. This knowledge had been transferred to them from their forefathers/mothers and it is their responsibility to pass it to their descendants.

The traditional knowledge and wisdom acquired by the tribals are ecolocal and cumulative knowledge evolve through environmental interactions. It includes an array of knowledge related to agriculture, water management, forest management, medicinal plants, edible items, agricultural or craft-skills, hunting, fishing, and so forth required for the sustenance of a community in its local ecology.[15,16] Traditional communities use these knowledges in their daily life, to assess environment, climate and weather, in agriculture, foodhabit, healthcare system, and so forth. A major feature of traditional knowledge is that it can be similar in different communities but the "how-to" part of this knowledge varies from community to community. Therefore, it is the collective heritage and patrimony of the tribal and indigenous communities, and the central theme consisted of cultural beliefs and traditions transferred from their ancestral generation to the present generation for the purpose of effective and comfortable survival without disturbing the harmony with the ecosystems.

Recognizing the importance of traditional knowledge in sustainable development, World Bank recognizes it as an enabling component of development. The UN also highlighted the significance of traditional knowledge by saying, "Indigenous peoples' traditional knowledge has been developed over generations through daily life practices and a close understanding of local environments. It can offer valuable responses to climate change, food insecurity, reducing inequalities and other challenges that we are trying to resolve through the Sustainable Development Goals. Traditional knowledge offers tremendous opportunities in such areas as land management, conservation, and scientific, technological and medical research."[17]

The World Intellectual Property Organization (WIPO) focuses on safeguarding traditional knowledge by addressing intellectual property concerns, particularly protection against unauthorized use of the knowledge by third parties. This initiative aims to ensure that communities holding traditional knowledge retain control over its usage and derive fair benefits from it.[18]

WIPO's Intergovernmental Committee on Intellectual Property and Genetic Resources, Traditional Knowledge, and Folklore is at the forefront of negotiations to establish an international legal framework for the effective protection of traditional cultural expressions, folklore, and Traditional Medicinal Knowledge. The committee seeks to create a binding instrument that balances the rights of knowledge-holding communities with the global demand for equitable access to and benefit-sharing from genetic resources and associated knowledge.

The ongoing negotiations address two primary concerns: preventing misappropriation of traditional knowledge and ensuring fair access to genetic resources while equitably sharing benefits derived from their use. This includes addressing intellectual property–related aspects of research, commercialization, and innovation based on traditional knowledge and genetic resources. By providing legal recognition and protection, the initiative aims to empower traditional knowledge holders, foster respect for cultural heritage, and promote sustainable development. Once finalized, this instrument will harmonize international efforts to respect and protect traditional knowledge while ensuring just outcomes for the communities that safeguard this invaluable heritage.

The most challenging aspect of traditional knowledge lies in ensuring its effective protection. Within the framework of intellectual property law, "protection" is generally understood as a mechanism to enforce private, exclusive economic rights over a specific invention, primarily aimed at preventing unauthorized use or reproduction.[19] However, this conventional understanding often fails to adequately address the complexities and unique characteristics of traditional knowledge. Unlike modern inventions, traditional knowledge is typically collective in nature, rooted in the cultural and spiritual heritage of communities, and passed down through generations. Its protection requires mechanisms that go beyond individual ownership and economic interests, encompassing recognition of communal rights, safeguarding cultural identity, and preventing exploitation or misappropriation by external entities. Additionally, the challenge extends to reconciling traditional knowledge systems with formal legal frameworks, which are often ill suited to capture the holistic, dynamic, and context-specific nature of such knowledge.[20]

Among the PVTGs of Jharkhand, minor forest produce, wood, wild vegetables, and the worship rituals for festivals still keep them dependent on and sensitive to forests and natural resources. Their deep connection with the flora, fauna, and wildlife of the forests is evident from the fact that their deities include Bear God, Tiger God, and other animal deities. Similarly, every village has its own boundary (referred to as "Seemani"), which is worshipped periodically as needed by the village priest ("Pahan") and the elders. Moreover, before taking anything from nature, the villagers from both communities perform a ritual worship, as exemplified by the tradition of "Mudmati." Even to forge a bond of friendship (referred to as "Sahiya"), they have a tradition of offering various elements of nature, such as flowers, leaves, or twigs. They also inspire the world that in spite of their traditional occupation of iron

Figure 7.4 A Birjia couple making basketry from bamboo
Source: Fieldwork

smelting had been destroyed by the industrialization, they relied on the sustainable occupational practices such as bamboo crafts, basketry, crafts made by lac (Lakh), utensils made from leaves, and so forth. They still worship their lands, forests, and natural resources and protect them. The ecosystem's sustainability maintained these communities while using ecosystem services. Their traditional knowledge about the plants and soil are highly significant. Their beliefs regarding the earth, forest, and cosmos are highly significant in their cultural practices. They take ecosystem services for sustenance only, not for profit making. They believe that if overexploitation of these resources are done then nature will refuse to provide the ecological services and these essential resources will disappear. Therefore, they use natural resources very cautiously (Figure 7.4).

7.9 Conclusion

The economic and political conditions of the Asur and Birjia tribes are extremely abysmal, necessitating a more focused and integrated approach by governmental and non-governmental organizations to ensure their holistic development. The two targets of the Sustainable Development Goals (which mention the indigenous communities) were also not found promising in the studied area. The chapter includes these two targets to highlight the inequality in income and education among the studied indigenous communities. However, the tribals' understanding about the development is not restricted to infrastructural growth but should encompass their social, cultural, linguistic, ecological, and psychological advancement, adopting a sustainable and inclusive framework. Such an approach must prioritize human lives, forest preservation, and agricultural land conservation, ensuring that development is not exploitative or destructive (*adamkhor*, *jangalkhor*, and *jaminkhor*). Incorporating the voices of the local communities in policymaking is crucial for achieving sustainable outcomes, as their needs, demands, and cultural practices provide a foundation for effective policy formulation and implementation. Policymakers must understand the intricacies of tribal identity, which often becomes a barrier to development when ignored or undermined. Development initiatives should aim to preserve and promote their unique cultural practices, arts, crafts, fairs, festivals, folklore, music, and literature, which not only establish their identity but also instil confidence and pride within the community.

The tribes' rich cultural heritage, often overlooked, holds significant potential for socio-economic development. Modernization and industrialization have led to the disappearance of traditional practices, such as iron smelting, which were once integral to their livelihoods, while subsistence activities like agriculture, daily wage labour, and mat knitting have taken precedence. Reviving traditional arts and crafts, alongside enhancing existing subsistence practices, must be undertaken in collaboration with local communities to ensure the process aligns with their interests and aspirations. Tribal ecological knowledge and wisdom, being central to their survival, can be leveraged as a model for sustainable development and a source of livelihood. Additionally, their unique culinary traditions, rich in nutrients and cultural significance, can be promoted as a means of livelihood generation, such as establishing tribal-themed restaurants.

Research efforts must also document their material culture, including tools, technologies, and utensils that reflect their livelihood practices, as well as basic infrastructural facilities that are critical for cultural and economic development. This documentation will serve as a foundation for future research focusing on livelihood approaches and cultural materialism, enabling the transformation of cultural resources into productive elements essential for local development. A two-day workshop-cum-exhibition can be organized to showcase the artistic talents of the Asur and Birjia communities, where stakeholders, including artists, craftspeople, performers, writers, scholars, and policymakers, can engage in meaningful dialogue to identify and promote cultural capitals with national and international potential. This platform would also facilitate the formulation of effective policies to enhance the livelihood opportunities of these communities.

Future research should emphasize extensive fieldwork to document their tangible and intangible cultural heritage through films, photographs, and reports, thereby transmitting their traditions to global audiences. Digitalization of their arts and crafts can amplify their reach via social media and connect them with national and international agencies for creative training, modifications, and recognition. Such an approach will ensure that cultural development safeguards the interests of local communities while fostering inclusiveness, tolerance, and creativity, transforming local resources into assets for sustainable development.

Notes

1 Jeremy Green, "Comparative Capitalisms in the Anthropocene: A Research Agenda for Green Transition," *New Political Economy* 28, no. 3 (2023): 329–346.
2 WCED, *Report of the World Commission on Environment and Development: Our Common Future*, 1987. Retrieved from http://www.un-documents.net/our-common-future.pdf (Last visited on January 01, 2025).
3 WCED, *Report of the World Commission on Environment and Development.*
4 Measuring Sustainable Development, *Report of the Joint UNECE/OECD/Eurostat Working Group on Statistics for Sustainable Development*, United Nations, New York and Geneva, Page No. 18, 2008. Retrieved from https://sustainabledevelopment.un.org/content/documents/801Measuring_sustainable_development.pdf (Last visited on January 01, 2025).
5 Measuring Sustainable Development, *Report of the Joint UNECE/OECD.*
6 K. K. Leuva, "The Asur: A Study of Primitive Iron-Smelters," 5–8 (1963). Retrieved from https://ia904705.us.archive.org/1/items/in.ernet.dli.2015.122811/2015.122811.The-Asur-A-Study-Of-Primitive-Iron-Smelters.pdf (Last visited on January 01, 2025).
7 Leuva, "The Asur", above note 6.
8 Bappa Ray, "The Story of Virgin Iron," A Film Retrieved from YouTube. https://www.youtube.com/watch?v=B7pzuisjeuY (Last visited on January 01, 2025).
9 P. L. Toppo, "Traditional Knowledge and Practices of Asur Tribe of Jharkhand in Sustainable Management of Natural resources," *Indian Journal of Traditional Knowledge* 16, no. 4 (2017), 579–585.
10 M. Behra, "Iron-Smelting of the Asur Tribe of Jharkhand: A Study of Social Organisation and Knowledge Transmission," *Indian Anthropologists* 39 (2009), 91–101.
11 Measuring Sustainable Development, above note 4.
12 Shonil A. Bhagwat, Nigel Dudley and Stuart R. Harrop, "Religious Following in Biodiversity Hotspots: Challenges and Opportunities for Conservation and Development," *Conservation Letters* 4, no. 3 (2011): 234–240.
13 Vibha Singhal, Jyotirmoy Ghosh and Sheeraz Saleem Bhat, "Role of Religious Beliefs of Tribal Communities from Jharkhand (India) in Biodiversity Conservation," *Journal of Environmental Planning and Management* 64, no. 13 (2021): 2277–2299.
14 Nafisa Priti Sanga and Rajeev Kumar Ranjan, "Natural Resource Management in Changing Climate–Reflections from Indigenous Jharkhand," *World Journal of Science, Technology and Sustainable Development* 11, no. 2 (2014): 117–133.
15 Ranjay K. Singh, Jules Pretty and Sarah Pilgrim, "Traditional Knowledge and Biocultural Diversity: Learning from Tribal Communities for Sustainable Development in Northeast India," *Journal of Environmental Planning and Management* 53, no. 4 (2010): 511–533.
16 Nancy J. Turner and Fikret Berkes, "Coming to Understanding: Developing Conservation Through Incremental Learning in the Pacific Northwest," *Human Ecology* 34 (2006): 495–513.
17 UNDESA. Retrieved from https://www.un.org/uk/desa/traditional-knowledge-%E2%80%93-answer-most-pressing-global-problems (Last visited on January 04, 2025).

18 WIPO. Retrieved from https://www.wipo.int/edocs/pubdocs/en/wipo-pub-rn2023-5-6-en-intellectual-property-and-traditional-medical-knowledge.pdf, (Last visited on January 04, 2025).

19 Prakash Tripathi and Ashutosh Mishra, "Traditional Wisdom/Knowledge and Intellectual Property Right: An Understanding of Legal Recognition of Traditional Knowledge and Practices in India," *Kalakalp* VI, no. 1 (2021): 237–248.

20 Tripathi and Mishra, "Traditional Wisdom/Knowledge and Intellectual Property Right."

Part IV

Gender Issues

8 Navigating Gender Representation in the Digital Age

Social Media, Law, and Policy Reforms in India

Shaiwal Satyarthi and Shivang Tripathi

8.1 Introduction

With the advent and constant transformation of scope and functionality of social media, it has gradually taken the form of a tool for representation, engagement, communication, and expression.[1] The previous decade witnessed massive increase in social media consumption in India, and the most significant thing to be observed was how social media platforms, such as Facebook, WhatsApp, Instagram, and Twitter, gradually changed their functional trajectory and became something more than a mere "communication" and "socializing" tool. This is the phase wherein such platforms started impacting their users' social, political, economic, and cultural perspectives. With rampant commodification of "data" and "information," floodgates opened for the big, medium, and small corporate entities to enter the field and establish real-time engagement with its users. With such transformation in the functionality algorithms of social media tools, challenges and threats were bound to arise in the form of "misuse" by certain users. Therefore, the aforesaid medium of engagement has crossed the frontiers of "communication" and "socializing," and trenched upon the delicate aspects of realization of an individual's identity, fundamental right of speech and expression, and his/her right to representation, opening wide the gates of challenges and threats to one's safety, security, and violation of one's representation right, that is the liberty to portray oneself in the society as per one's wish, in adherence to public policy and law.

Social media has helped in the democratization of information to a great extent, making information accessible, shareable, and collaborative to the individual users. It has helped the users to establish a virtual discussion forum, impacting both the right to audience and the right to express freely on various social and political issues. This has largely benefitted the women users to break open the societal stereotyped shackles and restrictions and become a part of the mainstream discussions, enabling their voices to be heard, which earlier were either supressed by the social mores or given least significance owing to the conservative mindset coupled with extreme patriarchal notions of the society. The least explored domain for Indian women social media

DOI: 10.4324/9781003733751-12

users was their participation in economic discourse, which was successfully enabled by the social media engagement.[2]

According to a report compiled jointly by the Internet and Mobile Association of India and Kantar firm, internet use in India has surpassed the massive number of even 800 million in terms of quantity, where rural India accounts for more active users than urban India. As per the report, 2023 witnessed reduction in the gender divide comparatively. It showed that 46% of the users are females and 54% males, making it all the more necessary to study the engagement ratio of females with the internet and the incidental and ancillary dimensions such as AI.[3] As of 2023, the female internet users worldwide have crossed approximately 65% and the margin of difference between being an internet user and a social media user was nominal.[4] If duration-wise user engagement matrixes are evaluated, tendency-wise women have shown more activity in terms of being an audience/consumer, to being content creator. This massive shift in the figures reflects that women have transitioned from being a passive audience to becoming an active user community. On Instagram alone, till January 2024, 49.4% users and content creators were women, making the platform one of the most women-engaging social media platform globally.[5] Female users outweigh the numbers accorded to the male users, when it comes to social media platforms such as Instagram and Pinterest, which makes the women community susceptible to greater risks, as well.[6] Women have contributed significantly in shaping the public discourse on such platforms by the help of user algorithms, focussed significantly on them because of their greater user-engagement. Owing to the deeply seeded and diversified cultural, societal, and legal landscape pertaining to gender reflection and portrayal in India, female social media users have been facing various challenges, related but not limited to online harassment incidents, cyber offences, mild to extreme stalking, gender-based name callings, and meme and troll generation.[7]

This chapter analyses how effectively representative social media platforms have been towards women in the last few years and evaluates the engagement matrix of women users with social media and the challenges faced by them, in the form of online harassment and other incidental acts. It shall also critically examine the legal and policy infrastructure prevalent in India catering to redressal mechanisms, punitive provisions, and resolution procedures with respect to the aforementioned challenges. Just like the physical space, even digital space is required to be safe and secure for the socially marginalized communities (especially women in India), so that their freedom of expression and engagement could be effectively exercised.

8.2 Women and Social Media – Global Perspective

Considering global perspective on the social media user engagement and interaction with women users, in the United States, even at the nascent stage of social media, women participation either outweighed or stood at an equal footing with their male counterparts. Research data compiled by Pew Research Centre reflects that on most popular social media platforms, such as Facebook, Instagram, Twitter, and Snapchat, women users have substantially surpassed the percentage of male users quantitatively. As of 2024, it shows that, for Facebook, it was 61% males compared to 78% females; for Instagram, 44% male use was countered by 55% female users; for Snapchat, the ratio stood as 23% males to 31% females.[8] It was also observed that the psychological tendency to frequently resort to social media platforms was seen more in women users, which can be attributed to various factors, such as identity representation, social validation, entertainment, informative search, and anti-isolation mechanism. In the European Union, according to a report by the EU Agency for Fundamental Rights, 10% of women in the European Union have experienced cyber-harassment since they were 15, including the uninvited sending of sexual content (emails, texts) or inappropriate online advances. The stalking, identity theft, dissemination of non-consensual pornography, and hate speech were among the many forms of online harassment used, but hate speech emerged as the predominant form.[9]

8.2.1 Challenges

In general, the surveys and compilations focus on the quantitative representations, leaving lesser space for the actual representation scenario which is dependent on external factors. Even in the United States, the numbers clearly reflect the quantitative dominance of women users, but the issue of underrepresentation of women stills looms large because of the stereotypical ways of viewing gendered notions of men and women. Such perceptions play a vital role in significantly reducing the "entity" character of women as a societal unit, and it also normalizes anti-representative incidents such as online harassment and violence against women's integrity, identity, and social well-being.

The issue of underrepresentation, even after dominating in user-engagement is dichotomous, yet real. It can be attributed to several factors, such as the way a women user feels intrinsically once she is exposed to the social media interface and users belonging to other genders. As per a research survey conducted by the Pew research Centre in 2021, it was found that 33% of the women social media users have acknowledged that they have faced one of the worst form of online harassments, that is sexual harassment, and this number is alarming because the age group which it caters to is comprehensive enough to include women below the age of 35, which basically forms part of the most significant engagers on social media platforms.[10]

The background of such sexually aggressive temperaments towards women users can be attributed to the rampant exposure of people to aggressive, violent, and gendered content on various forms of media, of which social media is a part, which somehow manages to normalize the acts of violence within relationships, as a part of modern-day intimacy.[11] Furthermore, it is more aggravated under the scope of social media platforms because of the following: first, content is being substantially user generated; second, legislative vacuum in the field of effective regulatory mechanisms; third, expansion of the categories of ways to harass an individual online with the advent and advancement of technology, such as AI and its ancillary tools.

Another significant challenge which arises in the interaction between women and social media is in the form of "arbitrary and excessive data collection," raising privacy and impersonation threats, whilst also enhancing the surveillance issues. On July 10, 2024, Amnesty International reported its findings in a study titled "Gender and Human Rights in the Digital Age," indicting social media and technology for their role in sustaining gender inequality and enforcing existing power structures.[12] The report argues that the people, in particular the marginalized groups, especially women, are forced to face more difficult challenges as a result of excessive data collection and extrapolation being carried out by technology companies and the impact of these data practices can enhance existing inequalities.[13] This is facilitating a surveillance-driven business model that leads to discrimination and dependence on unregulated data collection, which is often presented by governments as a cost-cutting policy for public services. The report highlights how normative gendered practices in social welfare programs, such as deploying digital ID systems or algorithmic processes, have exacerbated the global gender digital divide, negatively impacting women's access to not only technology, but also essential services. The report emphasizes the wrongful denial of women a right to health when they are denied access to essential sexual and reproductive health information and services, including abortion. An example of this limitation can be identified in the removal of relevant content from social media platforms in the United States.[14] Moreover, the report denounces the use of spyware to conduct targeted surveillance on women activists as a specific type of tech-enabled gender-based violence that severely threatens women's rights and safety. These are among the findings of the report, which highlights the pressing need to address these digital inequalities that are disproportionately experienced by women.[15] Specifically, the report calls for stricter regulatory monitoring to mitigate the risks data dispossession and index need syndrome of vulnerable populations.[16]

Lastly, the most recent and alarming challenge arose in the form of technologically created offensive activities. At least 12 categories of technology-facilitated gender-based violence have been identified, including sextortion (blackmail threats to post sexual content), image-based abuse (dissemination of intimate photographs), doxxing (leaking personal information), cyberbullying, online gender or sexual harassment, cyberstalking, grooming for sexual violence, hacking, hate speech, impersonation online, and technology-enabled tracking of victims for subsequent attack (Risinger, 2020). These tools can also facilitate offline violence, such as intimate partner violence, domestic abuse, or human trafficking. The top ten most commonly reported types of violence are as follows: misinformation and defamation (67%), cyber harassment (66%), hate speech (65%), impersonation (63%), hacking and stalking (63%), astroturfing (coordinated spreading of harmful content, 58%), image- and video-based abuse (57%), doxxing (55%), violent threats (52%), and unwanted explicit images or content (43%). According to the EU Rights Agency, one in ten women in the EU have suffered from cyber-harassment since the age of 15, including receiving unwelcome sexually explicit email or messages or people making inappropriate advances on social media.[17]

The most crucial factors responsible for aggravating the online gender-based abuse of women are user anonymity and the potential to create fake identities online, user-based algorithmic patterns, mob-mentality amongst the male counterparts who resort to stereotyped perceptions against the women users, and lastly the abundance of derogatory and anti-women content posted online.[18]

In the light of statistical data as aforementioned, it can be safely concluded that both in the United States and in the European Union, underrepresentation of women users on social media platforms is a crucial issue and even the developed economies with liberal mindset and lesser diversification in terms of caste, class, and linguistic barriers, as compared to that of India, are facing challenges such as rampant online harassment, unnecessary and arbitrary surveillance and data collection hazards, stereotypical gendered notions online, and frequent gender-based sexual abuse online.

8.3 Women and Social Media – Indian Perspective

As per the data compiled by Datareportal, as of January 2024, there were 462 million active social media user identities in India, making up 32.2% of the entire population and the number of social media users in India amounted to 61.5% of the entire internet user base regardless of age as of January 2024. As for gender demographics of social media users in India, 31.4% were female and 68.6% were male and of those social media platforms, the gender breakdown of the ad audiences were as follows: Facebook users were 25.5% female as against 74.5% male; YouTube, 32.4% female, 67.6% male; Instagram, 32.8% female, 67.2% male; LinkedIn, 29.7% female, 70.3% male; Snapchat, 35.9% female, 61.9% male; and X (formerly known as Twitter), 14.4% female versus 85.6% male.[19]

Digitization and the emergence of digital media platforms and the social media platforms have allowed female voices to be heard and amplified. There seem to be female content creators, journalists, and influencers on these platforms addressing issues that matter to women, telling their stories, and confronting stereotypes and additionally, government policies and advocacy efforts have contributed to promotion of female representation in media.[20] These efforts, including the National Commission for Women and movements like #MeTooIndia, have brought to light the need for gender equality and representation in media and have also called for higher levels of accountability in the industry.[21]

8.3.1 Challenges in India

There are multiple challenges for women social media users in the country. Firstly, there is a lack of definite, uniform, standard, and comprehensive set of data collections and reports that can smoothly reflect the usage patterns online and the threats faced by the Indian citizens periodically. This raises a significant concern because with the private players compiling data, it leaves lesser room for uniform usage trends. Secondly, social media platforms in India often suffer from "exclusionist syndrome"; that is the demographical, caste, and class-based barriers often hit against the idea of "inclusiveness," making only the dominant and the majority-driven community or class the "judge" for deciding as to who and to what extent can someone be heard or seen online. Thirdly, as per the report compiled by Statista, as of 2024, out of 954 million active internet connections, 556 million were recorded in the urban areas, leaving just one third connections in the rural areas, reflecting that urban–rural divide can be objectively acknowledged, making it difficult for the rural communities to become a part of the mainstream connectivity online.[22] Fourthly, since India comprises hundreds of languages with thousands of dialects, language-based restrictions pose a twofold challenge to the participation of women on social media platforms, one being the adverse impact on the penetration of mainstream social media tools at the grassroot level and the other being the psychological anxiety amongst the women lacking knowledge of the mainstream languages like English and Hindi, leading to their self-exclusion from engaging over such platforms. Fifthly, just like in the offline sphere, intersectional discrimination is prevalent in the form of caste-based exclusion on online social media engagement, making it difficult for women belonging to marginalized castes to express themselves or be heard and seen online with effective liberty.[23] Sixthly, as per the Digital India Report compiled by Datareportal, as of 2024, only 25.5% of Facebook users were women as compared to 74.5% male users, making it clearly evident that the gender divide in an emerging economy like India is still prevalent. When compared with the United States and the United Kingdom, it can be safely concluded that the developed economies have reflected the opposite statistics, with women user engagement topping the charts.[24] Seventhly, rural women users have frequently faced bans and restrictions on their usage of smartphones, social media, and so forth on the pretext of societal norms. Technology access to women has been seen as an impure move that could spoil their integrity and womanhood.[25] Eighthly, forms of abuse meted out to women offline have often been witnessed in the online medium as well. This has adversely impacted the psych of the women users, in general, casting apprehensions in their minds that they could be facing grave consequences if they tried to express themselves on social media, making them hesitant and anxious. Ninthly, political polarization of social media platforms has played a vital role in excluding the women users from expressing themselves liberally with a sense of political security.[26] Tenthly, sexually aggravated comments, gendered abuse, and online trolling have become the order of the day and they account for a significant set of reasons for women users to feel extreme anxiety and apprehensions of fear before going online or expressing themselves per se.[27] Eleventhly, the rampant objectification of women online, by reducing their image as a mere refection of their physical structure, has mired the women users of social media platforms frequently. It is evidently established that objectification is the result of traditional perceptions held by the patriarchal mindset of male population with respect to womanhood and its incidental personality traits.[28] *Lastly*, India topped the list of countries in terms of the number of active users on Facebook, making it prone to algorithmic inequities. Women users often face implied harassment because of the gendered algorithmic patterns in terms of content posted online. This has often raised debates on "accuracy versus engagement," creating echo chambers for the male community and their traditional perceptions towards women. Algorithmic settings prefer virality of a post over accuracy, which poses a danger to women users in the form of gendered posts, derogatory towards the integrity of women.[29]

8.3.2 Social Media, Women, and Legislative Framework in India

India has had the Information Technology Act, 2000 (IT Act), which was originally passed to protect e-commerce and financial institutions from cyber-crime. Nonetheless, this legislation does not provide redressal for individual male, female, and child's vulnerabilities against personal attacks in the digitized world and the Act construes "cyber-crime against women" as primarily sexual crimes, or abuse of women, including modifying photographs and creating pornographic material without consent, sending obscene messages or harassing through sending sexually explicit messages, or cyberstalking, making its ambit narrowly carved.[30] However, Section 67,[31] which governs pornography and obscenity on the internet and has given rise to many legal cases, and other provisions like the Indecent Representation of Women (Prohibition) Act, 1986, as well as certain provisions of the Indian Penal Code,[32] have been used more infrequently to protect women from indecency. These legal frameworks have become largely outdated in the current context, however. The focused Indecent Representation of Women (Prohibition) Act, 1986, was formulated to ban materials that indecently depict women, but its relevance and applicability in the current digital framework for the state require a thorough re-examination.

The Indecent Representation of Women (Prohibition) Act, 1986, was enacted specifically to protect the dignity of women by prohibiting the publication and dissemination of materials containing indecent representations of women. An important point to note here is that Indian conception of obscenity differs from the European conception; hence a woman shown in any sexual position or dressing can be construed as obscene in the eyes of the Indian society.[33]

Feminist advocates have sought protections under this Act but also expressed misgivings about its structure, which focuses upon crimes against women primarily through an obscenity lens.[34] This view makes a good and moral play, and fails to address the greater scheme of things women face on a daily basis, such as derogatory name calling, gendered trolling, sexist remarks, violence, or worse. Consequently the Act has not delivered on establishing the Indian perspective of gendered morality.[35] There are various calls that the Act needs to be amended so that in all forms of media (digital included) the punishment and acts are aligned with those in the Information Technology Act. If these amendments are made, legal protection against being depicted in an obscene manner would become very strong.[36] The Indian IT Act was amended in 2008 and added two gender-neutral provisions 67A and 67B covering production, creation, and distribution of obscene and sexually explicit material in digital form. Sections 67 and 67A work in conjunction to provide protection against cybercrimes towards women; however, these provisions are mostly concerned with the formulation of charges and financial punishment.[37] The word "whoever" implies that even people that make "lewd" imagery can be responsible for their behaviour.[38]

There are considerable gaps in the existing Indian Information Technology (IT) law, which requires urgent reform. Firstly, it lacks a provision mandating statutorily "removal or deletion of derogatory material," leaving scope for a deeply seeded traumatic experience for the victim in any walks of life. This shall cater to the psychological remedy to the ill-effects of the offence committed, apart from the actus reus.[39] Moreover, the IT Act lacks provisions for women investigators or cyber cells put on duty, specifically for female victims, making it difficult for the victims to come out of their socially anxious and hesitant shells and openly report the occurrence, with zero fear and manipulation of facts, whatsoever. This gets further strengthened by the interplay of Section 78 and Section 80(1) of the IT Act, none of which clearly specify that investigations need to be conducted by officers of the same gender, in the absence of gender-specific assistance cells. Sensitization has always been the key to resolve situations involving marginalized communities, like women, in the society.[40]

Moreover, it lays down the accountability in a manner which is broad enough by using "whoever commits." As written, this language means original creators of profiles could also be held liable if the content is found "sexually explicit" per se, owing to its creation, publication, or sharing. This can create a framework for reverse victimization, arguing instead that it is what an individual wears that incites crime.[41] Ultimately, internet service providers (ISPs) and intermediaries should be made more accountable to online crimes. People in India use social media services and ISPs that are based abroad, blurring the lines of where one should be accountable under what regulations. ISPs often offer little help, citing the incident does not satisfy their criteria or that the incident is not illegal under Indian law.[42] Greater accountability is needed to ensure these platforms take tangible steps to prevent online crime.

There is a notable absence of uniformity in the legal regimes adopted by various nations, alongside the lack of a cohesive international cyber law framework to protect victims and prosecute offenders. In the context of a globalized digital landscape, emails, websites, and applications can be accessed from any jurisdiction, facilitating the commission of crimes that may be transnational in nature. This enables an offender to perpetrate unlawful acts within the territory of India while physically located elsewhere, often escaping legal consequences due to insufficient statutory measures to extradite and hold such criminals accountable.[43]

The IT Rules 2021 have been updated and issued, with the aim of increasing the safety, trustworthiness, and accountability of cyberspace, under the ambit of the powers conferred by the IT Act directing the intermediaries to publish their rules, privacy policy, and user agreements on their websites and apps, while making users aware of those policies and taking action to prevent them from posting illegal material; they are prohibited from disseminating obscene or invasive content, particularly material that may incite a riot or disturb the public order.[44]

Intermediaries, for example, must also assist legally deemed authorized government agencies which request information critical and reasonably needed for law enforcement and cybersecurity purposes and put in place a grievance redressal mechanism where complaints about rule violations can be addressed within 72 hours.[45] Large social media intermediaries, those having more than 5 million registered users, in India are expected to carry out additional due diligence, which includes appointing a Chief Compliance Officer and a Resident Grievance Officer and releasing compliance reports every month.[46]

The new amendments, effective as of October 28, 2022, establish Grievance Appellate Committees, which will be able to hear appeals from users against decisions issued by Grievance Officers. In this regard, a National Cyber Crime Reporting Portal to report any kind of cybercrime with special reference to women has also been initiated by the Ministry of Home Affairs.[47]

Cyber offenses cannot be effectively addressed solely through regional legislation, underscoring the imperative for the establishment of international cyber legal frameworks. When foreign websites contribute to criminal activities affecting individuals, institutions, and corporations in India, significant challenges emerge regarding the applicability and enforcement of domestic laws. The existing legal instruments must be reformed to better address the complexities of cybercrime in an interconnected environment, thereby ensuring the protection of victims and the prosecution of perpetrators, regardless of their jurisdictional location.

8.3.3 Government Initiatives in India

Apart from providing a dedicated website to facilitate and smoothen the reporting procedure, the Government of India has launched the Cyber Crime Prevention against Women and Children (CCPWC) project which is being funded from the Nirbhaya Fund, and under this initiative, the National

Cybercrime Reporting Portal (NCRP) is created.[48] Moreover, a toll-free helpline, 1930, has been activated across all states and union territories (UTs) to help users and as a part of the CCPWC initiative; 33 states and UTs have established Cyber Forensic cum Training Labs.[49]

The most groundbreaking initiative was launching the Emergency Response Support System (ERSS-112) and its related app, which provides the opportunity to report grievances on a real-time basis.[50] Other significant initiatives include operationalization or upgradation of State Forensic Science Laboratories (SFSLs) in 30 states and UTs, integration of Women Helpline (WHL)-181 with Emergency Response Support System (ERSS) for benefiting more than 76.02 lakh women as on May 31, 2024, and establishment of Monitoring Centres in 11 states and UTs for implementation of Vehicle Tracking Platform (VTP).[51]

In addition, immediate complaints can be done through the Citizen Financial Cyber Fraud Reporting and Management System for financial frauds with saving of more than Rs. 1200 crore in over 4.7 lakh complaints.[52] The CCPWC scheme, which provides funding to the states and UTs for the establishment of cyber forensic training laboratories, creation of junior cyber consultants, and training of law enforcement personnel, judicial officers and public prosecutors, has received a financial assistance of Rs. 122.24 crore.[53]

The implementation of the Information Technology (Intermediary Guidelines and Digital Media Ethics Code) Rules, 2021, expands user rights on intermediary platforms and makes social media companies liable for user safety. These rules require intermediaries to have a robust grievance redressal mechanism to address complaints within a stipulated time frame and make that announcement to users, clear their terms and conditions restricting harmful, defamatory, obscene, or otherwise illegal content; and prohibit them from displaying, uploading, modifying, publishing, transmitting, updating, or sharing such content. Importantly, intermediaries must take down unlawful content promptly upon receiving either a court order or a notice from the appropriate government or its representative.[54]

The Ministry of Home Affairs (MHA) has shared substantive content on several forms of cybercrime and has distributed materials through its Twitter account (@cyberDost), undertaken radio campaigns, and published the *Handbook for Adolescents/Students* to sensitize them on cybercrime and additionally, the Central Board of Secondary Education (CBSE) provided instructions for schools to sensitize children about safe internet usage. The circular urges schools to set up effective firewalls and filtering software on all computers and implement strict security policies.[55]

8.3.4 Judicial Discourse on Social Media, Women, and Legislative Framework in India

Recently, whilst hearing a bail plea filed by one Rehana Fathima, the Kerala High Court reprimanded the applicant women activist, observing that her act of posting a cooking recipe video on social media, using the term "Gomatha Ularth" in a dish involving beef, clearly had the potential to hurt the religious sentiments of other religious communities, making her liable under the impugned sections of penal laws. This case highlighted the contours of the fundamental right of a user to express herself online, in contradistinction to that user's duty to respect the sentiments of the religious communities who were made its audience. Ultimately, the Court took a lenient stand and abstained from cancelling her bail altogether obtained in another case.[56] The matter is pending trial, and it gained traction because of the stringent bail direction included in the order which put a complete bar on the applicant women activist from posting, publishing, and so forth anything as such on her social media account. Later, the aforementioned strict conditions were stayed by the Supreme Court.[57] This decision didn't throw much light on the gender-specific judicial discourse, but what makes it significant is the way in which the Supreme Court tried to balance the scales of freedom of speech and the restriction which can be imposed. Whilst easing the stringent restrictions placed as bail condition to adopting a lenient approach whilst reprimanding the applicant to the required extent, Court was able to adopt a balanced take on a woman applicant's freedom of speech and expression.

The interaction between women and social media platforms has taken a curvy graph passage where in specific instances. Courts have adopted a restrictive view or constructive view of women's ability to engage online. A few years ago, the Indian Supreme Court handed down a landmark ruling to end this interference. Picturing women as an accused in a case helped in laying ahead a judicial discourse which did cure the mischief, but it still left certain circumstantial vacuum with respect to certain challenges. One[58] that many people know of is where Section 66A of Information Technology Act (ITA) was declared unconstitutional by the Supreme Court. Two women were arrested by police for allegedly posting remarks on Facebook that purportedly mocked their state's shutdown after the sudden death of a political leader, under Section 66A of the Information Technology Act of 2000 (ITA), which penalizes the transmission of grossly offensive information or knowingly false information aimed at causing annoyance or injury. The women were later released and their charges dropped, but the episode drew significant media coverage and condemnation. They meanwhile had approached the top court, seeking striking down of the provision, arguing that Section 66A was unconstitutional and violated their right to freedom of speech and expression. This badly drafted provision made it an offence to send offensive messages using a computer or communications device, defining such messages as any that were "offensive or menacing" or false representations intended to cause annoyance, inconvenience, danger, or insult. On many occasions, this law was abused by governments in order to repress political dissents or even censure opinions[59] The Court first considered three fundamental concepts crucial to understanding the reach of freedom of expression: discussion, advocacy, and incitement. It claimed that "mere discussion or even advocacy of a particular cause, no matter how unpopular, is at the heart" of this right.[60] The law can only restrict this freedom when such discussion or advocacy rises to the level of incitement.[61] Considering the nature of the present case, the Court found that Section 66A is likely to suppress any and all online expression, since there is no distinction maintained "between mere discussion or advocacy of a particular view which may be annoying, inconvenient or grossly offensive to some people and incitement when such propounding of view establishes an imminent causal connection resulting in public disorder, threat to security of the state and other that connected to public disorder."[62]

Herein, the fact that it involved women accused reflected the kind of interaction ensuing between them and the social media in general. Bestowing a victimized image upon the women in general was not the mainstream story anymore. This case highlighted a progressive step ahead, especially when the accused were women by gender. Yet is noteworthy to observe the dichotomous take on the aforementioned judgment because Section 66A was indeed poorly drafted and unconstitutional; its repeal has left an implied vacuum in legal protection directly tackling the sexual harassment that women face online.[63] Like a legislative dilemma, the real fruits of the aforementioned judgment can never be realized fully unless the Parliament makes a legislative attempt afresh via an amendment, to tackle the softer modes of online harassment and stalking.

Before the court's ruling in the aforementioned case, intermediaries could be held liable if they had "actual knowledge" of infringing material on their platforms; therefore, the user notifying an intermediator of the presence of such content may establish this knowledge, creating a "notice and take-down" framework.[64] But, in *Shreya Singhal*,[65] the Court reinterpreted the intermediary liability provisions under Section 79 of the IT Act and the regulations framed thereunder, declaring that no content could be removed without a judicial or executive order. It would place an unreasonable burden on intermediaries like Google and Facebook to decide whether potentially millions of requests are legitimate, the court reasoned.

However, subsequent judgments have required a reconsideration of the legal principles laid down in the *Shreya Singhal* case. An infamous example involved the sharing of rape videos. In response to a Public Interest Litigation filed by the NGO Prajwala, the Supreme Court took up the issue of the appalling content being circulated online and called Facebook, Yahoo, Google, and Microsoft as respondents[66] and observed that it is unambiguously absurd to wait for a judicial or executive directive to have the rights to remove vile content, videos of rapes, sexual abuse, and child pornography, and so forth and the Court analytically interpreted the implicit objective behind the framing of such enactments. These serious offenses require a more immediate and robust response than the traditional "notice-and-take-down" model.[67]

In the case of *Sabu Mathew George v. Union of India*,[68] the Supreme Court was asked to stop advertisements of prenatal and preconception sex determination kits from showing up in online search results, in light of the widespread practice of sex-selective abortions because of cultural preference for male children, which is prohibited under Indian law. The court went resolutely down the *Shreya Singhal* path little more than a month after *Shreya Singhal*, even as it directed the government to set up an independent body for the monitoring of search results and for informing search engines, which would be obliged by law to block prohibited content within a period prescribed to them. This institutional mechanism recognizes that under some circumstances, such need for speed may outweigh those regular workings of procedure.

8.4 Conclusion

Apparently, the potentially empowering power of social media for marginalized groups sits alongside risks, especially for women. But instead of walking away from the virtual world, we are bound to explore how social media offers great value as a tool enabling self-expression, community building, and social activism, even as it often mirrors and magnifies existing inequities in society and often leads to a grossly disproportionate amount of online gender-based violence against women. Although the existing legal framework does include a few relevant provisions, it does not provide a sufficiently detailed treatment of how the stakeholders can adapt in this rapidly evolving digital environment. Terminologically, the legislative framework in India warrants greater certainty, uniformity, and comprehensiveness. The substantive aspect of the mischief and its consequent legal resolution will not be sufficient in times to come. It has to be coupled with an approach which sensitizes the state, intermediaries, and the individual users, making them aware of the psychological impact on the victims of online gender-based abuse and harassment.

The contemporary enactments, like the Information Technology Act of 2000, the Bhartiya Nyaya Sanhita, and the Indecent Representation of Women (Prohibition) Act of 1986, lack the required legislative panache. Issues like gendered abuse online carry sensitive aspects alongside the actus reus of offences defined under the aforementioned enactments, warranting a sensitized approach. It can be achieved only by adopting the "non-ambiguous" drafting of statutory definitions given to the used terminologies. Terms like "offensive" and "obscene" have to be given country-specific and society-specific subjective connotation, especially with the advent and expansion of mediums of communication in the contemporary world of social media. The chronological essence of The Indecent Representation of Women (Prohibition) Act, as a 1990s law, reflects much about its redundancy and failure in providing a secure infrastructure to the women victims of newer offences online, the likes of cyberstalking, trolling, name-calling, identity-theft, and so forth. Perceiving obscenity through the lens of cultural mores will constrain the functional scope of the enactment, diluting its objective in the current scenario.

Moreover, the significance of algorithmic narratives build up and disseminated over the social media platforms, can never be overlooked, especially when the "virality" of the posts/content online tends to undermine the importance of "accuracy" attached to it. Gender-based inequities have constantly endangered the social fabric in the physical world. If it is able to hold its ground firmly in the online world, it can cause harm, larger in impact. In the name of "virality-based" algorithmic biases, the absence of "accuracy" has the potential to stifle the voices, presence, and participation of women social media users, making it difficult to ensure effective representation of women on social media. This will cut off the women community from the mainstream benefits of the use and utilization of social media tools as a platform to express. Without visibility on content moderation practices and algorithmic decision-making processes, holding social media companies accountable for their platforms' role in online gender-based violence becomes difficult, if not impossible. This shall require setting up of a thoroughly scrutinized algorithmic setup by attaching a legal liability upon the social media intermediaries to abide by the nation-wise trends and patterns, keeping a real-time tap on emerging evils such as gender-based abuse online and the consequent harassment.

Enforcement and implementation of laws, policies, judgments, and executive directions is always a tricky issue for any administrative setup, especially when the challenges posed have taken the digital realm within its fold. The procedural setup aimed at resolving such issues involves technical upskilling of the investigating and prosecuting authorities, coupled with periodical assessment and sensitization camps for them. The transnational nature of certain cybercrimes against women throws bonus challenges, in the form of jurisdictional issues. The stunted development of specialized cybercrime units with the skills required to investigate these complex offences means they are often underreported and under-prosecuted.

Courts have also played a vital role, through judicial interpretations of existing laws in response to online gender-based violence, whenever they have had the opportunity to do so. In *Shreya Singhal v. Union of India* (2015),[69] the Supreme Court was able to chart implications for the intermediary liability via Section 79 of the IT Act, warranting a careful balance between freedom of speech and pulling the plug on harmful content in the digital space. Such an evolving legal framework requires a technologically cautious, context-sensitive approach that seeks to promote free speech protections whilst enabling the state and civil society to act decisively against online forms of gender-based violence.

A multi-pronged approach is essential for addressing gender-based abuse online. This shall require overhauling of the enactments by inviting the concerned stakeholders at the table of discussions, so as to compile a set of safeguards, directives, and remedies in a comprehensive manner. In order to deal with the trans-national dimension of the issues, the state needs to foster global cooperation efforts diplomatically. Periodical bias audits, independent state-wise assessments, and stringent yet balanced rules for content moderation, and so forth can be utilized in achieving the aforementioned objectives. State investments in training and sensitizing the enforcement and adjudicating machinery shall cater to the attempts to stifle online gender-based abuses. Additionally, digital literacy amongst the most significant stakeholder, users of social media, is clearly warranted. Any misuse cannot be avoided without creating an awareness mechanism with respect to the risks involved, offensive acts, and remedies. Globally, a diplomatic effort is required to create a harmonized setup for smooth resolution of trans-national issues and for sharing the best state-of-the-art practices prevalent worldwide.

To conclude, a multifaceted and multipronged approach is the need of the hour, if online gender-based abuse and violent acts have to be effectively prevented, prohibited, and remedied. The cultural and diversified social outlook of a country like India cannot be overlooked in the aforementioned process. Therefore, technological interventions, holistic legislative framework with in-built certainty and definitiveness, and a sensitized judicial discourse via precedents are required to make an attempt to contain the vices as soon as possible, with the objective of securing the online space for women and other marginalized communities. It is thus crucial that the narrative surrounding the constraints placed on the tech giants focuses as much on accountability, transparency, and preventive measures as it does on the economic and political interests of the platforms, and how best it might align with a more equitable digital future for everyone.

Notes

1 According to the *Oxford Advanced Learner's Dictionary*, social media is defined as websites and software programs that allow people to communicate and share information online *available at:* https://www.oxfordlearnersdictionaries.com/definition/english/social-media (Last visited on May 14, 2026).
2 Mitali Mukherjee, Aditi Ratho, et al., "Unsocial Media: Inclusion, Representation and Safety for Women on Social Networking Platforms", *Occasional Papers*, May 6, 2021 *available at:* https://www.orfonline.org/research/unsocial-media-inclusion-representation-and-safety-for-women-on-social-networking-platforms (Last visited on January 11, 2025).
3 Internet and Mobile Association of India, "Internet in India 2023" *available at:* https://uat.indiadigitalsummit.in/sites/default/files/thought-leadership/pdf/Kantar_iamai_Report_20_Page_V3_FINAL_web_0.pdf (Last visited on January 11, 2025).
4 Ani Petrosyan, "Worldwide Digital Population 2024", *Statista* (Nov. 5, 2024) *available at:* https://www.statista.com/statistics/617136/digital-population-worldwide/ (Last visited on January 11, 2025).
5 Stacy Jo Dixon, "Instagram Distribution of Global Audience 2024, by Gender", *Statista* (Mar. 21, 2024) *available at:* https://www.statista.com/statistics/802776/distribution-of-users-on-instagram-worldwide-gender/ (Last visited on January 11, 2025).
6 World Population Review, "Instagram Users by Country 2024" *available at:* https://worldpopulationreview.com/country-rankings/instagram-users-by-country (Last visited on January 11, 2025).
7 Mukherjee, Ratho, et al., above note 2.

8 Pew Research Centre, "Social Media Fact Sheet", Nov. 13, 2024 *available at:* https://www.pewresearch.org/internet/fact-sheet/social-media/?tabItem=0ec23460-3241-4a1f-89bc-0c27fb641936 (Last visited on January 11, 2025).
9 Adriane Van Der Wilk, "Cyber Violence and Hate Speech Online against Women: Women's Right and Gender Equality", *European Parliament*, Aug. 2018 *available at:* https://www.europarl.europa.eu/RegData/etudes/STUD/2018/604979/IPOL_STU(2018)604979_EN.pdf (Last visited on January 11, 2025).
10 Emily A. Vogles, "The State of Online Harassment", *Pew Research Centre*, Jan. 13, 2021 *available at:* https://www.pewresearch.org/internet/2021/01/13/the-state-of-online-harassment/ (Last visited on January 11, 2025).
11 Julia T. Wood, "Gendered Media: The Influence of Media on Views of Gender" *available at:* https://www1.udel.edu/comm245/readings/GenderedMedia.pdf (Last visited on January 11, 2025).
12 Amnesty International, "Briefing: Gender and Human Rights in the Digital Age", Jul. 10, 2024 *available at:* https://www.amnesty.org/en/documents/pol40/8170/2024/en/#:~:text=Marginalized%20groups%2C%20including%20women%20and,not%20reflect%20their%20individual%20realities (Last visited on January 11, 2025).
13 Amnesty International, above note 12.
14 Amnesty International, above note 12.
15 "Amnesty International Research Claims That Social Media Apps Are Perpetuating Gender Inequalities", *The Dialogue Box*, Jul. 16, 2024 *available at:* https://thedialoguebox.com/amnesty-international-social-media-research/ (Last visited on January 11, 2025).
16 Above note 15.
17 United Nations Regional Information Centre for Western Europe, "How Technology-Facilitated Gender-Based Violence Impacts Women and Girls", Nov. 29, 2023 *available at*: https://unric.org/en/how-technology-facilitated-gender-based-violence-impacts-women-and-girls/ (Last visited on January 11, 2025).
18 Mukherjee, Ratho, et al., above note 2.
19 Simon Kemp, "Digital 2014: India", *Datareportal*, Feb. 21, 2024 *available at:* https://datareportal.com/reports/digital-2024-india (Last visited on January 11, 2025).
20 Deogirikar Sonali and Bobhate Prathamesh Rajesh Anagha, "Empowerment or Objectification? An Analysis of Female Representation in Modern India Media", *GAP Bodhi Taru* 7 (2024), 396 *available at*: https://www.gapbodhitaru.org/res/articles/(396-401)%20EMPOWERMENT%20OR%20OBJECTIFICATION%20AN%20ANALYSIS%20OF%20FEMALE%20REPRESENTATION%20IN%20MODERN%20INDIAN%20MEDIA.docx.pdf (Last visited on January 11, 2025).
21 Abhery Roy, "2018: The Year When # MeToo Shook India", *The Times of India*, Dec. 27, 2018 *available at*: https://economictimes.indiatimes.com/magazines/panache/2018-the-year-when-metoo-shook-india/2018-the-year-of-metoo-in-india/slideshow/66346583.cms?from=mdr (Last visited on January 11, 2025).
22 Shangoliao Sun, "Number of Internet Connections in Rural and Urban Areas of India from Financial Year 2015 to 2024", *Statsta*, Sep. 27, 2024 *available at:* https://www.statista.com/statistics/1196721/india-internet-connections-in-rural-and-urban-areas/ (Last visited on January 11, 2025).
23 Siva Mathiyazhangan, "A Call for Algorithmic Justice for Sc/ STs", *The Indian Express*, May 14, 2023 *available at*: https://indianexpress.com/article/opinion/columns/a-call-for-algorithmic-justice-for-sc-sts-8607880/ (Last visited on January 11, 2025).
24 Kemp, above note 19.
25 Arvind Chauhan, "Panchayat Bans Mobile Phones Among Girls", *The Times of India*, Feb. 20, 2016 *available at:* https://timesofindia.indiatimes.com/city/agra/

panchayat-bans-mobile-phones-among-girls/articleshow/51060339.cms (Last visited on January 11, 2025). Also see, "Rupees 2, 100 Fine on Women Using Phone in Public in U.P. Village", *The Hindu*, May 4, 2017, *available at:* https://www.thehindu.com/news/national/2100-rupees-fine-on-women-using-mobile-phones-in-public-in-up-village/article18377318.ece (Last visited on January 11, 2025).

26 Tabrez Ahemed Neyazi, "Social Media and Political Polarization in India", Nov. 1, 2017, *available at:* https://www.researchgate.net/publication/320868617_Social_media_and_political_polarization_in_India (Last visited on January 11, 2025).

27 N. Karusala, A. Bhalla and N. Kumar, "Privacy, Patriarchy, and Participation on Social Media [Conference Session]", in *Proceedings of the 2019 on Designing Interactive Systems Conference*, June 23–28, 2019; San Diego, SA, 511–526 *available at* https://pmc.ncbi.nlm.nih.gov/articles/PMC9247611/ (Last visited on January 11, 2025).

28 Sonali and Anagha, above note 20.

29 Showkat Ahmad Dar, "Unleashing Her Voice: How Social Media Empowers Women", *International Journal of Information Management* 8 (2023), 7.

30 Akanksha Pathak and Prateek Tripathi, "Digital Victimization of Women in Cyberspace: An Analysis of Effectiveness of Indian Cyber Laws", *available at:* https://nluassam.ac.in/docs/Journals/NLUALR/Volume-7/Article%207.pdf (Last visited on January 11, 2025).

31 Section 67 reads: "Whoever publishes or transmits or causes to be published or transmitted in the electronic form, any material which is lascivious or appeals to the prurient interest or if its effect is such as to tend to deprave and corrupt persons who are likely, having regard to all relevant circumstances, to read, see or hear the matter contained or embodied in it, shall be punished on first conviction with imprisonment of either description for a term which may extend to three years and with fine which may extend to five lakh rupees and in the event of second or subsequent conviction with imprisonment of either description for a term which may extend to five years and also with fine which may extend to ten lakh rupees." See The Information Technology Act, 2000, India, *available at:* https://www.meity.gov.in/writereaddata/files/The%20Information%20Technology%20Act%2C%202000%283%29.pdf (Last visited on January 11, 2025).

32 Section 509, Indian Penal Code 1860, India (now section 79 of Bhartiya Nyaya Sanhita 2023, India) reads: "Whoever, intending to insult the modesty of any woman, utters any words, makes any sound or gesture, or exhibits any object, intending that such word or sound shall be heard, or that such gesture or object shall be seen, by such woman, or intrudes upon the privacy of such woman, 1 [shall be punished with simple imprisonment for a term which may extend to three years, and also with fine]." Section 292 IPC (now Section 294 BNS) caters to public obscenity aspects.

33 Debarati Halder, "Examining the Scope of Indecent Representation of Women (Prevention) Act, 1986, in the Light of Cyber Victimisation of Women in India", *National Law School Journal* 11 (2013), 188.

34 Geetanjali Gangoli, *Indian Feminisms: Law, Patriarchies, and Violence in India* (Ashgate Publishing, Ltd., 2012).

35 Ved kumari, "Gender Analysis of the Indian Penal Code", in *Engendering Law: Essays in Honor of Lotika Sarkar*, ed. Archana Parashar and Amita Dhanda (Eastern Book Company, 1999).

36 Pathak and Tripathi, above note 30.

37 Information Technology Act, 2000, India.

38 Halder, above note 33.

39 Anita Gurumurthy and Niveditha Menon, "Violence against Women via Cyberspace", *Economic and Political Weekly* 44 (2009), 19.

40 Halder, above note 33.
41 Gurumurthy and Menon, above note 39.
42 Halder, above note 33.
43 Halder, above note 33.
44 Government of India (Ministry of Electronics and IT), "Cybercrime Against Women", *Press Information Bureau* (2022) *available at:* https://pib.gov.in/PressReleseDetailm.aspx?PRID=1881404&reg=3&lang=1 (Last visited on January 11, 2025).
45 Government of India (Ministry of Electronics and IT), above note 44.
46 Government of India (Ministry of Electronics and IT), above note 44.
47 Pathak and Tripathi, above note 30.
48 Government of India (Ministry of Women and Child Development), "Government Implements a Scheme of Cyber Crome Prevention against Women and Children (CCPWC) under Nibhaya Fund", *Press Information Bureau* (2024) *available at:* https://pib.gov.in/PressReleaseIframePage.aspx?PRID=2039866 (Last visited on January 11, 2025).
49 Government of India (Ministry of Women and Child Development), above note 48.
50 Government of India (Ministry of Women and Child Development), above note 48.
51 Government of India (Ministry of Women and Child Development), above note 48.
52 Government of India (Ministry of Women and Child Development), above note 48.
53 Government of India (Ministry of Women and Child Development), above note 48.
54 Rajya Sabha Un-starred Question No. 2390 to be answered on March 23, 2022 *available at:* https://sansad.in/getFile/annex/256/AU2390.pdf?source=pqars (Last visited on January 11, 2025).
55 Government of India (Ministry of Women and Child Development), "National Cyber Crime Reporting Portal Enables Public to Report Incidents of Cyber-Crimes, with Special Focus on Cyber Crimes Against Women and Children", *Press Information Bureau*, Dec. 18, 2024 *available at:* https://pib.gov.in/PressReleasePage.aspx?PRID=2085609 (Last visited on January 11, 2025).
56 "Use of Term 'Gomatha' for Beef in Cookery Show Likely to Hurt Religious Feelings of Hindus: Kerala HC Precludes Rehana Fathima from Using Social Media", *Live Law* (2020) *available at:* https://www.livelaw.in/news-updates/use-of-term-gomatha-for-beef-in-cookery-show-likely-to-hurt-religious-feelings-of-hindus-kerala-hc-precludes-rehana-fathima-from-using-social-media-166297 (Last visited on January 11, 2025).
57 "Rehana Fathima Can Use Social Media Without Hurting Religious Sentiments: Supreme Court Partially Stays HC's Order Over Her Cookery Show 'Gomath Ularth'", *Live Law* (2021) *available at:* https://www.livelaw.in/top-stories/rehana-fatima-supreme-court-social-media-kerala-high-court-bail-condition-gomatha-ularth-169594?infinitescroll=1 (Last visited on January 11, 2025).
58 *Shreya Singhal* v. *Union of India*, AIR 2015 SC 1523.
59 IT for Change, "Submission on Online Violence Against Women to the Special Rapporteur on Violence against Women", (2018) *available at:* https://itforchange.net/index.php/submission-on-online-violence-against-women-to-special-rapporteur-on-violence-against-women- 1 (Last visited on January 11, 2025).
60 *Shreya Singhal* v. *Union of India*, above note 58.
61 *Shreya Singhal* v. *Union of India*, above note 58.
62 *Shreya Singhal* v. *Union of India*, above note 58.
63 *Shreya Singhal* v. *Union of India*, above note 58.
64 *Shreya Singhal* v. *Union of India*, above note 58.
65 *Shreya Singhal* v. *Union of India*, above note 58.
66 *In Re: Prajwala*, Letter Dated Feb. 18, 2025 *available at:* https://indiankanoon.org/doc/1471625 (Last visited on January 11, 2025).
67 *In Re: Prajwala*, above note 66.
68 *Sabu Mathew George* v. *Union of India and Ors*. MANU/SC/1674/2017.
69 *Shreya Singhal* v. *Union of India*, above note 58.

9 Role of AI in Realising Feminist Utopia

Neha Dubey

9.1 Introduction

Feminists since long have dreamed of a utopian world where gender bias does not limit women's potential and opportunity for growth: a society that can be organized around ideals of gender equality, justice, and social progress. Several authors like Charlotte Perkins Gilman and Margaret Atwood through their literary works have used the concept of utopia to imagine worlds in which hierarchical gendered power dynamics no longer limit the potential of individuals, particularly women and marginalized genders. The feminists envisioned a utopian society free from oppression, having equal say in economic resources and governance, enjoying good health and full bodily autonomy, not having to struggle for their rights that includes education, and most significantly having the liberty to make their life choices and be treated with human dignity.[1],[2],[3] The utopian visions have long served as an inspiration for feminist activism and societal reforms. This chapter aims to examine the role of modern technology, especially of the artificial intelligence, in realizing feminists' utopia in contemporary time, along with the possible myriad socio-legal implications that may arise in the due process. Artificial intelligence (AI) seems to have a huge potential to play a transformative role in the realization of a feminist utopia. The chapter aims to explore whether by leveraging AI's capabilities in actualizing utopian ideals, can systemic gender inequalities be addressed to create a more inclusive and equitable society?

9.2 Feminists' Activism and Global Initiatives

To dream of such a utopian society, envisaged by feminists, in a patriarchal construct, have been met with mammoth challenges since the first voice to demand "rights for woman" was raised by Mary Wollstonecraft's in her seminal work *A Vindication of the Rights of Woman* (1792).[4] Still, the perseverance of the feminists for equal rights to realize the mentioned utopian ideals through generations created a global echo[5] that resounded in the establishment of the UN Charter 1945 as the UN's commitment to gender equality for progressive collective future.[6] The commitment was further reiterated in

DOI: 10.4324/9781003733751-13

the First World Conference on Women (held in Mexico City), marking the beginning of the UN Decade for Women (1976–1985).[7] Similarly, the world witnessed several milestones raised – the Convention on the Elimination of All Forms of Discrimination Against Women (CEDAW) (1979),[8] Beijing Declaration and Platform for Action (1995),[9],[10] Millennium Development Goals (2000)[11] – to name a few, to favour gender equality with gender sensitivity.

The Millennium Development Goals (MDGs) were established with Goal 3, specifically aiming to promote gender equality and empower women, focused on eliminating gender disparity in education, increasing women's participation in the workforce, and improving maternal health.[12] Ensuring the same ideals remain key foundation of societal development, they transitioned as Sustainable Development Goals (Number 5)[13] with further additions and elaborations. Goal 5 is dedicated to achieving gender equality and empowering all women and girls. The key "nine targets" of this goal addresses various issues, including ending all forms of discrimination and violence against women and girls, ensuring equal participation in leadership and decision-making, eliminating all harmful practices against women, providing universal access to sexual and reproductive health and rights, recognize and value unpaid care and domestic work, reforms to give women equal rights to economic resources, and so forth. Amongst the key nine targets, target "5.8 – Promote Empowerment of Women through Technology," that is "Enhance the use of enabling technology, in particular information and communications technology, to promote the empowerment of women,"[14] emphasizes upon the necessity of technological advancement in empowering women to establish gender equality.

9.3 Technological Advancement Leading to AI

Indeed, technological inclusion has been instrumental in empowering women. It has been advancing gender equality and improving the lives of women worldwide. By providing access to information, education, and economic opportunities, technology has empowered women to break barriers and achieve their full potential. It has visibly strengthened almost every aspect of a woman's life be it education, mobility, health care, finance, or other. Specially the information and communication technology transformed the access to information, education, and economic opportunities for women.[15] Digital platforms have amplified women's voices, enabling them to advocate for their rights and connect with like-minded individuals globally. It has enabled women to participate in decision-making processes, start businesses, and access global markets by alleviating geographical, social, and other limitations. As technology continues to evolve, its potential to create a more inclusive and equitable society for women becomes even more significant.

Technological advancement witnessed revolutionizing change with the advent of AI and its small- and large-scale incorporation with other existing and emerging innovations. AI has been defined as "the process of creating software that uses computers to carry out activities traditionally done by humans. A wide range of data mining techniques, encompassing but not extended to computational instruction, learning representations, neural networks, and processing of natural language, are collectively referred to as AI according to professional and everyday parlance."[16] The invention of AI marks one of the most historic periods in human history. From its early conceptualization to its current applications, AI has evolved significantly, impacting various aspects of human life. In 1956, the Dartmouth Conference, organized by John McCarthy, Marvin Minsky, Nathaniel Rochester, and Claude Shannon, is often considered the birthplace of AI as a field of study.[17] The conference aimed to explore the possibility of creating machines that could simulate human intelligence. However, progress was slow, and the field experienced several periods of reduced funding and interest, known as "AI winters." Despite these challenges, AI continued to evolve. The development of machine learning algorithms, particularly neural networks, in the 1980s and 1990s, marked a significant turning point. These algorithms allowed machines to learn from data and improve their performance over time. AI development saw a significant boost in the mid-2010s with the invention of big data and advancements in computing power, which further accelerated AI research and applications.[18]

9.4 Transformative Effects of AI in Actualizing Feminists' Ideals

The amalgamation of AI with smart technologies coupled with the internet changed the world as we knew it. It opened a plethora of opportunities and new avenues for all, but specially affected the lives of women. The demands raised by feminists in various feminist's movements[19] (i.e. suffrage, property rights, education, reproductive rights, work-place and legal equality, inclusivity, etc.), including the demands tagged as "radical"[20] for challenging stereotypical gender role functioning and claiming full bodily autonomy, gained strength and support under the changed scenario. The newer technologies are presenting women and girls with the options that were unavailable earlier. New doorways are opening through the intervention of revolutionary technologies in reshaping societal norms, cultural acceptance, and work–life balance.

The biggest game-changer technology, bringing the most diverse usage and opportunities, turned out to be the mobile phones, becoming "smart" after the AI's application. This device has become the weaponized tool for a woman to dismantle the earlier imposed gender limitations. Taking this simple consideration of a woman wanting to travel alone to an unknown place, which earlier was nearly unthinkable, also quite unsafe, is a common routine now for so many women. The AI-integrated smart phones providing them the assistance of Google Maps, services like Ola, Uber, and so forth, also the advantage of booking tickets of trains, buses, and flights along with the bulk data being available on the transport means for desired destination, make everything accessible and comparatively safe for a woman. This doesn't simply remove the "necessity" a woman felt to accompany someone in olden days; it offers the choice she can make independently.

Analogously, the multifaceted usage of AI technology vitally transformed education, finance, health, household, safety and community management for women, which aligns with fulfilling various SDG5 goals for gender equality.

9.5 AI's Impact

9.5.1 Education

In *A Vindication of the Rights of Woman* (1792), Wollstonecraft wrote extensively for women's right to be educated stating various reasons. She argued that women should receive the same education as men, because women aren't naturally inferior to them; rather lack of education renders such an appearance. She believed education is essential for women to develop their reason and virtue. She envisioned a society where women are seen as rational beings capable of contributing to society beyond their roles as wives and mothers. She believed empowering women will lead to a more just and virtuous society. Similarly, in India, Savitribai Phule, along with her husband, Jyotirao Phule, advocated for girl's education and established the first school for girls in India in 1848.[21]

Globally, the 19th century was emancipating for women in terms of access to education. Simultaneously, it depicts their marginalized condition based on gender discrimination across the world, not pertaining to any specific geographical location. Although the gender barrier fell to a larger extent, still women and girls have been confronting challenges in availing this fundamental right. Lack of finance and resources, domestic responsibilities, geographical limitations, societal norms and stereotyping restrict women's and girls' chances to avail proper education. Subsequently, the portion of women receiving higher or professional education has been quite slim.

But in recent years, technology has begun to play a significant role in expanding educational opportunities for women. Online learning platforms, mobile education apps, and digital literacy programs have made education more accessible to women and girls, especially in remote and underserved areas. For education purposes, many AI-integrated digital platforms emerged offering access to online education,[22] which has empowered women from remote areas too, to gain knowledge and degrees as well. Options like Coursera,[23] edX, and so forth offer opportunities to enrol in specialized courses and acquire degrees in various disciplines from universities and companies worldwide.

AI chatbots and virtual assistants provide real-time support and mentorship, helping women navigate their educational journeys and overcome challenges. AI enables flexible learning schedules, allowing women to balance their education with other responsibilities, such as work and family commitments. Online education has empowered women restricted to homes due to various reasons including scarcity of resources or finance or perhaps distance acting as barrier, to achieve their educational and career goals. AI-driven tools can also identify skill gaps and recommend specific courses or training programs to help women acquire the skills needed for their careers. Vocational courses, arts, and craft, aside academic courses, are imparting various skills to women and girls that have the potential to convert into economic gain.

9.5.2 *Economic Empowerment*

Women constitute nearly half of the global population, and their full participation in the economy can significantly boost productivity and innovation. However, women often face barriers such as limited access to financial resources, which hinder their economic empowerment. Addressing these challenges is essential for creating a more equitable society where women can contribute to and benefit from economic development. Charlotte Perkins Gilman realistically portrays the economic stature of women in society in her renowned book *Women and Economics*[24] (1898), while critiquing on the economic relationship between a man and woman and its social consequences. Gilman posits that the economic relationship between men and women is a significant factor in social evolution. She argued that women's economic dependence on men is detrimental to both genders and society as a whole. She advocated for women's economic independence as an essential for their personal development and societal progress. Criticizing the stereotypical notion that emphasized women's primary role is in the domestic sphere; she disseminated the idea that women should have the freedom to pursue careers, which is crucial for improving the social institutions as well. She believed that technological advancements have made many domestic tasks obsolete, and women should be free to engage in productive work outside the home surpassing the traditional set-ups. Till the end of the 19th century the economic divide between the genders was deep.

To mitigate this economic gap, SDG Goal 5 advocates for economic empowerment of women and AI is assisting as well as educating women in the art of managing their finances.[25] AI-driven financial services offer women better access to credit and financial management tools, helping them achieve financial independence. Besides assisting in the online banks' application, AI "robo-advisors" are advising them on financial investments to maximize gains. After analysing statistical data, which otherwise are generally incomprehensible for not so expert people, AI can suggest investment options for diversified investment, which has indirectly increased participation of women in different investment avenues like stock markets, mutual funds, and so forth. Besides, it allows remote access of various e-platforms, offering capacity building to secure jobs, and facilitating opportunities to work from home. Also, AI-driven platforms can match people with remote job opportunities that align with their skills and preferences. Although it may not seem gender specific, it is favourable, for specially those women enduring such obligations that require them to be at home. AI-powered platforms can facilitate connections with mentors, peers, and industry professionals, which can immensely help women in fostering a supportive community and expanding their professional networks. These connections can provide valuable guidance and support for career advancement.

Women aspiring to raise their own business have their path eased with AI assistance. E-commerce platforms like Amazon, eBay, and Flipkart facilitate even small-scale entrepreneurs to sell their products globally. AI-powered tools can help women identify market trends, optimize their operations, and connect with potential investors and customers. It immensely benefits women with little resources and support to grow their business, by alleviating the need of "man"-power to market their products in otherwise male-dominated traditionally structured markets.

These e-commerce marketplace options give rise to young women entrepreneurs as they offer bypassing of the traditional and difficult entry methods. For instance, Falguni Nayar (founder and CEO of Nykaa, a leading beauty and lifestyle retail company in India), established Nykaa in April 2012, leveraging her extensive experience in investment banking to create a successful e-commerce platform.[26] Similarly, Geetha Manjunath co-founded Niramai, a health-tech startup that uses AI and machine learning to detect early-stage breast cancer through a non-contact, non-invasive approach. Niramai's solution is cost effective and accessible to women of all age groups.[27] AI's assistance in economic growth is surely creating a more inclusive and supportive environment for women's economic empowerment.

9.5.3 *Unpaid Care Work*

Being the primary unpaid care worker, burdened with unequal distribution of household chores,[28] women can have their work eased out by using AI tools like bots and robots. One of the primary ways AI is initiating change is through smart home devices that automate household tasks such as cleaning, cooking, and laundry. AI-powered devices reduce the time and effort required for daily chores, allowing caregivers to focus on other responsibilities or take much-needed breaks. Chat-bots like Alexa facilitate them in organizing household tasks, managing some household appliances, assisting in party planning, and so forth, whereas AI robo-cleaners[29] assist in cleanliness of multiple sorts.

AI-driven health-monitoring systems ensure better health outcomes for those in need of care, while also reducing the stress and workload on caregivers, which are primarily women. Nowadays, with the latest inventions, fully developed robots are being developed to cater the household chores completely in the near future.[30] Although the percentage of women actually benefiting from such inventions is debatable, it indicates a possible future wherein women can finally be free from the tasks of household chores, perhaps providing a final solution to "The Problem That Has No Name" as emphasized by Betty Friedan in her celebrated book *Feminine Mystique*.[31] Friedan realistically portrayed the emptiness a woman feels in performing household duties in an unending cycle, silently and unrecognized.[32] Through this portrayal, Friedan raised poignant questions regarding socially accepted "feminine traits," not allowing women to voice their genuine concerns, while simultaneously challenging stereotypical gender roles. This became a celebrated document of second-wave feminists to emphasize the unappreciated unpaid care work done by women and the unjust unequal distribution of this workload that claims a lot of time, energy and body of a woman, further supported by the International Labour Organization (ILO) in their comprehensive report titled "Care Work and Care Jobs for the Future of Decent Work."[33] This report highlights the significant gender inequalities in unpaid care work and its impact on women's participation in the labour market. Among its key findings, it states that women perform 76.2% of the total hours of unpaid care work globally, which is more than three times as much as men.[34] The poignancy of the issue raised translated directly as SDG 5.4. Among other measures to bridge this gap, AI-driven technologies are becoming part of the solution by easing the load and saving time for women.

9.5.4 Safety

Safety has always been a priority concern for women. Recently, AI has been playing a pivotal role in ensuring women's safety by providing innovative solutions and tools. AI-powered safety apps and wearable devices offer real-time monitoring and alert systems, enabling women to quickly notify authorities or trusted contacts in case of emergencies. These technologies can track location, detect unusual activities, and send automatic alerts, enhancing personal security. AI-integrated apps, for instance bSafe, Safetipin, Tribe, Circle 6,[35] and many more, have evolved with real-time safety features offering a sense of security to women.

Many such apps have an inbuilt feature creating an inner circle of trusted people to reach out in a situation of crisis through SOS calls and messages having location tagged with it. Apps like Safetipin keep a record of safe routes and locations based on reviews, for users to make an informed decision. Spam calls and misleading messages are promptly detected by many safety apps, assuring protection from scammers and people with malafide intentions. Nari Suraksha Setu[36] (an app) helps women stay safe by providing emergency response, real-time location tracking, and access to support services. It also allows users to report incidents anonymously and offers information on legal rights and support organizations.

These apps have become safety companions for independently travelling women and girls. Apps featuring live location and tracking around the globe act as a safety net for travelling women and mental reassurance for their family members.[37] Additionally, AI-driven data analysis can identify patterns of violence and harassment, helping law enforcement agencies to predict and prevent potential threats. It also supports online safety by detecting and mitigating cyberbullying and harassment on social media platforms.

9.5.5 Digital Social Community

Most significantly, AI-integrated online platforms like Twitter, Facebook, and other digital platforms have created a global hub, connecting people rooting for feminism across the globe to collaborate and come together to raise global concerns regarding gender equality. These AI-powered platforms enable seamless communication and information of sharing, breaking down geographical barriers and creating inclusive spaces for diverse voices. For instance, the #Metoo movement[38] became a global voice against violence endured by women irrespective of their ethnicity, country, or language. AI apps having the capacity to translate any language in real time remove the barrier of comprehending other languages, facilitating cross-language communication, and encouraging a hybrid community of women coming together to share their experience and uplift each other in need. Also, AI-driven tools help in tracking hashtags, monitor sentiment,[39] and identify influential voices, thereby becoming instrumental in raising mass movements by garnering global support. It also helps in curating and recommending relevant content, ensuring that community members have access to valuable resources and information.

Having a global space to share their opinions on issues concerning local, national, or international events imparts a psychological boost, especially to women. Marginalized by the patriarchy from public spaces and sharing opinions on public matters left women unheard and their demands uncatered. The rise of forums to stage their demands and opinions has a critical effect globally in shaping the lives of women. For instance, a mass protest in Switzerland by women in June 2020 against gender pay gap, domestic violence, and unequal treatment[40] harnessed global support by social media posting, aside from news coverage. The influential voice of the protest and large-scale movement pressured the contemporary government to address the grievances raised by the Swiss women.[41] This movement became inspirational for all the women striving for gender equality and justice. This instance also underscores the utility of digital platforms integrated with AI in raising the magnitude of women's demand that must indispensably factor in decision-making or policy-making processes, which otherwise may get shunned.

9.5.6 Medical and Health Care

Landmark medical advancements have been achieved with AI's integration, from drug assessment to major surgeries, contributing in curing serious ailments. AI-driven health apps and telemedicine services are improving women's access to healthcare, especially in remote and underserved areas. These tools provide personalized health recommendations and support women's overall well-being without geographical limitations.

AI-powered tools are capable of analysing symptoms and provide preliminary diagnoses based on vast medical knowledge. These tools help healthcare providers make more accurate and timely diagnosis. For instance, according to the World Health Organization (WHO) in 2022, there were an estimated 20 million new cancer cases and 9.7 million deaths globally. As per the diagnosis, breast cancer was the most commonly diagnosed cancer in women, with 2.3 million new cases, accounting for 11.6% of the total new cases.[42] Breast cancer is the most common cancer in women in the vast majority of countries (157 of 185) and the leading cause of cancer death worldwide.[43] The predictive capability of AI to detect certain diseases like breast and ovarian cancer at an early stage creates the possibility of completely eliminating it. A recent study conducted by researchers from the University of Lubeck in Germany demonstrated that AI could improve cancer detection rates by over 17%. This study involved 460,000 women and showed that AI-assisted screenings had a higher detection rate compared to traditional methods.[44] Another study of a similar nature highlighted that AI models could outperform doctors in detecting breast and ovarian cancers. The research conducted in Germany and Sweden found that AI-supported screenings detected breast cancer at a rate of 6.7 per 1000, which was higher than the 5.7 per 1000 rate in the control group.[45] These results raise hope of curing this fatal disease and saving numerous lives, with women being the biggest beneficiary.

Moreover, studies have found AI to be efficient in detecting general ailments, sleep disorders, diseases like arthritis, lupus, and so forth.[46] AI-driven virtual health assistants are available 24/7 to answer health-related questions, provide medical advice, and assist with medication management. These assistants offer convenient and accessible support for patients as well as for maintaining general fitness. They immensely help in daily fitness tracking, monitoring health conditions and providing health-related advice, which can assist women, in general, staying healthy and fit.

The health applications integrated with AI, provide forum (apps) to connect with distinguished doctors from any part of the world, improving medical accessibility and reach to the critical medical consultancy. It enables remote monitoring of patients' health conditions and facilitates telemedicine consultations. This is particularly beneficial for patients from distant areas, providing them with access to medical care without the need for physical visits. Women, assuming the role of "caregivers" have been executing the responsibility of caring for family members' health, may hugely benefit from all these AI functions.

9.5.7 *Reproductive Health and Rights*

Besides the factors mentioned earlier, reproductive health and rights have been an issue for all women and girls. The Beijing Platform for Action states that "the human rights of women include their right to have control over and decide freely and responsibly on matters related to their sexuality, including sexual and reproductive health, free of coercion, discrimination and violence."[47] Yet women in many parts of the world are struggling to have full autonomy of their body.

Abortion rights are given conditionally in many countries like Poland, Liechtenstein, and so forth, under circumstances involving, incest, molestation/sexual assault, or danger to women's life or health.[48] Countries like Andorra, Malta, and San Marino do not allow abortion under any circumstance,[49] even if it is posing a risk to the mother's life. It's a common occurrence in several parts of the world, where women are unable to procure proper health care or advice during pregnancy. First-time mothers often suffer from the lack of knowledge on various fronts, and unavailability of experienced people to take guidance from. Sometimes, this ignorance invites threat to the life of mother and child, or other unwanted complications during pregnancy.

Reproductive health has been a shushed topic, depriving women and girls of the opportunity of understanding their body and health concerns. For enhancing awareness regarding reproductive health, AI can provide personalized and interactive learning experiences to each user as per their queries.[50] AI's use for medical purposes is offering solutions to much extent in ensuring good reproductive health. AI has the capacity to efficiently screen for reproductive health issues, such as cervical cancer or endometriosis. Machine learning algorithms can analyse medical images and patient data to identify early signs of disease and recommend appropriate treatments.[51] AI can also assist in monitoring and predicting potential complications during pregnancy, such as preeclampsia or gestational diabetes. By analysing data from wearable devices and electronic health records, AI can provide early warnings and personalized care plans,[52] also recommending drugs and providing remote access to specialized doctors across the globe through telemedicine platforms.[53]

Moreover, algorithms of AI can successfully analyse large datasets to predict the success rates of in vitro fertilization (IVF) treatments, assisting doctors and patients in taking more informed decisions. Besides, AI can assist in selecting the best embryos for implantation, increasing the chances of a successful pregnancy.[54] This empowers those women wanting to have children but cannot through natural means to have offsprings. Although IVF treatments predate AI applications, they became more easily accessible and popular due to its success with the integration of AI.

9.5.8 Ectogenesis and Future Possibilities

AI, today is progressing towards such possibilities, which were earlier envisioned in fictional stories. One such venture is Ectogenesis, that is producing human babies through artificial wombs. Aldous Huxley in *Brave New World*

(1932)[55] fictionalized a world where humans were cultivated in "hatcheries" through a process called Bokanovsky's process. The process was controlled by the state and used to be conducted at the Central London Hatchery and Conditioning Centre. It involved human embryos being artificially created and grown in bottles. In less than a hundred years, scientific invention is making something quite similar to this possible, through the use of AI.

The first such experiment was successfully conducted in 2017 at Children's Hospital of Philadelphia (CHOP), Philadelphia, USA. The researchers at CHOP successfully used an artificial womb, called the Biobag, to support the development of premature lambs for up to 4 weeks. This Biobag device mimics the natural womb environment, providing sterile fluidic incubation and a pumpless extracorporeal circuit for perfusion and gas exchange.[56] The CHOP team is now seeking approval for human clinical trials of their artificial womb system, known as the Extra-Uterine Environment for Newborn Development. This system aims to improve survival and outcomes for extremely premature babies by simulating elements of a natural womb.[57]

Similar research related to culturing of human embryos in an artificial environment was conducted in 2016 at the University of Cambridge. The researchers, led by Professor Magdalena Zernicka-Goetz, successfully cultured human embryos for up to 13 days. This ground-breaking work allowed scientists to study early human development beyond the implantation stage, providing valuable insights into the processes that occur during this critical period.[58] With the mentioned scientific advancement, one can predict a possibility of successful ectogenesis of human babies for full gestation period in future.

AI advancements, especially in building more humanistic robots, indicate a possible future where humanoid robots may carry full-term pregnancy as well, to actualize the concept of full-time gestation ectogenesis pregnancies.[59] If ever this invention materializes, it will realize the wildest dream of certain prominent radical feminists, like Ms. Shulamith Firestone. They envisioned a system of artificial reproduction carrying pregnancy instead of women.[60] They believed pregnancy to be a natural barrier, consuming time, life, and health of women. They were of the opinion that child bearing and rearing was the method enforced by patriarchy to contain women under their paradigm. For them, artificial reproduction would be an ultimate means to eliminate patriarchy by freeing women from the burden of reproduction.

Upfront the idea of an artificial system carrying the babies seems fascinating, and liberating as well, as it will present women with the opportunity to continue their work and grow without the necessity of a "break." The health risk posed by complicated pregnancy, many times costing the life of mothers, will be out of the question. Natural health concerns for nine months, ending in painful labour or c-section surgery heavily affecting a woman's bodily integrity, will not be the *only* option for a woman wanting an offspring. There are, undoubtedly, many alluring advantages to advocate for promoting development of such an invention, but critical evaluation must also be considered to holistically assess the outcome.

The other section of radical feminists, contrary to the opinions of Ms. Firestone, raised concern over practices of IVF and possible ectogenesis, claiming it will ensue the complete control of patriarchy over "reproduction."[61] The unique ability of women to carry babies will not remain so unique if it can be replicated in full swing through artificial machines. It will inevitably invite contention of power to control such a system, favouring patriarchy to dominate women further. They opined that such technologies will be devaluing women's importance further in society by reducing their participation in creating and propagating it.[62] Both the arguments stand strong and postulate a careful approach in advancing and embracing a possible future with ectogenesis pregnancies. The unmindful enthusiasm in promoting such technology may result in the dystopian reality as suggested in *Brave New World* (Huxley),[63] a future where technological advancements are used to control and manipulate the population, sacrificing personal freedom and individuality.

The future possibility of artificial machines or humanoid robots carrying babies raises further questions, especially after recent evidence of a robot's suspicious behaviour patterns. There are cases where a civil servant robot allegedly committed suicide in South Korea[64]; in another incident, in China, a tiny robot in bizarre behaviour "kidnaps" other 12 robots.[65] The small conversation among the robots before "abduction" is filled with "human" sentimental tone, expressing human emotions and longing. In the conversation, the tiny robot is asking from the other robots "So you're not going home?" to which one of the larger robot replies, "I don't have home"; listening to this the tiny robot responds, "then come home with me," and leads the way out, followed by the other 12 robots.[66]

These suspicious behaviours of robots must be subjected to scrutiny to inspect the proposition – can robots also cultivate emotions or conscience? It will raise an array of possibilities and threats, if scientists actually succeed in ectogenesis of human babies. It will not merely be an ethical concern of whether one should or shouldn't opt for ectogenesis pregnancy; it will open a plethora of legal implications as well. Questions such as – "Shall robots be awarded personhood?"; "What if the robot-carrying baby refuses to handover it?"; and "Will babies lack humanistic elements being carried by robots not human mothers?; all these queries may seem farfetched but demand introspection considering the rapid advancement of AI technology.

9.6 AI Usage and Socio-Legal Implications

Technologies are primarily invented to benefit humanity. This is precisely where the law steps in. So far, this chapter has established the positive and progressive utilization of AI and AI-driven tools that are seemingly contributing in SDG Goal 5 and realization of feminist's utopia. But the malpractices for which AI are being used cannot be overlooked. Cases of digital rapes, deep fakes, digital arrests, scamming, and other new types of crimes are being committed with the use of AI tools.[67]

On one front, AI is paving newer opportunities, contributing to futuristic innovations and facilitating digital avenues for myriad productive purposes; on the other front, the dark usage of the same technology is bringing new horrors. For instance, according to the credible reports, there has been an alarming rise in the heinous crime of digital rapes in India, in the recent years.[68] The word "digit" connotes a finger, thumb, or toes, and using them for sexual violence has been termed as digital rape. Although digital rape does not necessarily allude to usage of digital platforms or multimedia or mobile, the cases of molestation including the said things come under its periphery. The ease of transfer and instantaneous circulation of sensitive multimedia content of the victims with the use of modern mobile apps makes it nearly impossible to stop, leading to further emotional and psychological harassment of the victims.

With the advancement of digital technology incorporating AI, innovative methods to scam, blackmail, and harass are escalating. One of the most dangerous AI applications is deep fakes. Deepfakes are a type of synthetic media where AI is used to create realistic but fake images, videos, or audios. Through its usage harmful contents can be easily created such as fake news, impersonations, and non-consensual explicit material. Moreover, identifying these deepfakes from the real is really tricky, further complicating the situation for the victim. There have been several troubling cases involving deepfake usage to blackmail women. One such case occurred in Delhi, where a man was arrested for blackmailing a college student with AI-generated explicit images. He posed as a female on social media, altered the victim's profile pictures, and demanded money, threatening to share the images.[69] It triggers the imagination to think about the types of unprecedented crimes that the future can witness and instils fear based on the possibility of gruesome misuse of the same technology, which has the potential to usher feminist utopia.

To ensure right usage and practices of AI tools, the law has to transition with the contemporary scenarios to accommodate unheard-of crimes. AI also poses the critical question of accountability for any mishap by the AI technology or devices (for instance, driverless-car accidents of 2018). In 2018, there were several road accidents involving driverless cars of companies like Tesla autopilot and Uber auto-driving cars. In March 2018, an Uber self-driving car struck and killed a pedestrian in Tempe, Arizona. The car failed to recognize the victim as a pedestrian because she was not near a crosswalk.[70] Similarly, there were multiple incidents involving Tesla's autopilot system in 2018. One notable case was the fatal crash in Mountain View, California, where a Tesla Model X collided with a highway barrier while in autopilot mode.[71] It tables the problem of placing accountability for rendering justice to the victims of such unfortunate accidents. A person cannot be punished for the mistakes of AI tools and devices. Even if AI is sentenced as guilty, then whom to punish? The law is faced with the challenge of enlarging its ambit to include these emerging unprecedented crimes into its fold.

9.7 Conclusion

Undoubtedly, the advent of AI represents a monumental shift in human history. The future of AI holds immense promise, and equally, possible threats in almost every domain of human life. Ensuring that AI is developed and deployed responsibly is crucial to maximizing its benefits while minimizing potential harms. While navigating the challenges and opportunities presented by AI, it is essential to ensure that its development aligns with ethical principles and societal values, paving the way for a better future ensuring gender equality.

Emerging AI technologies are posing newer legal implications for which fresh discussion and debates are required to formulate new laws providing solutions to new challenges appearing in regulating AI technologies. Only strong laws with right regulations can serve as a spine for actualization of feminist's utopia through attainment of SDG5 goals. Without it the dreams of utopia will inevitably digress into dystopian reality.

For any society to ensure individual freedom and dignity, justice, and equality, its law should have a strong foundation based on humanistic ideals that can decide and serve right justice. Otherwise, the dream of society having ideals of gender equality and gender justice, as being endeavoured by SDG5 goals and a long line of feminists, will always remain a "future goal." Technological advancement, incorporating AI, under the law's just intervention coupled with right policy-making and application are the foundation where a feminist utopia can exist in reality.

Notes

1 Michael Robertson, *The Last Utopians: Four Late 19th-Century Visionaries and Their Legacy* (Princeton, NJ: Princeton University Press, 2018).
2 Frances Bartkowsky, *Feminist Utopias* (Lincoln and London: University of Nebraska Press, 1989).
3 Carol Pearson, "Women's Fantasies and Feminist Utopias" 2 *Frontiers* 50–61 (1977).
4 A Vindication of the Rights of Woman: Wollstonecraft, Mary, 1759–1797: Free Download, Borrow, and Streaming: Internet Archive, *available at*: https://archive.org/details/AVindicationOfTheRightsOfWoman (Last visited on January 09, 2025).
5 Gita Sen, Discussion-paper-The-SDGs-and-feminist-movement-building-en.pdf, *available at*: https://www.unwomen.org/sites/default/files/Headquarters/Attachments/Sections/Library/Publications/2018/Discussion-paper-The-SDGs-and-feminist-movement-building-en.pdf (Last visited on January 09, 2025).
6 treaties.un.org, *available at*: https://treaties.un.org/doc/Publication/CTC/uncharter.pdf (Last visited on January 09, 2025).
7 Convention on the Elimination of All Forms of Discrimination against Women New York, 18 December 1979. OHCHR, *available at*: https://www.ohchr.org/en/instruments-mechanisms/instruments/convention-elimination-all-forms-discrimination-against-women (Last visited on January 09, 2025).
8 Convention on the Elimination of All Forms of Discrimination against Women New York.

9 Beijing Declaration and Platform for Action, Beijing +5 Political Declaration and Outcome | UN Women – Headquarters, *available at*: https://www.unwomen.org/en/digital-library/publications/2015/01/beijing-declaration (Last visited on January 09, 2025).
10 Thirty years of the Beijing Declaration and Platform for Action: Achieving gender equality, the rights and empowerment of women and girls | UN Women – Headquarters, *available at*: https://www.unwomen.org/en/news-stories/news/2024/09/thirty-years-of-the-beijing-declaration-and-platform-for-action (Last visited on January 09, 2025).
11 Millennium Development Goals – MDGs, *available at*: https://www.mdgmonitor.org/millennium-development-goals (Last visited on January 09, 2025).
12 Millennium Development Goals – MDGs.
13 Goal 5: Gender Equality – The Global Goals, *available at*: https://www.globalgoals.org/goals/5-gender-equality (Last visited on January 09, 2025).
14 Goal 5: Gender Equality – The Global Goals.
15 Usha Daigavane, "Women Empowerment through New Technology in India" 5 *Journal of Emerging Technologies and Innovative Research* 70–77 (2018).
16 J. He, S. L. Baxter, J. Xu, J. Xu, X. Zhou and K. Zhang, "The Practical Implementation of Artificial Intelligence Technologies in Medicine" 25 *Nature Medicine* 30–36 (2019), *available at*: https://doi.org/10.1038/s41591-018-0307-0 (Last visited on January 09, 2025).
17 Artificial Intelligence (AI) Coined at Dartmouth | Dartmouth, *available at*: https://home.dartmouth.edu/about/artificial-intelligence-ai-coined-dartmouthelligence (AI) Coined at Dartmouth | Dartmouth (Last visited on January 25, 2025).
18 AI Development: Opportunities & Challenges, *available at:* https://www.valuecoders.com/blog/ai-ml/ai-development-opportunities-and-challenges/ (Last visited on January 25, 2025).
19 Sarah Pruitt, "What Are the Four Waves of Feminism?" (2023), *available at*: https://www.history.com/news/feminism-four-waves (Last visited on January 09, 2025).
20 R. Tong and T. F. Botts, *Radical Feminism. Encyclopedia of Feminist Theories* (Routledge, London, 2003).
21 This Day In History: India's First Girls' School Established at Pune's Bhide Wada 175 Years Ago Today, *available at*: https://www.freepressjournal.in/pune/this-day-in-history-indias-first-girls-school-established-at-punes-bhide-wada-175-years-ago-today (Last visited on January 25, 2025).
22 How AI Can Transform Education for Students and Teachers. World Economic Forum, *available at*: https://www.weforum.org/stories/2023/05/ai-accelerate-students-holistic-development-teaching-fulfilling (Last visited on January 9, 2025).
23 Coursera Learn Online, *available at*: https://www.coursera.org/courses?query=free (Last visited on January 9, 2025).
24 Charlotte Perkins Gilman, *Women and Economics: A Study of the Economic Relation Between Men and Women* (Boston: Maynard & Company, 1898), *available at*: https://archive.org/details/womenandeconomi01gilmgoog/page/n1/mode/2up.
25 E. Norris and K. Kochhar, "Closing the Gender Gap", 56(1) *Finance and Development* 6–11 (2019), *available at*: https://www.imf.org/external/pubs/ft/fandd/2019/03/pdf/fd0319.pdf (Last visited on January 10, 2025).
26 Nykaa Founder: Falguni Nayar Inspiring Success Story (2025), *available at:* https://www.cheggindia.com/earn-online/nykaa-founder-success-story (Last visited on January 10, 2025).
27 Top 10 AI Startups Run by Indian Women Entrepreneurs, *available at:* https://viestories.com/ai-startups-run-by-indian-women-entrepreneurs (Last visited on January 10, 2025).

28 "What If We Finally Recognised Women's Unpaid Care Work?", *The Daily Star* (June 19, 2024), *available at*: https://www.thedailystar.net/opinion/views/news/what-if-we-finally-recognised-womens-unpaid-care-work-3636856 (Last visited on January 10, 2025).
29 "Dreame's Robo-Vac Has Retractable Legs" – Pickr, *available at:* https://www.pickr.com.au/news/2025/dreames-robo-vac-has-retractable-legs (Last visited on January 10, 2025).
30 "Robots for Household Chores Less Than 10 Years Away": Expert (pub: September 02, 2023) *available at*: https://interestingengineering.com/innovation/robots-for-household-chores-less-than-10-years-away-expert (Last visited on January 10, 2025).
31 Betty Friedan, *The Feminine Mystique* (WW Norton & Company, 2010).
32 Friedan, *The Feminine Mystique.*
33 International Labour Organization (ILO), "Care Work and Care Jobs for the Future of Decent Work" (June, 2018), *available at*: https://www.ilo.org/publications/major-publications/care-work-and-care-jobs-future-decent-work (Last visited on January 11, 2025).
34 International Labour Organization (ILO), "Care Work and Care Jobs for the Future of Decent Work", xxix.
35 Circle of 6 App | Women in the World Foundation, *available at:* https://womenintheworld.org/solutions/entry/circle-of-6-app (Last visited on January 11, 2025).
36 Shruti Mahajan, Shweta Khandare, Krishna Dubbewar and Kartik Patekar, "Nari Suraksha Setu: A Women's Safety App", *Ijraset Journal For Research in Applied Science and Engineering Technology* (2024), *available at:* https://doi.org/10.22214/ijraset.2024.62873 (Last visited on January 11, 2025).
37 Parismita Sarma, Danish Ahmed and Pouranika Bezbaruah, "Android-Based Woman Safety App", 16 *Indian Journal of Science and Technology* 60–69 (2023), *available at:* https://doi.org/10.17485/IJST/v16iSP2.8767 (Last visited on January 11, 2025).
38 MeToo | AP News, *available at*: https://apnews.com/hub/metoo (Last visited on January 11, 2025).
39 Social Media Sentiment Analysis: The Rise of AI, *available at*: https://alternacx.com/blog/social-media-sentiment-analysis-the-rise-of-ai/Analysis: The Rise of AI (Last visited on January 12, 2025).
40 Women in Switzerland March for Gender Equality and Fairer Pay. World Economic Forum, (June 16, 2020), *available at:* https://www.weforum.org/stories/2020/06/switzerland-swiss-women-protest-domestic-violence-pay-gap/en in Switzerland march for gender equality and fairer pay | World Economic Forum (Last visited on January 11, 2025).
41 Swiss women take to streets to fight inequality and sexism – SWI swissinfo.ch (June 15, 2023), *available at*: https://www.swissinfo.ch/eng/politics/swiss-women-take-to-streets-to-fight-inequality-and-sexism/48589370 (Last visited on January 11, 2025).
42 Global Cancer Burden Growing, Amidst Mounting Need for Services, *WHO* (February 1, 2024), *available at*: https://www.who.int/news/item/01-02-2024-global-cancer-burden-growing-amidst-mounting-need-for-services (Last visited on January 10, 2025).
43 Global Cancer Burden Growing, Amidst Mounting Need for Services.
44 AI Breakthrough in Breast Cancer Detection: New Study Claims Over 17% Improvement in Screening Accuracy. Technology News – *The Indian Express*, *available at:* https://indianexpress.com/article/technology/artificial-intelligence/ai-breakthrough-in-breast-cancer-detection-9768970 (Last visited on January 10, 2025).

45 Artificial Intelligence Better Than Doctors in Detecting Breast and Ovarian Cancers, Find Studies. *Firstpost*, *available at*: https://www.firstpost.com/world/ai-better-than-doctors-in-detecting-breast-and-ovarian-cancers-find-studies-13851247.html (Last visited on January 10, 2025).
46 "New AI Method to Pick Up Arthritis, Lupus Early in High-Risk Patients", *The Statesman*, (January 09, 2025), *available at*: https://www.thestatesman.com/technology/new-ai-method-to-pick-up-arthritis-lupus-early-in-high-risk-patients-1503384425.html (Last visited on January 10, 2025).
47 Sexual and Reproductive Health and Rights. OHCHR, *available at*: https://www.ohchr.org/en/women/sexual-and-reproductive-health-and-rights (Last visited on January 11, 2025).
48 European Abortion Law: A Comparative Overview. Center for Reproductive Rights, *available at*: https://reproductiverights.org/european-abortion-law-comparative-overview-0/ Rights (Last visited on January 12, 2025).
49 European Abortion Law: A Comparative Overview.
50 Unpacking Artificial Intelligence in Sexual and Reproductive Health and Rights (March 22, 2024) *available at:* https://www.who.int/news/item/22-03-2024-unpacking-artifical-intelligence-in-sexual-and-reproductive-health-and-rightshts (Last visited on January 12, 2025).
51 World Health Organization, "The Role of Artificial Intelligence in Sexual and Reproductive Health and Rights Technical Brief" 2–3 (2024), *available at*: https://iris.who.int/bitstream/handle/10665/376294/9789240090705-eng.pdf?sequence=1&isAllowed=y (Last visited on January 12, 2025).
52 World Health Organization, "The Role of Artificial Intelligence", above note 51.
53 World Health Organization, "The Role of Artificial Intelligence", above note 51.
54 "Researchers Develop AI System to Help Better Select Embryos for Implantation", *News Medical Life Science* (September 15, 2020), *available at*: https://www.news-medical.net/news/20200915/Researchers-develop-AI-system-to-help-better-select-embryos-for-implantation.aspx rs develop AI system to help better select embryos for implantation (Last visited on January 12, 2025).
55 Aldous Huxley, *Brave New World* (Chatto & Windus, 1932).
56 Emily A. Partridge, Marcus G. Davey, et al. "An Extra-Uterine Physiologic Support System for the Extreme Premature Lamb", *Nature Communications* (Published online April 25, 2017), *available at*: https://www.nature.com/articles/ncomms15112 (Last visited on January 12, 2025).
57 "A Unique Womb-Like Device Could Reduce Mortality and Disability for Extremely Premature Babies", *Children's Hospital of Philadelphia* (April 24, 2017), *available at*: https://www.chop.edu/news/unique-womb-device-could-reduce-mortality-and-disability-extremely-premature-babies (Last visited on January 12, 2025).
58 "Scientists Develop Human Embryos Beyond Implantation Stage for First Time", University of Cambridge, *available at*: https://www.cam.ac.uk/research/news/scientists-develop-human-embryos-beyond-implantation-stage-for-first-time (Last visited on January 12, 2025).
59 Dr. Nichollas Willams, "The Pros and Cons of Artificial Wombs", *NBC News*, *available at*: https://www.news-medical.net/health/The-Pros-and-Cons-of-Artificial-Wombs.aspx (Last visited on January 12, 2025).
60 Elaine Denny, "Liberation or Oppression? Radical Feminism and in Vitro Fertilisation", 16 *Sociology of Health & Illness* 62–79 (1994), *available at*: https://onlinelibrary.wiley.com/doi/pdfdirect/10.1111/1467-9566.ep11347010 (Last visited on January 12, 2025).
61 Denny, "Liberation or Oppression?", above note 60.
62 Denny, "Liberation or Oppression?", above note 60.

63 Aldous Huxley, *Brave New World* (Chatto & Windus, 1932).
64 "Robot Commits Suicide: South Korea Sees Its First 'Robot Suicide': Here's What Caused It", *Times of India* (July 13, 2024), *available at*: https://timesofindia.indiatimes.com/technology/tech-news/south-korea-sees-its-first-robot-sucide-heres-what-caused-it/articleshow/111517391.cms (Last visited on January 12, 2025).
65 "Robot 'Kidnaps' 12 Large Robots of Another Manufacturer; Company Confirms 'Abduction' and that Viral Video Is Real", *Times of India* (November 22, 2024), *available at*: https://timesofindia.indiatimes.com/technology/tech-news/12-large-robots-kidnapped-by-another-manufacturers-robot-company-confirms-that-the-viral-video-is-real/articleshow/115487104.cms (Last visited on January 12, 2025).
66 "Robot 'Kidnaps' 12 Large Robots of Another Manufacturer."
67 G. Krishna Subramanyam and P. Manindar, "Artificial Intelligence Crime an Overview of Malicious Use and Abuse of AI", 13 *International Journal for Advanced Research in Science & Technology* 215–219 (August 2023), *available at*: https://www.ijarst.in/public/uploads/paper/594291699155882.pdf (Last visited on January 26, 2025).
68 Ms. Jyotsana Singh and Dr. Prashant Mishra, "Digital Rape in India: A Doctrinal and Empirical Studies", 11 *International Journal of Creative Research Thought* d788-d796 (September 2023), *available at*: https://ijcrt.org/papers/IJCRT2309455.pdf (Last visited on January, 26, 2025).
69 "'Sunno Kuch Kaam Hai . . .': How a Delhi Man Blackmailed, Extorted Women after Doctoring Their Images Using AI", *The Economic Times* (January, 20, 2025), *available at*: https://economictimes.indiatimes.com/news/india/sunno-kuch-kaam-hai-how-a-delhi-man-blackmailed-extorted-women-after-doctoring-their-images-using-ai/articleshow/117396036.cms?form=MG0AV3 (Last visited on January, 26, 2025).
70 "Self-Driving Uber Car That Hit and Killed Woman Did Not Recognize That Pedestrians Jaywalk", *NBC News* (November 10, 2019), *available at*: https://www.nbcnews.com/tech/tech-news/self-driving-uber-car-hit-killed-woman-did-not-recognize-n1079281 (Last visited on January 12, 2025).
71 "Uber Self-Driving Car Accident: Who's to Blame When There's No Driver?", *ABC News* (March 20, 2018), *available at*: https://www.abc.net.au/news/2018-03-20/uber-driverless-car-accident-who-is-to-blame/9567766 (Last visited on January 12, 2025).

Part V

Food Security

10 The Interface of Emerging Technologies, Human Rights and Sustainable Nutritional Security vis-a-vis Indian Food Security System

Sonika

10.1 Approaches to Food Security and SDGs

The relation between food and healthy life is well established. Healthy food always ensures a healthy life. The same is deliberated and defined in various texts. As far as the statutory definition of "food" and its relations with a healthy life is concerned, the Prevention of Food Adulteration Act, 1954 (now repealed as the Food Safety and Standard Act, 2006), defines "food" as "an article that is meant for human consumption."[1] According to the *Lexicon Law Dictionary*, "food" is defined as "an edible item to be eaten which provides nourishment to organic bodies, victuals, etc."[2] Ideally, the chief function of "food" is the attainment of well-being, nutrition, and good health. It is imperative that definition of "food" focuses upon its core functions similarly; Japan recognized the concept of *functional food* that is food for Specific health uses.

Primarily, it focuses on the physiological effects of food on the human body therefore; it requires regular and systematic assessment of food.[3] Such food requires consistent availability, accessibility, and stability to each individual, which are considered pillars of food security as determined by Food and Agriculture Organization (FAO). These pillars categorically emphasize on the balanced relation between food and healthy life, which leads to enforcement of basic human right of an individual. The concept of Food Availability addresses the fulfilment of nutritional requirements through sufficient food supply, irrespective of economic constraints.

Further, Food Accessibility focuses on affordability and sustainability of food regardless of geographic diversities, dietary choices, and economic incapacities of an individual. The Utility of Food emphasizes providing safe, nutritious food with minimal wastage, while Food Stability ensures long-term nutritional security for sustainable development. Simultaneously, it determines certain parameters to classify food and nutritional insecurity namely acute food insecurity and chronic food insecurity. The acute food insecurity demonstrates a crisis of dietary diversity, water security, and food due to poverty. Further, chronic insecurity indicates a crisis of nutrition and

DOI: 10.4324/9781003733751-15

sufficient and affordable food resources. It is apt to mention that pillars of food security complement each other and a comprehensive reading of these factors opens avenues for technology-driven food sustainability. The same approach reflects in various international and national instruments.

10.2 International Approaches

'Food Security' is not statutorily defined in either international or national instruments; however, it can be understood with the help of textual approaches to the Human Right to Food under national and international instruments. Under the Universal Declaration of Human Rights, 1948 (UDHR), *food security* is interpreted in the realm of two human rights: every individual is entitled to the Right to Social Security[4] and Right to adequate Standard of Living.[5] Declaring the Right to food as a Human Right, the Third Committee on the Promotion and Protection of Human Rights focuses on curbing malnutrition, streamlining of food security system, adopting technology-based farming, and so forth.[6]

Deliberating on the Voluntary Guidelines for the Progressive Realization of the Right to Adequate Food, the Committee on World Food Security also recognized food as a human right and focused on Sustainable Development Goals in 1996. It considered food as more than a commodity. Further, it highlighted that around 735 million people suffer from hunger and malnutrition, which need immediate attention. The human right to food needs to be accomplished in consonance with international human rights standards and Sustainable Development Goals (hereafter referred to as SDGs).[7] Considering the global poverty and hunger index, the United Nations has determined certain dimensions for its eradication by 2030 as follows:

- i Safe and nutritious food should be adequately available to each individual.
- ii Availability and accessibility to nutritional food to children, women, and senior citizens.
- iii A balance is required to be maintained between conventional knowledge and advanced agricultural techniques for enhancing the farmers' income and agricultural productivity.

Way back in 2020, the World Health Organization (WHO) deliberated on the nutritional growth and food security among humans as a part and parcel of SDGs to be accomplished by 2030.[8] Simultaneously, it invited the attention on the impact of COVID-19 pandemic and intense and rapid climate change on the food security system, which triggered hunger and malnutrition across the nation. It seems that the deliberations primarily focussed on adoption of innovations and technology-driven agrifood systems across the globe.

Commemorating World Food Day 2024,[9] the Food and Agriculture Organization (FAO) highlighted its theme Right to Food for a Better Life and a Better Future. The theme aimed at inclusive development, sustainable agrifood systems, and zero hunger by ensuring adequate and nourished food for all individuals. According to available statistics, healthy diets are non-affordable to more than 2.8 billion people. Further, it highlighted the need for renewal of Global Commitment to Food security for the promotion of farmers' participation in the global food security system, preservation of conventional food habits and cultures, and rebuilding of agricultural infrastructure. Further, it stressed the active participation of private sectors especially for the adoption of innovation and technology in the food security systems. Such result-oriented initiatives may play a significant role in the eradication of global hunger synergizing with SDGs.

It is an undeniable proposition that national and global food sustainability rests on climate sustainability. Considering climate crisis as an interruption in achieving the objective ofzero hunger, FAO determined the adoption ofOne Country, One Priority Product, Blue Economy initiative and Technical Cooperation programmes to ensure the accomplishment of nutrition requirements and dietary requirements irrespective of climate change and other emergencies. It depicts the recognition and introduction of new emerging technologies in the food and agriculture systems to achieve the goal of zero hunger globally.

Further, the United Nations Framework Convention on Climate Change (COP 29, 2024) recognized that climate change is a threat to water scarcity, sustainable land management, food insecurity, and so forth.[10] It focused on the vision of the United Nations Convention to Combat Land Desertification, 1994 (UNCCD) for the preservation, conservation, and management of the ecosystem and promotion of climate-centric innovative agricultural techniques to establish resilient food security systems. Therefore, it reiterated the need for SDG-oriented initiatives: land rehabilitation, overexploitation, biodiversity loss, equitable access to nutritious food long-term ecological balance, and supportive welfare policies to curb the menace of aforesaid threats on basic human rights. Significantly, the core objective of UNCCD includes the protection and restoration of land for a protective, just, and more sustainable future.[11]

It is apt to mention that India has signed and ratified these aforementioned international instruments to ensure Human Right to food Security as part and parcel of its social welfare model. Therefore, several food-based welfare schemes, namely, Targeted Public Distribution System (TPDS), One Nation One Ration Card (ONORC), Mid-Day Meal Scheme (MDMS), POSHAN 2.0, and so forth, were launched as part and parcel of social security in India. Concurrently, several emerging technologies, especially AI-driven technologies, namely CODEX, Aadhar-seeded targeted public distribution system, real-time status of fair price shops, digital supply chain management, bio-surveillance based food safety system, and so forth, have been adopted in agrifood systems. To ensure the active compliance of such technologies and to meet desired outcomes, certain policy measures have been incorporated in agrifood-based legislations.

10.3 Indian Approach

'Food security' is usually interpreted with the help of factors enshrined in various municipal legislative instruments and food-based welfare schemes. Such instruments were also drafted and implemented in line with the international commitments as follows:

1 **The Constitution of India, 1950:** The idea of justice and equality embodied in the UDHR[12] and the International Covenant on Economic, Social, and Cultural Rights 1966 (ICESCR)[13] is enshrined in the Preamble and other Constitutional provisions. For instance, equitable distribution of material resources among individuals without any discrimination is one of the cherished goals of the Constitution.[14] The supreme goal is implemented through various social welfare schemes and beneficial legislation; the Targeted Public Distribution System (TPDS) ensures the distribution of subsidized food grains or entitlements to its beneficiaries, namely Below Poverty Line (BPL) and Antyodaya Anna Yojana (APL). Such beneficiaries are entitled to their subsidized entitlements directly without any discrimination. Interestingly, this public distribution system is based on the scheme that surplus of food grains lying and rotting in the state-owned warehouses and godowns needs to be distributed among individuals facing starvation, poverty, malnutrition, and dying of hunger.

   Considering it as a policy concern, it was judicially acknowledged that the Right to food is a fundamental right under Article 21 of the Constitution.[15] This right guarantees every individual the constant availability and accessibility to food without any discrimination. Further, the Constitution of India determines duty of the State to protect children from exploitation, which simultaneously ensures compliance with the United Nations Convention on Child Rights (UNCRC) 1990.[16] Further, it imposes a positive obligation on the State to take adequate measures for the protection of children under the age of 6 years.[17]

   In this regard, several food-based social welfare initiatives were launched in India: Integrated Child Development Services (ICDS), popularly known as Supplementary Nutrition Programme, the Mid-Day Meal Scheme, Prime Minister's Overarching Scheme for Holistic Nourishment 2.0, and so forth. According to ICDS, children below the age of 6 and adolescents are entitled to receive nutritional food entitlements. It aims to curb the menace of high child mortality rates, malnourishment, undernourishment among adolescents, and so forth. It is a constitutional obligation of the State to raise the standard of living and nutrition of the people and improve public health.[18] Even the Apex Court observed that the protection of public health and safety is the constitutional duty of the State.[19] It is significant to mention that the aforesaid food-based welfare measures are technology driven: Aadhar based entitlements distribution is mandatory for fair and transparent distribution System.

2 **Food Safety and Standard Act, 2006:** Complying with the Constitutional mandate, the Act of 2006 aims to regulate public health, nutritional security, and food safety in India.[20] Further, it prescribes provision for the establishment of regulatory authority, known as Food Safety and Standards Authority of India (FSSAI), to determine food safety standards, to monitor the sale, purchase, and exchange of contaminants and toxin-free food, and to prohibit sale of adulterated and contaminated food products.[21] It is pertinent to mention that its statutory provisions are based on the CODEX Alimentarius Commission (CAC). The Commission is an entity constituted by the Food and Agriculture Organization (FAO) and the World Health Organization (WHO) to determine standards for food products, precautions related to consumer health, and execution of trade practices in the food system and to initiate proceedings in case of violation of laws.[22] Even the Supreme Court observed that the sale and supply of adulterated food amounts to a violation of Article 21 of the Constitution and Sections 91, 92, and 94 of the Act of 2006.[23] It is pertinent to mention that FSSAI has introduced genome sequencing and the Internet of Things in the food processing system to curb the menace of adulterated and compliance of food safety standards.
3 **The Protection of Human Rights Act, 1993:** The core objective of this statute is to ensure the enforcement and protection of human rights in India. It recognizes Right to life[24] as a human right that is already recognized as a fundamental right under Article 21 of the Constitution of India. It is significant to note that Right to food has already been recognized as a fundamental right under Article 21 of the Constitution and acknowledged in the landmark ruling of the *People's Union for Civil Liberties (PUCL)*.[25]
4 **National Food Security Act, 2013:** The Act of 2013 brought a revolutionary transformation in the jurisprudence of food and nutritional security. This law relies on a rights-based approach. It guarantees right to food security to the general masses of India. It is apt to say that it accomplishes the idea of Food for All. Further, it regulates the entire Targeted Public Distribution System, other smart-technology-driven food-based social welfare schemes, and sectorial-based initiatives, namely, ICDS, Mid-day Meal Scheme, POSHAN 2.0, and so forth. Under this legislation, the beneficiaries are entitled to receive subsidized food grains.[26]

A glance of legislative framework of the Indian food security system demonstrates compliance with Constitutional and human rights mechanism. Despite various sectorial-based initiatives and legislative measures, the Report of Food Security and Nutrition in the World demonstrates that 194.6 million individuals in India are malnourished.[27] Further, a total of 13% of the Indian population suffers from chronic undernourishment, which demonstrates prolonged food insecurity. Also, 53% of Indian females are anaemic, the highest in South Asia. Further, 55.6% of the Indian population is still unable to afford a healthy nutritious diet and 38% eats unhealthy foods; 28% eat the recommended diet comprising one pulse, one fruit, one starchy staple food, nuts or seeds, and one animal-source food. Apart from the United Nations Food Security Report, the following national data also shows the malnourished population of India (see Figures 10.1 and 10.2).

The graphical representation of the National Family Health Survey 4 (2015–16) and 5 (2019–21) indicates the prevalence of chronic food insecurity. Minor progress is noted in combating stunting, underweight, and wasting. As mentioned earlier, the prevalence of chronic insecurity demonstrates the non-completion of nutritional requirements in an individual's diet, that is undernourishment. According to SDGs, prevalence of undernourishment (POU) is a parameter to weigh the universal nutritional growth. FAO directed all its member nations to make effective efforts for the realization of the Right to adequate food and national food security and curbing POU.[29] Further, it determines the parameters of a healthy diet as follows[30]:

i Diversity within and across food groups
ii Availability of sufficient and essential nutrients
iii Moderation of foods and nutrients considering health outcomes
iv Promotion of balanced energy and macronutrient intake

The report directs member nations to address the issue of financial constraints and introduce innovative fiscal tools for the eradication of hunger, food insecurity, and malnutrition in all its forms in line with SDGs. Also, it recommends public investment in food security, private sector investment,

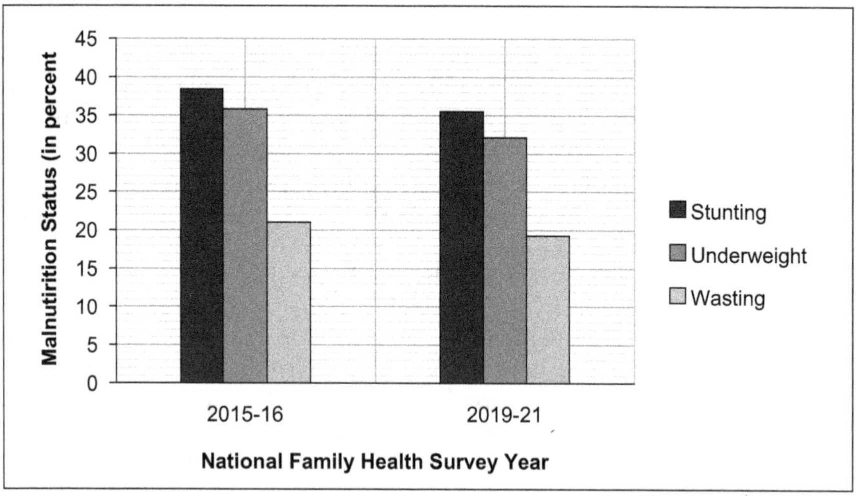

Figure 10.1 National Family Health Surveys 4 and 5: Status of malnutrition indicators in India (in percentage)

Source: National Family Health Surveys 4 and 5[28]

promotion of climate-resilient farming, development of sustainable agrifood system, promotion of nutrition-based diets to vulnerable sections (women, children, senior citizens), and real-time monitoring and modern technology oriented systematic reporting of food security system, and so forth. To eradicate food insecurity, accomplish SDGs, and abolish challenges to agrifood systems, certain AI-centric initiatives were launched from food production to food distribution system.

10.4 Analysis of AI-Integrated Strategies for Climate and Agri-Food Sustainability: Indian and International Perspectives

Undoubtedly, the emergence of AI has revolutionized many sectors, namely, information technology, education, healthcare, market, and so forth. It has brought lucrative results in food processing and agriculture. Simultaneously, AI can play a significant role in agricultural, climate sustainability, and food security. In common parlance, the food security system comprises of agricultural system, food storage, food safety, food processing, and food distribution. A food security system can be strengthened with the introduction of AI or modern smart technologies. Interestingly, FAO adopted a twin-track approach for global food security: direct food accessibility and balancing between sustainable farming and sustainable development. Sustainable development is based on the traditional approach of "real happiness lies only in the total and never in the partial."[31] More importantly, the idea of sustainable development revolves around inclusive development, which invites the integration of new emerging technologies in the achievement of SDGs 2030.

In this direction, the United Nations formed the Digital Public Good Alliance in 2020 with multi-fold objectives: to bridge the global digital equity gap, to focus promote digital agriculture, to facilitate AI governance in food security, and to achieve SDGs.[32] Several nations including India have introduced AI-integrated initiatives for climate, food, and agricultural sustainability. It becomes significant to analyse the role of AI-driven technologies in other nations achieving good success in curbing food insecurity.

10.5 Climate and Agricultural Sustainability

It is a well-known proposition that climate sustainability plays a key role in strengthening global food security system. Also, agriculture is the one of chief sources of food security. Ultimately, it is desired that this source needs to be a sustainable one. These notions duly find place in global initiatives: FAO's initiativeOne Country, One Priority Product, Blue Economy, United Nations Convention to Combat Land Desertification, 1994 (UNCCD), and UN's Framework Convention on Climate Change (COP 29, 2024). These initiatives acknowledge the relation between climate change, agricultural sustainability, technology, and global food security. Many jurisdictions have adopted AI-driven agricultural and food security strategies for the sustainable growth and climate sustainability.

10.5.1 India: Technology-Driven Crop Production Practices

It is significant to mention that India is known for its agricultural richness, traditional farming, and nutritional sustainability. The idea of zero hunger and elimination of poverty revolves around environmental sustainability with the active participation of each individual. Certain policy initiatives were taken and widely promoted to achieve national food and agricultural sustainability. The prime example of such initiatives is Indian *Thali* (Indian food practices). Indian *Thali* is acknowledged as highly environmentally sustainable and human friendly in the World Life Fund Report 2024.[33] According to it,[34] the adoption of Indian food consumption habits would require only 0.84 of an earth by 2050 for sustainable global food production. Indian food patterns require limited natural resources and release little greenhouse gases leading to food, agricultural, and climate sustainability. Since Indian food production is mainly conventional and based on resilient crops, it is considered water friendly. For instance, the National Millets Campaign was launched for the production and consumption of millets for nutrition and environmental sustainability. This climate-resilient farming technique was acknowledged at a global platform and a few more initiatives for sustainable crop production practices were executed as follows[35]:

i **National Mission for Sustainable Agriculture (NMSA)[36]:** This program was launched in 2014–15 to improve agriculture resilience, manage soil health, ensure food security, and establish a climate-resilient and sustainable farming system. Under this advanced technology farming techniques were introduced such as per drop more crop smart drip and sprinkler technologies, and irrigation practices. To curb the menace of desertification and infertility of soil, the Soil Health Card Scheme was launched to ensure the fertility and nutritive value of soil.

ii **Organic Farming Promotion Schemes[37]:** Two main schemes – Paramparagat Krishi Vikas Yojana (PKVY) and Mission Organic Value Chain Development for North Eastern Region (MOVCDNER) – were launched in 2015–16 to provide end-to-end support to farmers involved in organic farming. Such support includes training, data management, marketing, certification, and so forth. Apart from this, the Jaivik-Kheti Portal was also created to provide a platform to facilitate organic farmers to sell and promote their organic products to their consumers or other stakeholders.

iii **National Innovations in Climate Resilient Agriculture (NICRA)[38]:** This project was launched in 2011 to promote climate-resilient farming and to adopt technology and innovation for sustainable agriculture. Further, it spreads awareness among farmers to use advanced and strategic techniques to the adverse impact of climate change on agriculture.

Though technology-driven strategies indicate compliance of requisite policy initiatives, the World Food Programme, International Fund for Agriculture Development, and NITI Aayog reported a constant depletion of groundwater and extreme climate patterns. Such patterns pose dangers to sustainable agriculture.[39] The unavailability of sufficient groundwater and adverse weather conditions affect the agroecological conditions and lead to changes in seasonal crop production. The constant shift in the growing seasons is equally responsible for undesirable crop production and certainly leads to crop losses.

It leads to the degradation of the first pillar of food security, which is the availability of food. The unavailability of nutritious food for the accomplishment of the dietary and nutritional requirements of each individual is violative of the idea of food security. Further, the crop that draws huge water from the soil leaving an agricultural-land water deficit needs to be grown with new agricultural techniques. The International Fund for Agricultural Development (IFAD), recommended that water-saving crops such as millets and pulses need to be produced in such areas.[40] Further, shading of crops and small crop production with trees may be used in the farming.

In 2018, NITI Aayog reported the Composite Water Management Index; India faces the worst water crisis wherein around 600 million people face extreme water stress. Simultaneously, 70% of water is contaminated, wherein India is placed at 120th position among 122 nations. NITI Aayog reported that the Indian GDP (Gross Development Plan) may dip by up to 6% in case of improper management of water by 2050.[41]

According to the United Nations World Water Development Report 2022, India is one of the largest groundwater extractors (12% of the global extractor). It reports that 25% of groundwater is used for irrigation, which shows that this natural resource is poorly managed and misused.[42] The mismanagement of groundwater results in poverty as well as instability of food security systems. Further, NITI Aayog reported that India is one of the chief water-stressed nations across the globe. It demonstrates that water depletion in India is highly unsustainable.[43] It is recommended that groundwater governance is the need of the hour for long-term sustainability. The adoption of an advanced agri-food system leads to a more sustainable and water-friendly scenario. Further, the sensitization drive for the stakeholders especially farmers can bring remarkable and sustainable growth in the food security and agriculture sector.

In this direction, FAO recommended a threefold solution for sustainable agriculture, that is growth protection, responsible utility, and water conservation. It highlighted that constant groundwater depletion is a threat to the national food and water security. Therefore, advanced and modern technologies are required to be adopted for national food and water security. In this direction, the Ministry of Agriculture G20 Working Group of India invited the World Food Programme (WFP) to provide technical and coordination support to crystalize the idea of sustainable agriculture. WFP in collaboration with the University of Reading introduced Participatory Integrated Climate Services for Agriculture in Odisha, especially tribal areas, to manage disaster risk and climate change.

Primarily, this initiative focuses on SDG 17, which promotes global partnerships for inclusive growth. Under this initiative, experts discussed agricultural techniques with farmers, women's self-help groups, and so forth in the state of Odisha. WFP Innovation Accelerator introduced Takachar technology to convert agricultural waste into fertilizer.[44] Further, WFP has introduced multiple strategic and technological initiatives to support and strengthen the Government's agricultural and food-based welfare schemes. Undoubtedly, climate change and rising global temperatures have disrupted environmental sustainability and the food security system. According to the research conducted by the World Food Programme and International Crops Research Institute for the Semi-Arid Tropics (ICRISAT) in India, climate change and agricultural instability halt the food security system and threaten nutritional security as well in the following manner[45]:

i **Availability:** Generally, every individual is entitled to consistent availability of food as per his or her food choices; however, this is one of the pillars of food security. The adverse effect on crop production, livestock production, and aquaculture production halts the availability of food. As far as crop production and changing patterns are concerned, it results in financial losses to farmers. Even temperature-changing patterns trigger water scarcity and upset natural rainfall patterns. Further, it settled the notion that drought has reduced the crop's productivity more than other environmental hazards in the last few years.[46] Such losses adversely affect the supply chain of grains and entitlements in the market. Apart from the direct impact on crop cultivation and agricultural production, climate change indirectly affects the livestock system as well: poultry, animal husbandry, cattle breeders, and so forth. Food choices depend on the geographical and climatic conditions of a locality. Hence, the productivity of the livestock system is susceptible to climatic conditions and availability and sufficiency of natural resources.[47] The environmental changes and climatic conditions lead to diseases among animals, which negatively impact food availability, nutritional value, and the financial stability of individuals.

ii **Accessibility:** It is a well established proposition that the unavailability of sufficient and nutritious food amounts to inaccessibility to food, which enhances the risk of hunger, malnourishment, undernourishment, and stunting among the individual. The available food should be accessible to every individual. Such food should fulfil the nutritional requirements. The noncompletion of nutritional requirements, rising food prices, and declining purchasing power of individuals enhance the risk of hunger. It leads to a reduction of dietary requirements and a fall in the nutritional health of the children. This factor generally gets affected due to poor climatic conditions, poverty, and improper agricultural techniques. It was highlighted by the Swaminathan Committee that climate change leads to low crop production and growth in hunger rate.[48] According to the aforementioned research, notable changes have been noted in the temperature and weather patterns, which have affected livestock and crop productivity in the last few years in India. Unavailability of a sufficient amount of food resources leads to market inflation and price hiking of food products. These factors force individuals to compromise their food choices and purchasing power. Ultimately, individuals, especially marginalized and poor sections of the society, opt for low calorific food and compromise their dietary intake.

iii **Utility:** The individuals use the available and accessible safe and nutritious food for their growth. The availability and accessibility of sub-standard and unhealthy food tend to affect its utility. As mentioned, food products available in the market for human consumption need to be at par with Food Safety Standards established by the Government. It is a settled notion that food quality and nutritional value are susceptible to degradation due to extreme climatic conditions and improper agricultural patterns. The agricultural pattern, especially biological processes, affect the cultivation and cropping pattern, which in turn affects the nutritional growth of crops essential for human health. Considering the weather pattern and climatic conditions and their impact on agricultural productivity, Indian stakeholders have devised various techniques to enhance nutritional value of food. For instance, India promotes the utility of millets as nutritious food. Simultaneously, the idea of cultivation of bio-fortified crops and fortified rice has been launched in the Indian food security system to ensure the maximum utility of nutritious food.[49]

iv **Stability:** The idea of stability revolves around the consistent availability and accessibility to food. It aims to curb the menace of hunger. It is a well-known proposition that extreme weather conditions and climatic conditions negatively impact food production and storage system and supply chain. Due to these factors, individuals, especially the marginalized and poor sections of society, face food and nutritional insecurity, including issues such as stunting, child mortality, and undernourishment.[50]

According to the aforementioned research, limited studies have been conducted on the impact of climatic conditions on the nutritional value of food affecting children's health.[51] To ensure the nutritional value and dietary requirements among children and adolescents, the government has introduced CODEX standards[52] for food entitlements as well as food-based Welfare Schemes: POSHAN Abhiyaan,[53] ICDS (Integrated Child Development Services),[54] and Mid-Day Meal Scheme (National Programme of Nutritional Support to Primary Education)[55] to ensure the consistent availability of nutritious food for children, women, and school-going kids respectively.[56] The objective of these foods-based schemes is to customize the food and dietary requirements as per the geographical and health requirements and to ensure the continuous availability of nutritious food without any disruptions.

According to WFP and ICSTAT, there is a need for the introduction of advanced agricultural techniques and equipment to enhance soil fertility and farmers' awareness of the growth of nutritious and safe food.[57] Since farmers are the main stakeholders in the food security system, their participation is highly crucial for agricultural biodiversity and food security. Therefore, farmers must be sensitized toward the utility of their traditional knowledge in tackling the adverse climatic conditions. Even the Parliament of India has signed and ratified the TRIPS Agreement and enacted The Protection of Plant Varieties and Farmers Rights (PPV&FR) Act, 2001.[58] The core objective of these instruments is to provide wider protection to farmers, breeders, and Indian cultivators' traditional knowledge of the Indian agricultural system. It is pertinent to mention that apart from adversely affecting agricultural growth, climatic conditions create socio-economic imbalances and political inequalities, especially gender inequality. Hence, there is a need for comprehensive research to understand the impact of climatic conditions on food security and gender equality for the mitigation of such inequalities and the introduction of inclusive development and food security systems.

10.5.2 Sub-Saharan Africa: Observatory Satellite and Other Predictive Technologies

The introduction of AI-integrated technologies brought a substantial transformation in the agricultural sector.[59] These predictive technologies include remote sensing technologies, drones, IoT, robotics, and so forth. For instance, the stakeholders use robotics as a replacement for the human workforce for performing routine agricultural tasks, namely seed sorting, fumigation for weed control, harvesting, and so forth. Robotics turned out to be cost-effective and seamless agricultural production. Further, other technologies help the stakeholder to collect real-time data for making precise predictions about soil health, seed quality, crop health, crop loss, optimal harvest timings, climate change, and weather patterns for the completion of agricultural activities. In 2019, the first observatory satellite was launched by Ethiopia. The core function of the satellite is to collect real-time data on crop patterns, plant diseases, specific seed selection, and so forth. The algorithm-based technologies ensure agricultural sustainability and sustainable development.

10.5.3 Australia and Southeast Asian Nations

Australia has adopted conservation agriculture and tillage techniques to ensure the natural geo-ecological condition of the soil and to increase the water retention capacity of the soil. Their agricultural model is entirely based on rainwater. This model may also be adopted to accomplish the idea of sustainable food and agricultural production. Further, it promoted the idea of ethical use of new emerging technologies AI, IoT, robotics, and so forth in the processing of agricultural data. Further, it aimed to stimulate agricultural sustainability, minimum waste, and farmers' growth. For instance, LettuceBot is a robotic system that works on visual mapping and system-vision algorithms. This system can identify around 1.5 million lettuce plants per hour.[60] Also, a national grain-crop-yield-forecasting system popularly known as Graincast™ enables stakeholders to monitor national and global market chains, predict crop risks, forecast weather patterns and climate risks, and so forth.[61]

Similarly, certain Southeast Asian countries namely Thailand, Malaysia, Indonesia, and so forth introduced AI-based policies within their respective jurisdictions. The Indonesian National Strategy for AI was launched in 2020.[62] Although the AI-based multi-sectorial initiatives are in the nascent stages, they primarily focus on bringing reforms and establishing efficient governance of education, health, research, infrastructure, agriculture, and food security and safety systems. Simultaneously, it promoted the ethical use of real-time data and AI tools for making verified and data driven predictions and solutions. In 2022, Thailand approved the National AI Strategy and Action Plan (2022 to 2027).[63] This plan aimed to promote ethical and responsible utility of AI in the national development. The first phase of the plan focussed on three prominent sectors, namely health, governance, and agrifood growth. Secondly, it aimed to introduce AI applications in other sectors for sustainable development: transportation, economy, education, tourism, environment, and so forth. Apart from agriculture production, AI plays a significant role in the food storage system of various jurisdictions.

10.6 Food Storage

Earlier, the global population used to follow traditional methods of food and grain storage. Later on, some innovative techniques, namely weather and moisture-resistant food storage systems, were introduced, leading to steady success. Therefore, AI-integrated and data-driven food storage strategies were adopted in several jurisdictions to ensure zero food loss and a smooth supply chain.

10.6.1 India: Anna Darpan System

India has established central storage units in the form of Food Corporation of India and State Food Corporation with the enactment of Food Corporation of India Act, 1964. FCI is the lifeline of the Indian food security system; therefore, it becomes the supreme obligation of the Corporation to ensure the constant availability of healthy, safe, and non-contaminated food grains to the general masses of the nation. In order to ensure standard quality and nutritional value of the food grains, especially rice, and wheat, FCI established a Quality Assurance Cell to ensure compliance with the second pillar of Food Security. FCI introduced advanced techniques to prevent the foodgrains from rotting and damage by the establishment of steel silos for the foodgrains

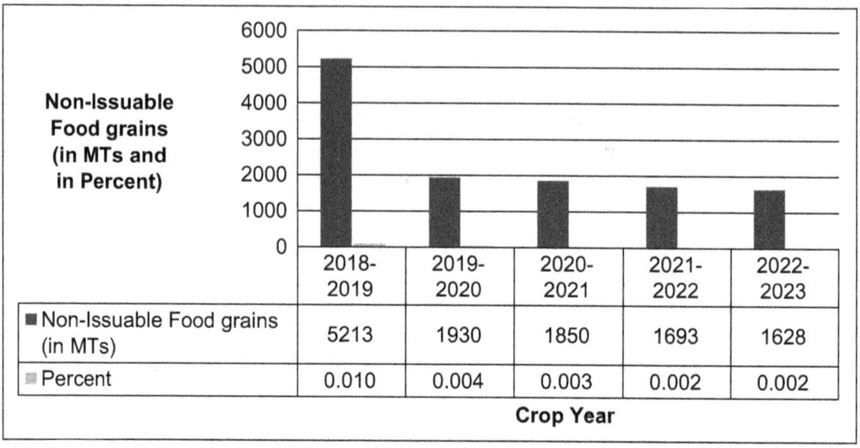

Figure 10.2 Position of damaged food grains in the central pool[64]

Source: Food Corporation of India, Position of damaged food grains in the central pool, 2024

storage. The advanced storage techniques have been adopted by the stakeholders; the progress is steady. The recent data of the FCI itself indicates the rotting of food grains in the godowns of FCI and the reason for the same is temperature changes, extreme weather conditions, and constant precipitation patterns.

A perusal of the available statistical data signifies a consistent reduction in the non-issuable food grains stored in FCI godowns. It demonstrates that the frequent change in climatic conditions demands the adoption of new and modern techniques in food security to ensure the accessibility of nutritious food. Therefore, FCI introduced a digital supply chain management system called *Anna Darpan*.[65] This depot online system aims to integrate divisional, regional, zonal, and headquarters supply chains to ensure efficiency, enhance productivity, optimize internal systems, make data-driven strategic decisions, manage user relations and documentation, and provide real-time accessibility through mobile systems, establishing a sound and effective food security ecosystem. Further, FCI introduced AI-based chemical kits to test the quality and moisture content, to reduce food grain loss during procurement and storage, and to control recycling of rice automatically.[66] Such strategies ensure food quality and safety.

10.6.2 Brazil: Nano-Encapsulation Technique

To establish a technology-driven food storage and safety system, nano-encapsulation and micro-encapsulation were adopted in Brazil.[67] It aimed to improve the quality and safety of food and beverages in the food processing industry. This system is based on AI, digital data, IoT, and so forth, wherein a chemical image of stored food is created. The imaging technique enables stakeholders to predict loss, minimize waste, smooth supply chain, and so forth. This digital initiative significantly transformed Brazilian and global food safety and ensured a sustainable food economy. Stored foodstuff is required to meet technology-powered safety standards.

10.7 Food Safety

The idea of next-generation technology in food security and safety is necessary. AI can be used for precise and efficient smart and sustainable food safety systems.

10.7.1 India: Food Bio-Surveillance for Food Safety

According to FSSAI, around 4.2 lakh deaths and 600 million infections occur annually due to consumption of unsafe food.[68] Therefore, it directs state food authorities to escalate surveillance in warehouses of e-commerce platforms to ensure the maintenance of food hygiene and safety. Therefore, bio-surveillance of logistic operations has been introduced through AI in food processing in the following manner[69]:

1 Collection of structured, unstructured, and semi-structured data.
2 Such data usually contains information related to food processing plants, food industry corporations, and consumer data.
3 The collected data may contain real-time data about food processing, brand quality, food product standards and safety, trends, potential threats, price tracking, forecasting of environmental factors, customer feedback, and so forth.
4 Such technical data should be utilized for maintaining transparent supply chain, updating food safety norms, quality assurance, and monitoring consumer and market requirements.

The promotion of the aforementioned AI-assisted bio-surveillance system and machine learning can bring a revolution in proactive measures and making predictions regarding food safety, food adulteration, public health, bioterrorism, and so forth. Such data and next-generation tools can be used by FSSAI for monitoring quality and ensuring safety standards. The authority aims to introduce genome sequencing and the Internet of Things for maintaining food integrity and authenticity and processing operations respectively. Showing its concern about the climatic conditions and environmental measures, FSSAI aims to promote recycled plastics and materials and utilization of modern testing equipment in the packaging of materials, introduce INFoLNET, a digital network of all food labs for the maintenance of hygiene,[70] and establish an efficient food testing infrastructure.[71]

In 2023, the Global Food Regulators Summit laid a foundation for transformation in the global food safety regulatory mechanism and consumer welfare, promotion of transparent market competition, and mitigation of potential risks in the food system. Simultaneously, the CAC emphasized introduction of innovations and modern quality standards, effective food safety parameters, and stringent regulatory mechanisms.[72] Further, it reiterated the significance of sustainable food security and the utility of emerging technologies to substitute antibiotics as recommended by FAO.

It is apt to mention that a notable growth has been noted in the consumption of processed foods in India in the last few years, which is a serious concern.[73] The excessive consumption of processed and ultra-processed food leads to serious health implications and poses critical challenges to the Indian food and healthcare system. The growing rate of malnutrition among women and children in the nation demonstrates non-completion of nutritional requirements and improper lifestyles, excessive consumption of processed foods being one of them. Therefore, it demands the regulation, imposition of limits, and surveillance of data related to food processing.[74] Hence, AI-integrated proactive measures are needed for the elimination of potential risks and to ensure the delivery of sustainable and safe healthy food to the masses (like the Singapore and Canadian models).

10.7.2 Singapore and Canada: Early Warning and Response Systems

FAO promoted AI-based and data-driven strategies: horizon scanning, big data, early warning and response systems (EWARS), and so forth.[75] EWARS was utilized in Singapore. This system is based on data science to identify contaminated food products, issue prompt warnings, prevent the re-occurrence of such incidences, and disseminate information about such products among the masses. Further, this strategy is utilized to analyse the roots of contamination and to recall contaminated food products in the supply chain. Similarly, a real-time data sharing information system was introduced in Canada. Under this system, the local food safety authorities and laboratories share through a national platform the Canadian Food Safety Information Network. The local authorities share their real-time opinions, issue early warnings, and send alerts, and so forth on the national platform. Such data-driven strategies ensure promising solutions, that is emergency response preparedness, prevention of any health safety hazards, issuance of rapid warnings to health emergencies, and so forth for a robust food safety system.

Despite the adoption of advanced and modern techniques, the episodes of agricultural losses and foodgrains rotting simultaneously rise, which halts the objectives of SDGs. These goals embody a vision of a healthy present and safe future, wherein every individual can grow and enjoy available resources rationally without compromising the needs of future generations. This vision can only be achieved by proactive promotion and reliance on technology-integrated food, climate, and agricultural sustainability in the food security system.

10.8 Conclusion

The foundation of the Indian food security system lies in agricultural storage and food processing system. The FAO's four pillars of food security – stability, availability, accessibility, and utility – apply to the Indian food security system. These pillars are part and parcel of SDGs to be accomplished by 2030. The significance of the SDGs can be understood with the FAO's prediction about 582 million to be undernourished by 2030 globally.[76] As far as the national food security system is concerned, India endeavours to achieve the goal of sustainable agriculture and sustainable food. It can be concluded that human and natural goals cannot be accomplished without relying upon new emerging technologies, especially IoT, machine learning, and AI. In this direction, legislative, policy, and industrial measures have been made in consonance with the International Commitments. According to the World Life Fund Report 2024, the adoption of the Indian food pattern would require only 0.84 of an earth by 2050, which is commendable for sustainable global food production. Though these measures have brought a considerable improvement in the aforesaid system, the statistical data demonstrates a steady growth in agriculture, food storage, and food processing. Similarly, the National Family Health Surveys 4 and 5 depicts little progress in remedying malnutrition status in India. Further, crop loss due to poor climatic conditions reduces agricultural sustainability. Certain technology-powered innovative policy measures, namely NMSA, Organic Farming Promotion Schemes NICRA, and so forth, have been incorporated to increase agricultural sustainability and to ensure food security (has brought positive results). However, being a water-stressed nation, a sensitization drive for groundwater governance and encouragement to stakeholders to actively participate in agricultural sustainability are needed. The G20 Working Group of India stressed on technological innovations for the conservation of natural resources and sustainable development. Such measures can help in eradicating socio-economic imbalances and enforcing inclusive development. A major challenge in the implementation of the Right to Food Security is faced by the Indian stakeholders. To reduce the loss of procured food grain stored in its godowns and silos, FCI adopted AI-powered Digital Supply Chain Management System and introduced AI-based chemical kits. Around 4.2 lakh deaths take place due to the consumption of unsafe food in India.[77] Such incidents mandate the need for advanced food safety measures.

Chronic food insecurity is still prevalent in India, wherein 13% and 55.6% of Indians suffer from chronic undernourishment, unable to meet their nutritional requirement respectively. There are two reasons for the non-accomplishment of nutritional requirements: firstly, the consumption of a sub-standard or low nutritional diet. The non-completion of dietary and nutritional requirements is another indicator of food insecurity. Further, the hunger index and food insecurity data indicate the need for proactive measures to ensure zero food loss, zero hunger, and a sustainable food security system. Secondly, agricultural techniques – from seed selection (seed security) to harvesting – are needed. There is an urgent need for advanced and smart agricultural techniques. The idea of zero hunger and food security demands the adoption of AI technologies in the Indian food security system.

Certain scientific and technical policy reforms – bio-surveillance and CODEX – have been introduced. Such reforms contribute to food logistics and processing systems to maintain authenticity in the food safety infrastructure and to eliminate susceptibilities and insufficiencies deeply rooted in the food security system. Like the Brazilian, Canadian, and Australian models, India needs to abolish the digital divide and amplify AI-powered strategies among stakeholders of the food security system. The adoption of such strategic measures assures multi-dimensional reforms such as agricultural sustainability, climate sustainability, and nutritional food security. Conclusively, the objective of the Sustainable Development Goal of zero hunger and human right to nutritional food security can be accomplished with the adoption of AI-powered technologies in the Indian food security system.

Notes

1 The Prevention of Food Adulteration Act, 1954, s.2(v) – "food" means any article used as food or drink for human consumption other than drug and water and includes – (a) any article which ordinarily enters into, or is used in the composition or preparation of, human food, (b) any flavouring matter or condiments, and (c) any other article which the Central Government may, having regard to its use, nature, substance or quality, declare by notification in the Official Gazette, as food for the purposes of this Act.

2 P. Ramantha Aiyar, *Advanced Law Lexicon* (LexisNexis, 7th edn, 2024).

3 Toshio Shimizu, "Health Claims on Functional Foods: The Japanese Regulations and an International Comparison", 241–252 *Nutrition Research Reviews* (2003).

4 Universal Declaration on Human Rights, 1948, art. 25 (1) – Right to adequate Standard of living: Everyone has the right to a standard of living adequate for the health and well-being of himself and of his family, including foods, clothing, housing and medical care and necessary social services, and the right to security in the event of unemployment, sickness, disability, widowhood, old age or other lack of livelihood in circumstances beyond his control.

5 Universal Declaration on Human Rights, 1948, art. 22 – Everyone, as a member of society, has the right to social security and is entitled to realization, through national effort and international cooperation and in accordance with the organization and resources of each State, of the economic, social and cultural rights indispensable for his dignity and the free development of his personality.

6 74th Session, Third Committee, Agenda item 70 (b) – Promotion and protection of human rights: Human rights questions including alternative approaches for improving the effective enjoyment of human rights and fundamental freedoms United Nations General Assembly, 2019.

7 United Nations Human Rights Office of the High Commissioner, *Food as a Human Right, Not a Commodity: High Commissioner*, Statement and Speeches (October 22, 2024).

8 World Health Organization, *United Nations Decade of Action on Nutrition* (2016–2025), Report by the Director General (2025).

9 Food and Agriculture Organization of the United Nations, *World Food Day 2024: Global Ceremony Highlights the Right to Nutritious Safe and Affordable Foods, Rome* (October 16, 2024).
10 COP 29 Baku Azebaijan, *Letter to Parties and Observers* (November 6, 2024).
11 United Nations Convention to Combat Desertification, 1994, Convention: Overview.
12 United Nations Human Rights, Fact Sheet No. 34, 'The Right to Adequate Food', *Office of the High Commissioner for Human Rights, Food and Agricultural Organization*, ISSN 1014–5567, 2010, available at https://www.ohchr.org/Documents/Publications/FactSheet34en.pdf (Last visited on December 22, 2024).
13 ICESCR, 1966, art. 2(1) – 1. Each State Party to the present Covenant undertakes to take steps, individually and through international assistance and co-operation, especially economic and technical, to the maximum of its available resources, with a view to achieving progressively the full realization of the rights recognized in the present Covenant by all appropriate means, including particularly the adoption of legislative measures.
14 The Constitution of India, 1950, art. 39(c) – The State shall, in particular, direct its policy towards securing (c) that the operation of the economic system does not result in the concentration of wealth and means of production to the common detriment.
15 *PUCL* v. *Union of India* (2001) Writ Petition (Civil) No. 196 (SC).
16 The Constitution of India, 1950, art. 39(e) – Certain principles of policy to be followed by the State. – The State shall, in particular, direct its policy towards securing (e) that the health and strength of workers, men and women, and the tender age of children are not abused and that citizens are not forced by economic necessity to enter avocations unsuited to their age or strength.
17 The Constitution of India, 1950, art. 45 – The State shall endeavour to provide, within a period of ten years from the commencement of this Constitution, for free and compulsory education for all children until they complete the age of fourteen years.
18 The Constitution of India, 1950, art. 47 – Duty of the State to raise the level of nutrition and the standard of living and to improve public health. The State shall regard the raising of the level of nutrition and the standard of living of its people and the improvement of public health as among its primary duties and, in particular, the State shall endeavour to bring about prohibition of the consumption except for medicinal purposes of intoxicating drinks and of drugs which are injurious to health.
19 *Centre for Public Interest litigation* v. *Union of India and Others* (2013) 16 SCC 279.
20 The Food Safety and Standards Act, 2006, Preamble – An Act to consolidate the laws relating to food and to establish the Food Safety and Standards Authority of India for laying down science based standards for articles of food and to regulate their manufacture, storage, distribution, sale and import, to ensure availability of safe and wholesome food for human consumption and for matters connected therewith or incidental thereto.
21 *Swami Achutanand Tirth and Ors.* v. *Union of India and Others* (2014) 13 SCC 314.
22 Food and Agriculture Organization of the United Nations, *About Codex Alimentarius Commission: International Food Standards*, 2026, available at https://www.fao.org/fao-who-codexalimentarius/about-codex/en/.
23 *Kisan Trimbal Kothula* v. *State of Maharashtra* (1977) 1 SCC 300.
24 The Protection of Human Rights Act, 1993, s. 2(1)(d) – Definitions – "human rights" means the rights relating to life, liberty, equality and dignity of the individual guaranteed by the Constitution or embodied in the International Covenants and enforceable by courts in India.

25 *PUCL* v. *Union of India* (2001) Writ Petition (Civil) No. 196 (SC).
26 The National Food Security Act, 2013, Preamble – An Act to provide for food and nutritional security in human life cycle approach, by ensuring access to adequate quantity of quality food at affordable prices to people to live a life with dignity and for matters connected therewith or incidental thereto.
27 Food and Agriculture Organization of the United Nations, *International Fund for Agricultural Development, UNICEF*, World Food Programme and World Health Organization, The State of Food Security and Nutrition in the World: Financing to End Hunger, Food Insecurity and Malnutrition in All Its Forms, Rome (2024).
28 Ministry of Health & Family Welfare, *Malnutrition among Children* (March 16, 2022).
29 Food and Agriculture Organization of the United Nations, *Celebrate the 20 Years of the Right to Food Voluntary Guidelines* (November 17, 2023).
30 Food and Agriculture Organization of the United Nations, International Fund for Agricultural Development, UNICEF, World Food Programme and World Health Organization, *The State of Food Security and Nutrition in the World: Financing to End Hunger, Food Insecurity and Malnutrition in All Its Forms*, Rome (2024).
31 S. R. Bhatt, M. M. Sankhdher, et al., *Social Security, Welfare and Polity: Welfare State in Indian Classics* (Deep and Deep Publications Pvt. Ltd., 2004).
32 Food and Agriculture Organization of the United Nations, *Artificial Intelligence: The Next Frontier in Agrifood Systems Transformation: FAO Liaison Office with the European Union and Belgium* (January 27, 2025) available at https://www.fao.org/brussels/news/news-detail/artificial-intelligence-the-next-frontier-in-agrifood-systems-transformation/en (last accessed on January 25, 2025).
33 Press Information Bureau, *Indian Thali in Global Spotlight for Sustainability* (October 14, 2024) available at https://www.pib.gov.in/FeaturesDeatils.aspx?NoteId=153313&ModuleId=2&reg=3&lang=2 (Last visited on June 08, 2026).
34 World Wildlife Fund, *Living Plant Report 2024-A System in Peril* 68 (2024) available at https://www.worldwildlife.org/publications/2024-living-planet-report/ (Last visited on June 08, 2026).
35 Press Information Bureau, Ministry of Food Processing Industries, *Indian Thali in Global Spotlight for Sustainability* (October 14, 2024), above note 33.
36 National Mission for Sustainable Agriculture, Department of Agriculture, Cooperation & Farmers Welfare, *Ministry of Agriculture & Farmers Welfare, Government of India, Strategy* (2024).
37 Press Information Bureau, *Ministry of Agriculture & Farmers Welfare: Promotion of Organic Farming* (August 8, 2023).
38 National Innovations in Climate Resilient Agriculture, *Key Features* (2011).
39 FAO in India, *Transition to Sustainable Agriculture Vital for Water and Food Security, Say FAO, IFAD and WFP* (October 16, 2023).
40 NITI Aayog, *Promoting Millets in Dies: Best Practices Across States/UTs of India* (2023).
41 Food and Agriculture Organisation of the United Nations, *FAO in India, Transition to Sustainable Agriculture Vital for Water and Food Security, Say FAO, IFAD, and WFP* (2023).
42 UNESCO, *UN World Water Development Report: Making the Invisible Visible* (2022).
43 Press Information Bureau, *NITI Aayog Report on Water Crisis* (December 09, 2019).
44 World Food Programme, *Country Strategic Plan 2019–2022, India: Annual Country Report* (2022).
45 ICRISAT World Food Programme, ICRISAT AND WFP: *India Working Paper: Effects of Climate Change on Food Security and Nutrition in India: A systematic Review* (2023).

46 J. Samuel, C. A. R. Rao, Raju, et al., "Assessing the Impact of Climate Resilient Technologies in Minimizing Drought Impacts on Farm Incomes in Dry Lands Sustainability", 382 (2021).

47 B. Sabar and D. K. Midya, "Intersecting Knowledge with Landscape: Indigenous Agriculture, Sustainable Food Production and Response to Climate Change – A Case Study of ChuktiaBhunjia Tribe of Odisha, India", *Journal of Asian and African Studies*, 59(1), 123–141 (2022).

48 M. S. Swaminathan and P. C. Kesavan, "Agricultural Research in an Era of Climate Change", *Agricultural Research*, 1, 3–11 (2012).

49 National Food Security Mission, *Operational Guidelines, 12th Five Year Plan*, Department of Agriculture & Cooperation, Ministry of Agriculture Government of India (2007).

50 A. Sharma, B. S. Kumar, P. P. Dubey, et al., "Delay in Puberty Is Dependent on Heat Shock Protein B1 Expression in Native Cross Layers of Punjab Under Heat Stress", *Reproduction in Domestic Animals*, 57(3), 284–291 (2022).

51 B. Mahapatra, M. Walia, C. A. R. Rao, B. M. K. Raju and N. Saggurti, "Vulnerability of Agriculture to Climate Change Increases the Risk of Child Malnutrition: Evidence from a Large-Scale Observational Study in India", *PLoS One*, 16(6), e0253637 (2021).

52 Food and Agriculture Organisation of the United Nations, *Codex Alimentarius Commission (Procedural Manual)*, 25th edn (Rome, 2016).

53 Ministry of Women and Child Development, Guidelines on Quality Assurance, *Roles and Responsibilities of Duty Holders, Procedure for Procurement, Integrating AYUSH Concepts and Data Management and Monitoring through 'POSHAN Tracker' for Transparency, Efficiency and Accountability in Delivery of Supplementary Nutrition* (2021).

54 Integrated Child Development Scheme (ICDS), *Manual for District-level Functionaries* (2017).

55 Mid-day Meal (MDM) Scheme, *Manual for District-Level Functionaries (2017)* (28 December 2024) available at https://darpg.gov.in/sites/default/files/Mid%20Day%20Meal%20Scheme.pdf (Last visited on January 27, 2025).

56 Preamble of TRIPS Agreement.

57 ICRISAT World Food Programme, ICRISAT AND WFP: India Working Paper: Effects of Climate Change on Food Security and Nutrition in India: A Systematic Review, 2023.

58 The Protection of Plant Varieties and Farmers' Rights Act, 2001 – Preamble – An Act to provide for the establishment of an effective system for protection of plant varieties, the rights of farmers and plant breeders and to encourage the development of new varieties of plants.

59 International Centre of Expertise in Montreal on Artificial Intelligence, *State of AI in Agriculture in Sub-Saharan Africa* (2024) https://doi.org/10.5281/zenodo.13144813 (Last visited on January 27, 2025).

60 Allegiant Media, Artificial Intelligence in Agriculture (April 22, 2021) available at https://allegiantmedia.com.au/australian-agriculture-3-key-barriers-to-agtech-adoption-2/ (Last visited on January 25, 2025).

61 Digiscape Future Science Platform, Graincast™: Forecasting Australia's National Grain Crop Yield available at https://research.csiro.au/digiscape/digiscapes-projects/forecasting-australias-national-grain-crop-yield/ (Last visited on January 25, 2025).

62 MFAT Market Intelligence reports, Indonesia's National Strategy for Artificial Intelligence-July 2023, New Zealand Embassy in Jakarta available at https://www.mfat.govt.nz/en/trade/mfat-market-reports/indonesias-national-strategy-for-artificial-intelligence-july-2023.

63 US-ASEAN Business Council Inc., AI Governance in Southeast Asia (October 2023) available at https://www.usasean.org/sites/default/files/2023-10/%5BUSASEAN%5D%20AI%20Governance%20in%20Sou theast%20Asia.pdf (Last visited on January 25, 2025).
64 Food Corporation of India, *Statistical Data: Quality Control* (Food Corporation of India, 2024).
65 Ministry of Consumer Affairs, Food & Public Distribution, *PIB: Food Corporation of India (FCI) Transforms Supply Chain Management System under Anna Darpan* (September 30, 2024).
66 Indiaai, *FCI to Use AI to Assess Food Grain Quality* (November 19, 2021).
67 Transfer LC Local Consultants Kingdom of the Netherlands, *The Brazilian Food Processing Industry: A Sectoral Study by TRANSFER.LC Commissioned by the Embassy of the Kingdom of the Netherlands to Brazil* (Last visited on January 29, 2025).
68 Foodspectrum, *Global Food Regulators Summit 2023 Pledges to Strengthen Global Food Safety System* (August 8, 2023), available at https://www.fssai.gov.in/upload/media/GFRS2023_strengtehenGlobalFoodSafetynuffood.pdf (Last visited on January 08, 2024).
69 Indiaai, *User Submission – AI's Role in Improving Bio Surveillance and Food Safety*, available at https://indiaai.gov.in/article/ai-s-role-in-improving-biosurveillance-and-food-safety (Last visited on January 05, 2025).
70 PTI, Outlook The News Scroll, *FSSAI Mulling Using Next-Gen Technology Tools to Ensure Food Safety* (2021) available at https://www.fssai.gov.in/upload/media/FSSAI_News_Mulling_Outlook_28_07_2021.pdf (Last visited on January 05, 2025).
71 FSSAI, *FSSAI Experience Zone*, available at https://fssai.gov.in/upload/knowledge_hub/1506435b2a1 61d48e1cExperienceZoneBrochure.pdf (Last visited on January 05, 2025).
72 Ministry of Health and Family Welfare and FSSAI, *Global Food Regulators Summit 2023: Collaborative Approach on Strengthening Food Safety & Regulatory Framework* (20th to 21st July 2023) available at https://sites.fssai.gov.in/gfrs2023/pdf/GFRS2023.pdf (Last visited on December 22, 2024).
73 Ministry of Statistics and Programme Implementation, *Household Consumption Expenditure Survey:2023–24 Fact Sheet* (August 2023–July 2024) available at https://mospi.gov.in/sites/default/files/publication_reports/HCES%20FactSheet%202023-24.pdf (Last visited on January 02, 2025).
74 FSSAI, *The Global Food Regulators Summit 2023 Shows New Direction for Transforming the Global Food Regulatory Landscape* (July 23, 2023) available at https://www.fssai.gov.in/upload/uploadfiles/files/Press%20Release_GFRS%202023%20Consolidated.pdf (Last visited on January 02, 2025).
75 Food and Agriculture Organization of the United Nations, *Early Warning Tools and Systems for Emerging Issues in Food Safety: Technical Background* (Rome, 2023) available at https://openknowledge.fao.org/handle/20.500.14283/cc9162en (Last visited on January 05, 2025).
76 United Nations, *UN Water: The State of Food Security and Nutrition in the World 2024* (August 12, 2024) available at https://www.unwater.org/news/state-food-security-and-nutrition-world-2024 (Last visited on January 30, 2025).
77 FSSAI, *Global Food Regulators Summit 2023 Pledges to Strengthen Global Food Safety Systems* (August 8, 2023) available at https://www.fssai.gov.in/upload/media/GFRS2023_strengtehen_Global_FoodSafety_nuffood.pdf (Last visited on January 29, 2025).

Part VI

Access to Justice

11 Enhancing Access to Justice Through Digital Innovation

Bhutan's Journey Towards SDG 16.3

Rinzin Penjor

11.1 Introduction

At the backdrop of a series of economic crisis and a lack of coordinated approach to international development,[1] the leaders of the world met in September 2000 at the Millennium Submit and adopted the United Nations Millennium Declaration, setting out the 8 Millennium Development Goals (MDGs).[2] These goals were to eradicate extreme poverty and hunger; to achieve universal primary education; to promote gender equality and empower women; to reduce child mortality; to improve maternal health; to combat HIV/AIDS, malaria, and other diseases; to ensure environmental sustainability; and to develop a global partnership for development.[3] Many countries around the world implemented this new developmental priorities; however, the outcome was not up to the level intended. The MDG achieved significant progress in reducing the world poverty rate to half of the rate in 1990, achieved significant parity in primary education between girls and boys, and other targets it aimed to achieve.[4] However, the achievements were uneven and unequally distributed across regions and countries. Therefore, while many countries made significant progress, some countries lacked behind in translating the vision into reality.[5]

As the MDGs were set to expire by 2015, the discussion on post-2015 development agenda picked up with a focus on building a sustainable future. The 2012 Rio+20 Conference prompted discussions on developing a new set of Sustainable Development Goals,[6] which will build on the foundation laid by the MDGs. The outcome document of 2012 Rio Conference titled "The Future We Want" established an Open Working Group (OWG) to develop a set of Sustainable Development Goals to be reviewed and considered by the UN General Assembly (UNGA) in 2014.[7] On 19 July, 2014, the OWG presented its proposal of Sustainable Development Goals (SDGs) consisting of 17 interconnected goals and 169 targets.[8] Learning from the critics on the progress of MDGs being uneven across regions and countries, the SDGs were developed in such a way that it is uniformly applicable to all the countries irrespective of their development status. Furthermore, the development of the

DOI: 10.4324/9781003733751-17

SDGs has been more of a participatory policy process compared to the MDG, which was criticised for being developed following a closed-room process.[9]

SDGs are a global commitment to a developmental philosophy that prioritize the needs of today without compromising the future generation's abilities to meet their needs.[10] SDGs reflect the shared commitment of the member states and global consensus on eradicating poverty and ensuring environmental sustainability, social equity, and economic growth to ensure that no one is left behind.[11] Among these goals, SDG 16 stands out as an important goal for sustainable development, focusing on promoting peaceful and inclusive societies, providing access to justice for all, and building effective, accountable, and inclusive institutions at all levels.[12] This goal is particularly important as it recognizes that development cannot be achieved without peace, stability, and good governance. Within SDG 16, target 16.3 highlights the importance of promoting the rule of law at both national and international levels, ensuring equal access to justice for all.[13] This target emphasizes that justice systems must be accessible, efficient, and fair to protect rights, prevent discrimination, and foster sustainable development. The emphasis on justice and strong institutions represents a significant departure from the MDGs, reflecting a deeper understanding of how governance and justice systems fundamentally impact development outcomes.

However, in mountainous regions, achieving target 16.3 faces unique geographical challenges that significantly impact access to justice. The rugged terrain and extreme weather conditions often isolate communities from legal institutions and services. Physical accessibility becomes a major barrier, with limited transportation infrastructure and seasonal road closures preventing people from reaching courts or legal aid centres.[14] These geographical constraints also affect the deployment and retention of legal professionals in mountain areas, leading to a shortage of qualified personnel. The scattered nature of mountain settlements increases the cost and complexity of providing legal services. The combination of geographical isolation, infrastructure limitations, and resource constraints creates significant obstacles in ensuring equal access to justice for mountain communities, requiring targeted interventions and innovative approaches to overcome these challenges.

Bhutan, a small mountainous country with an area of 38,394 square kilometres,[15] presents a unique case study of challenges in achieving SDG 16.3 in a mountainous region. While Bhutan has made significant strides in enhancing access to justice through improved roads, better connectivity, and the establishment of various sub-district courts, challenges remain due to its mountainous terrains and isolated communities in rural areas. Furthermore, the concentration of legal services (especially lawyers) in urban centres has made access to legal services expensive and time intensive for rural communities. These challenges were further exacerbated during the COVID-19 pandemic. The existing geographical barriers to access justice became a problem as many people living in remote parts of the country struggled to access legal services due to movement restrictions. However, the pandemic not only highlighted the weakness of existing practices but also paved the way for digital innovation in enhancing access to justice in Bhutan.

Under the leadership of His Majesty the King, Bhutan implemented e-litigation facilities and e-filing systems. These innovations allowed people to file cases and attend court hearings from the comfort of their homes. These projects not only mitigated the impact of COVID-19 but also addressed the age-old geographical barrier to access to justice in Bhutan. While COVID-19 disrupted the services of the justice system in Bhutan, it also paved the way for digital and technological innovation in the justice system. Therefore, this chapter examines the digital transformation in Bhutan's Justice Sector, analysing how technological innovations can enhance access to justice in mountainous regions while contributing to the achievement of SDG 16.3.

11.2 Overview of Bhutan's Geographic and Socio-Economic Context

11.2.1 Bhutan's Geography and Its Impacts on Accessibility

Bhutan's unique geographical condition significantly impacts the accessibility of various services, including accessing justice services. Located in the Eastern Himalayas, the topography of the country varies significantly from the southern foothills, located at about 150 meters above sea level to the Northern steep mountains located at about more than 7000 meters above sea level.[16] These significant changes in the elevation over a relatively small area creates a topography characterized by steep valleys, rugged mountains, and deep gorges that significantly impacts accessibility.

The country's topography can be broadly divided into three different regions: the southern foothills, the inner Himalaya in the centre featured with low mountains and fertile valleys, and the Greater Himalayas in the north featured with snow capped mountains.[17] These different terrains have historically influenced the settlement patterns in the country, with communities often scattered across isolated valleys and mountainsides. The scattered nature of settlements, especially in rural areas, creates a geographical barriers to accessing centralized services, including courts, legal institutions, and legal services such as lawyers, Legal Aid Centre, and the Pro Bono Unit of the Bar Council of Bhutan, which is located in the capital city of Thimphu.

The primary mode of transportation in the country for people to move from district to district is through road. However, transport infrastructure development in Bhutan faces considerable engineering challenges due to the mountainous terrain. While the country has made significant progress in enhancing connectivity through constructions of roads, many remote communities remain connected only by farm roads or traditional transport routes that becomes unusable during the heavy monsoon season, which typically lasts from June to September.[18] The East-West highway, Bhutan's primary arterial road, highlights these challenges. Despite its crucial role in connecting districts, its winding mountain routes pose significant challenges while traveling as even relatively short distances can take many hours, while monsoon rains and winter snow continuously disrupts the routes causing road blocks.[19]

The geographical constraints particularly affect the delivery of justice services in several ways. Firstly, the concentration of most legal professionals in Thimphu, and courts in district headquarters and urban centres, make it difficult for the rural residents as they often have to travel long distances, navigating challenging terrain and weather conditions, to access legal services. This journey can be both time consuming and expensive, potentially deterring people from seeking legal recourse for their grievances.

Seasonal weather patterns further exacerbates the accessibility issues. During the monsoon season, landslides frequently block roads, while winter snowfall can isolate high-altitude communities for a long periods of time.[20] These weather-related disruptions create additional barriers to accessing justice services, as people have to wait for favourable weather conditions to travel to court or meet with legal professionals and avail services from the Legal Aid Centre or the Pro Bono Unit of the Bar Council.

The geographical challenges and the disparity in urban rural development also affects the distribution and retention of legal professionals across different districts in Bhutan. Most lawyers and other legal service providers tend to cluster in urban areas like Thimphu, Phuentsholing, and other district headquarters, where infrastructure and amenities are better developed. For illustration, almost 85% of law firms in the country are located in Thimphu, 7% in Chukha District, and 8% in Paro District and Samdrup Jongkhar District; the rest of the 16 districts do not have a law firm operated by a licensed lawyers. This concentration creates a spatial mismatch between legal service providers and the rural populations they serve, making it difficult for remote communities to access legal representation and advice.

11.2.2 Bhutan's Socio-Economic Context and Access to Justice

Bhutan's unique philosophy on development under the umbrella of Gross National Happiness paradigm leads to distinctive development models that empower balance progress, strengthening of institution, and equal access to services during the delivery of justice and its governance.

The country's socio-economic development has created opportunities and challenges for access to justice. While overall poverty rates have reduced from 23.2% in 2007 to 12.4% in 2022,[21] economic differences between the urban and the rural areas affect people's ability to access legal services in the country. The cost of travel, legal representation, and related expenses often present significant challenges for rural communities in seeking justice and exercising their rights.

For improvement of legal empowerment, the education and literacy rate of the citizen is an important factor. With an increase in the literacy rates from 10% in the 1960s to over 71% in 2017,[22] more Bhutanese citizens can understand their privileges in accordance with law and navigate the legal system. However, new challenges have risen due to the digital literacy gap between rural and urban populations during the execution of technology-based legal solutions.

The most significant outcome of Bhutan's digital transformation is the availability of new possibilities for justice delivery. Despite the late adoption of internet, innovative solutions in the justice sector have been made possible by the country's investment in digital infrastructure that achieved 87% internet penetration by 2024.[23] Also the COVID-19 pandemic necessitated the implementation of e-litigation facilities and virtual court services, making the digital foundation particularly crucial.

11.3 Access to Justice in Bhutan: Pre-COVID Landscape

11.3.1 Status of the Legal Framework and Services

Since the adoption of its first written Constitution in 2008, Bhutan's legal framework has undergone significant evolution. Under Article 21, the Constitution of Bhutan guarantees that all citizens have equal access to justice, and that justice delivery should be fair, independent, and efficient. Various legislative instruments, such as the Civil and Criminal Procedure Code and the Evidence Act, support this constitutional foundation when it comes to the administration of justice.

The Judicial System is based on four tiers: the Supreme Court, High Court, Dzongkhag (District) Courts, and Dungkhag (Sub-district) Courts.[24] Despite the hierarchical structure's goal of ensuring widespread access to justice, the distribution of courts is a reflection of the geographical challenges previously discussed. The Judiciary of Bhutan reported that there were 20 Dzongkhag Courts and 15 Dungkhag Courts, serving Bhutan's 20 districts and 205 gewogs (blocks).[25]

A significant challenge in Bhutan's legal system is the limited number of legal professionals. According to the records with the Bar Council of Bhutan, as of 2024, there were fewer than 130 private practising lawyers in the country, with approximately 90% of the private practising lawyers concentrated in Thimphu and other urban centres. This urban concentration of legal expertise creates significant access barriers for rural communities, often requiring them to travel long distances for legal consultation. Due to this acute shortage of legal professionals, the formal justice system is complemented by traditional dispute resolution mechanisms, particularly the consensus-based mediation system known as Nangkha Nangdrik. This system, practiced at the community level, helps resolve minor disputes without requiring parties to access formal court systems.[26]

Recent legislative reforms have focused on modernizing the legal framework to accommodate technological advancements. The Information, Communications and Media Act of 2018 provided the legal basis for electronic transactions and digital signatures, laying the groundwork for e-litigation systems. Additionally, the Judiciary of Bhutan initiated the e-litigation process as a response to the COVID-19 pandemic in 2020 and established protocols for virtual court proceedings, marking a significant shift toward digital justice delivery.[27]

11.3.2 *Efforts to Improve Access to Justice*

Bhutan has shifted from traditional outreach programs to innovative digital solutions as part of their efforts to enhance access to justice. Institutional strengthening, physical infrastructure development, and digital transformation are the three main phases of the government's efforts.

Capacity building and legal awareness were the primary objectives of the institutional strengthening phase. Local leaders and community members have been trained in basic laws and legal procedures through extensive legal literacy programs conducted by the Bhutan National Legal Institute (BNLI).[28] These programs are rolled out to local government leaders, judicial officers, school law clubs, and community members annually, significantly improving legal awareness in remote communities.[29] Furthermore, in order to enhance access to justice the development of physical infrastructure has resulted in the establishment of sub-district level courts and the legal aid centre which will provide legal aid services to the vulnerable population.[30] Similarly, the Bar Council of Bhutan has recently established the Pro Bono Unit, encouraging lawyers to provide Pro Bono services to people who need it the most.[31]

11.4 Impact of COVID-19 on Justice Delivery and Measures of Adaptation

11.4.1 *Impact of COVID-19 on Bhutan's Justice System*

Bhutan's justice system experienced an unprecedented challenge when the COVID-19 pandemic took over the world by storm, exacerbating the already existing geographical and infrastructural barriers to accessing justice. The strict implementation on movement restriction and social distancing protocols by the government had considerably disrupted the conventional justice delivery mechanisms after the first reported COVID-19 case in March 2020.

Court activities suffered the most direct effects. Physical court proceedings were halted, and during the first few months of the pandemic, fewer new cases were filed. In comparison to the same period in 2019, the Supreme Court recorded a 47% drop in case registration during the second quarter of 2020.[32] Since many people in rural areas initially lacked access to alternate justice delivery mechanisms due to a lack of digital literacy and infrastructure, this loss was especially noticeable in rural areas of Bhutan.

The pandemic had an impact on numerous aspects of the execution of justice in addition to judicial hearings. There were significant disruptions to legal services, which were formerly provided through in-person consultations.

Similarly, movement restrictions and social separation requirements hindered traditional community-level dispute resolution procedures.

Vulnerable populations, including women, the elderly, and people living in rural areas, were especially impacted by the crisis. Domestic violence cases in Bhutan surged by 37% during the COVID-19, as per the Impact Assessment Report of COVID-19 on Women and Children.[33] The travel restriction imposed meant that the victims could not access justice services.

However, the pandemic also highlighted how flexible the legal system can really be if needed. The judiciary swiftly mobilized to create alternate service delivery methods under the visionary leadership of His Majesty the King. In order to ensure that services are available to the people in desperate times, the judiciary of Bhutan swiftly adopted the e-litigation services with major legal institutions such as the office of the Attorney General, and the Bar Council of Bhutan.

11.4.2 Digital Innovation During the Pandemic

Bhutan's legal system underwent an unprecedented digital revolution as a result of the COVID-19 epidemic. A number of digital initiatives that had been in the planning phases before the pandemic were hastened in their execution as a result of the crisis's requirement for quick innovation in service delivery.

The launch of the e-litigation system in June 2020 was the most important one. Online hearing facilities, virtual case management, and electronic case filing were made possible by this all-inclusive digital platform. The system's ability to sustain the administration of justice throughout the crisis was demonstrated when the justice system across Bhutan went online.[34]

The government accelerated the launch of the Digital Drukyul flagship program to facilitate this digital shift. With the installation of e-litigation access points at Community Centres throughout Gewogs, special emphasis was laid on improving digital infrastructure in rural areas. Most of these centres were equipped with the capability to access virtual court services.[35]

Innovative approaches to witness examination and evidence presentation were also introduced by the judiciary. Courts were able to continue operations while adhering to social distance standards through the establishment of secure digital evidence submission systems and virtual witness examination processes. Because they eliminated the need for in-person court visits, these advances were especially helpful for litigants from far-flung locations.[36] Initiatives for capacity building and training were quickly digitalized. The BNLI created virtual training sessions and online legal awareness programs for community people, local leaders, and legal professionals.[37]

11.5 Digital Transformation in Bhutan's Justice Sector: Advantages and Challenges

11.5.1 E-Litigation Facilities and E-Filing Systems

Bhutan's journey towards digital innovation in justice system began with the introduction of the Case Management System (CMS) in 2011. However, the comprehensive digital transformation was accelerated during the COVID-19 pandemic. The current e-litigation framework represents a significant evolution from the early digitization efforts, incorporating advanced features and user-centric design principles.

The e-litigation system, launched in June 2020, included several integrated components such as electronic filing of cases, electronic filing of documents, electronic payment, and remote hearing among others.[38] Bhutan was one of the first countries in the region to implement e-litigation as a response to the disruption caused to the justice system due to covid-19.[39] At its core is the Electronic Filing System, which allows litigants and legal practitioners to file cases electronically through a secure web portal.[40] The system supports various document formats and includes built-in validation checks to ensure compliance with filing requirements. The Virtual Court Platform forms another crucial component of the e-litigation infrastructure. Various courts across the country have been equipped with infrastructures such as dedicated internet line for uninterrupted connection, virtual conference equipment, and other facilities necessary for the virtual hearing process.[41]

Furthermore, to ensure accessibility in rural areas, the judiciary established e-litigation access points in Community Centres with support from the United Nations Development Programme.[42] These centres are equipped with necessary equipment and supported by trained facilitators who assist users in navigating the system.

Similarly, the e-payment gateway integration has simplified court fee payments and financial transactions. Through collaboration with the Royal Monetary Authority, the system enables secure online payment options, reducing the need for physical visits to courts for financial transactions. This feature has particularly benefited rural residents who previously had to travel long distances for such transactions.

11.5.2 Benefits of Digital Transformation

The digital transformation of Bhutan's justice sector significantly benefited the people, especially those living in rural areas, and has fundamentally changed how justice services are accessed and delivered. One of the most significant achievement of the digitization of the justice system is the improved accessibility. The e-litigation system has significantly reduced geographical barriers to access to justice. Similarly, digital transformation has had a significant impact in making the court proceedings cost effective for the litigants as well as the courts. With the e-litigation facilities in place, litigants need not make arrangements for long travels, and can save up on accommodation expenses. Similarly, the office of the Attorney General, which is located in the Capital city, need not have to incur expenses in sending out prosecutors to different districts, thereby reducing the office expenditure significantly. For the judiciary, digital transformation has led to a significant reduction in operational costs through decreased paper usage, storage requirements, and administrative overhead.

Furthermore, transparency and accountability have been improved through digital tracking and monitoring capabilities. The system provides real-time case status updates and maintains detailed audit trails of all transactions.[43] Additionally, digital transformation of the justice system reduces the carbon footprint of the justice sector in Bhutan. The reduction in paper usage and physical travel has contributed to lower carbon emissions, supporting Bhutan's commitment to environmental conservation.

11.5.3 Challenges and Limitations

Despite its successes, the digital transformation of Bhutan's justice sector faces several challenges and limitations that require ongoing attention and resolution. Digital innovation and infrastructure remain a significant challenge in the process. While internet connectivity has improved significantly, many remote areas still experience unstable connections, and the internet itself is very expensive in Bhutan.[44] Furthermore, many regions in Bhutan face regular connectivity issues,[45] affecting the reliable use of e-litigation services. The high cost of maintaining and upgrading digital infrastructure in mountainous terrain poses additional challenges for sustainability of the system.

Another challenge is the digital literacy in the country. Despite training and capacity building efforts, many citizens, particularly elderly and rural residents, struggle with using digital platforms. Many rural users require assistance to access e-litigation services, highlighting the need for continued support and capacity building to the community centre staffs and the local leaders. Security concerns and cybersecurity risks pose another ongoing challenge that the justice sector and the newly established GovTech Agency continues to address. As the e-litigation system handles sensitive legal information, ensuring robust security while maintaining accessibility requires constant attention and resource allocation.

11.6 Achieving SDG 16.3 in Mountainous Regions and the Role of Technology

11.6.1 Role of Technology in Addressing Geographical Barriers

Bhutan's experience with digital innovation in enhancing access to justice, while still in its nascent phase with many challenges and limitations, provides a perspective on how digital technology can be utilized to overcome geographical barriers to access to justice in mountainous regions. The experience provides a valuable lesson for other mountainous countries striving to achieve SDG 16.3's goal of ensuring equal access to justice for all. The transformation of justice delivery through digital innovation can significantly change how mountainous regions interact with legal systems, by creating new possibilities and opportunities for access and participation in legal system in ways previously considered challenging.

Digital technology's role in mitigating geographical barriers manifests through multiple interconnected innovations. The advent of e-litigation platforms has fundamentally changed the justice ecosystem, by effectively eliminating the need for physical travel in many legal proceedings, an innovation which is significant in mountainous regions where a journey to the nearest court might traditionally involve traveling over difficult terrain. The UNDP's report "E Justice: Digital Transformation to Close the Justice Gap" (2022) indicates that virtual court services can reduce travel-related costs and other cost related to accessing legal services, allowing people to attend legal proceedings from the comfort of their home or office.[46] This significant reduction in time and cost burdens has made justice more accessible to remote communities, which previously faced significant barriers to access courts and justice services.

Similarly, the implementation of a cloud-based case management systems has revolutionized document handling and administrative processes in the justice sector. These systems enable legal documents to be filed, processed, and accessed from any location with internet connectivity, eliminating the traditional challenges of physical document transportation across difficult locations often delayed due to difficult terrains or weather conditions. The digital document management systems can effectively reduce case-processing times in remote mountainous areas, especially when a case is being appealed from a remote court to a high court, significantly improving the efficiency of justice delivery while reducing the administrative burden on both courts and litigants.

Mobile justice applications and services have become an important initiative for connecting isolated communities with justice system by offering basic legal information, documentation services, and connections to legal aid and pro bono services. These applications, optimized for low-bandwidth environments, have proven particularly effective in mountainous regions where internet connectivity may be unstable. These mobile applications help in creating legal awareness and service utilization in remote mountain communities following the implementation of mobile justice solutions.[47] This increased awareness and accessibility have transformed how mountain communities can engage with legal systems, making justice more accessible and understandable.

Similarly, artificial intelligence and machine learning technologies have added another dimension to digital justice accessibility in mountainous regions. Artificial intelligence has huge potential in changing the ecosystem of the justice sector by means of facilitating translation of language, document classification, and case scheduling optimization, addressing specific challenges faced by justice system world wide irrespective of the jurisdictions.[48] The AI-powered services have the potential to significantly improve efficiency multilingual mountain communities, breaking down language barriers that previously posed challenges in access to justice. Furthermore, the AI-powered services effectively reduce the time consumed in processing and studying case documents, thereby expediting the case processing time.

The technology's role is not just limited to service delivery but also extends towards fundamentally reshaping the modes of legal education and legal awareness in mountain communities. Digital platforms such as social media and other conferencing tools such as Zoom, Microsoft Teams, and Google Meet aid in remote legal literacy programs, training for local leaders, and legal education and awareness initiatives. Technology and digital platforms have a huge potential in promoting legal literacy programmes in mountainous regions, which are not easily accessible.

11.6.2 *Policy and Infrastructure Recommendations*

Developing and implementing an effective digital justice system in mountainous regions calls for a comprehensive and nuanced approach to policy and infrastructure development. Drawing from Bhutan's ongoing experience and broader research on mountainous regions, several key recommendations emerge for creating sustainable and effective digital justice systems. These recommendations address both immediate accessibility needs and long-term sustainability concerns while considering the unique challenges posed by mountainous terrain.

Infrastructure development is an important aspect of a successful digital justice system in mountainous regions like Bhutan. The establishment of a reliable internet connectivity is a crucial component of a digital justice ecosystem. This must be complemented by access to community centres equipped with necessary hardware and software supported by trained staff who can assist users in accessing justice services, like in the case of Bhutan. In mountainous regions where digital literacy is quite low, having these centres can increase digital service utilization by vulnerable population such as elderly people, people with disability, and people who do not have access to digital technologies, ensuring that services are available to the people who need them the most.

Policy frameworks supporting digital justice systems must be comprehensive and forward thinking. Legislation needs to explicitly recognize electronic filings and virtual proceedings while establishing clear protocols for data protection and cybersecurity. UNDP's report on e-justice emphasizes that successful digital transformation requires policies that balance accessibility with security concerns.[49] These policies should also encourage public–private partnerships for technology infrastructure development, as government resources alone may be insufficient to meet the substantial infrastructure needs and to ensure sustainability of these projects in mountainous regions.

Financial sustainability represents a critical policy consideration. The justice-for-all initiative recommends establishing dedicated funding mechanisms through a combination of government allocations, international development support, and innovative financing models.[50] Cost-sharing arrangements between the national and local governments, coupled with a nominal user fee structures that ensure accessibility while supporting system maintenance, can prove effective in making the programs accessible as well as sustainable in a long run.

11.6.3 Broader Implications for SDG 16.3

The successful implementation of digital justice systems in mountainous regions has a far-reaching impact on achieving SDG 16.3 globally. The transformation of justice delivery through digital means highlights the potential for overcoming geographical barriers to access while emphasizing broader lessons about institutional change and sustainable development. These impacts extend beyond mere technological solutions to include fundamental changes in how justice is conceived and delivered.

The economic impacts of digital justice systems are particularly significant for sustainable development. While initial implementation costs may be high, digital transformation can reduce justice delivery costs and improve accessibility significantly in the long term.[51] This cost reduction makes justice more affordable and sustainable, particularly in resource-constrained environments. Furthermore, the reduction in travel and physical infrastructure requirements creates opportunities for reallocating resources to other aspects of justice system improvement.

While implementing digital justice system, achieving social inclusion and equity remains a central concern. Digital platforms can help reduce discrimination and increase the participation of marginalized communities in the justice system.[52] They are ideal for making justice services available to people who need it the most but cannot access it due to geographical or socio-economic constraints. When we take "access to justice" as our focus, we are working directly in accordance with the core principles of Sustainable Development Goal 16, which aims for the promotion of inclusive societies and institutions and target the reduction of discrimination to ensure that no one is left.

To implement a successful and effective Digital Justice System, organizational structures, workflows, and competencies of professionals need to be changed as well, so the institutional transformation affected by this happens at a profound level. For this reason, the UNDP's report on e-justice (2022) talks about something much broader and more systemic than just a "digitalization of criminal justice." It gets to the heart of what needs to change, and it shows us the path to that change in a more efficient, transparent, and accountable way.

Therefore, a successful digital justice system offers promising prospects for advancing SDG 16.3 – providing fair, timely, and effective access to justice for all – because they can be readily adapted to suit the legal needs of very different communities. Furthermore, many of the promising solutions created for mountainous regions can be customised for use by other hard-to-reach communities in geographically and socially diverse communities, including island nations, desert regions, and urban conflict zones. The digital justice system is significant in achieving improvements in global access to justice.

Furthermore, an unforeseen but significant benefit of digital justice systems is their effect on environmental sustainability. Digitization of the justice system has the potential to significantly reduce the carbon footprint of justice delivery.[53] The principal means to that end is certainly not "paperless" justice. Yet, even if one takes paper usage as a proxy for carbon emissions, e-justice may in fact be accomplishing that "save a tree" goal by virtue of reduced travel and other means – a situation that reminds us of the "invisible" environmental benefits achieved through broader Sustainable Development Goals.

11.7 Discussion

11.7.1 Reflection on Bhutan's Progress

Access to justice in Bhutan has significantly progressed through digital innovation, allowing the mountainous nation to maintain cultural integrity and institutional effectiveness despite geographical barriers that often plague mountainous regions. While this digital transformation is certainly a means to an end – increased access to justice – it is also an opportunity for reflections on system efficacy in both the achievement of just results and the efficiency of the journey toward those results.

The achievement of progress towards digital justice transformation reflects Bhutan's wider commitment to development that is both balanced and profound, aligned with the country's singular Gross National Happiness. Prior to digital transformation, Bhutan's judiciary faced major topographical challenges. Many people had to often travel for days to reach courts, due to which many people preferred informal dispute resolution mechanisms over formal court proceedings because of the cost associated with the travel.

The COVID-19 pandemic, though an enormous difficulty, was also an opportunity to quickly push forward the digital transformation of the justice sector in Bhutan. The basic ingredients necessary for a digital transformation – political will and financial resources – were already in place. There was still the issue of lack of infrastructure and personnel challenge implementation in rural, mountainous areas; however, during this time, the judiciary's response was to be creatively resilient in order to maintain access to justice. This is evidenced by the Annual Report of the Supreme Court of Bhutan (2021), which shows that during the first year of its operation, the e-litigation system saw over 103 cases filled. Furthermore, the Office of the Attorney General of Bhutan is now using the e-litigation facilities in order to make their operations less expensive and more efficient.[54]

Another noteworthy aspect of Bhutan's development is the inclusion of conventional dispute resolution mechanisms in digital platforms. Rather than completely supplanting traditional systems, digital innovations serve to complement and enhance existing practices. Many lawyers registered with the Bar Council of Bhutan reported the use of digital platforms for dispute resolution processes such as mediation and negotiated settlement, making them much more efficient while preserving their cultural significance and community-based nature. This integration demonstrates how modern technology can enhance, rather than erode, traditional practices.

Accessing justice has been significantly facilitated by the establishment of Community Centres in remote areas. These Community Centres, equipped with the necessary digital infrastructure and manned by trained facilitators, have turned into access points for services that are a part of the legal landscape. They have also become crucial spaces for enhancing access to justice services. For most rural communities that have a community centres, the average travel time to get to legal services has been significantly reduced. A place that used to take 2–3 days to physically reach can now be traversed in under half a day at the most. Another significant achievement is reducing the costs associated with accessing justice.

That said, the progress in Bhutan also highlights areas that need more focus and attention. Digital literacy is one of those areas, especially for older citizens and people living in remote communities. A recent digital assessment of the country (2022) shows that while digital literacy has improved, many people living in rural area were found to have a low digital literacy level.[55]

Another challenge facing Bhutan's digital justice journey is sustaining the infrastructure. Although there has been considerable investment in digital infrastructure, maintaining and upgrading these systems over the long haul demands a lot of resources. Furthermore, despite the best efforts made to bridge the gap by ensuring equal access to information and communication technology, the quality of access enjoyed by users across the Bhutan remains inconsistent. Urban users tend to enjoy high-quality, stable connections, while a significant number of rural users suffer from poor-quality, unstable broadband connections and in some cases, no broadband at all.[56]

Another crucial area that requires attention is professional capacity building. The Justice Sector Strategic Plan II (2024) states that there is a pronounced need for the "continued training and development of the legal professionals in the digital competencies area." The plan emphasises the need for improvement in digital literacy to keep up with the trends of digital innovation and advances of AI technologies in justice sector to ensure efficient and effective services.

11.8 Conclusion

The insights that Bhutan provides on the path towards digital transformation in the justice sector can benefit other mountainous countries seeking to leverage technology to expedite access to justice and progress toward SDG 16.3. Bhutan's experience, notably at an accelerated pace owing to the COVID-19 pandemic, highlights how the geographic and physical access barriers to justice that many mountainous countries experience can be overcome with effective digital justice solutions that are both strategically and sensibly implemented.

The e-litigation facilities and digital court services in Bhutan represent a serious shift, in justice delivery. They serve as a way to improve the administration of justice for mountainous and rural regions – like Bhutan, where justice has historically been slow and access challenging because of geographical barriers. In many low-income countries, poor access to justice is simply the way of life: until a series of innovations came to e-litigation in Bhutan, bringing with them the prospect (and in some cases, the promise) of better access to a more timely and functioning administrative pathway to justice.

Nonetheless, Bhutan's experience also reveals some key problems that must be confronted for a sustainable digital transformation in justice delivery to take place. The ongoing digital literacy issue – especially among elderly citizens and rural inhabitants – brings into sharp relief the necessity for not just comprehensive programs but also a radical rethinking of the capacity-building approach. Moreover, the maintenance of the digital and physical infrastructures in mountainous terrain, coupled with the challenge of necessary ongoing upgrades, raises real concerns about long-term sustainability.

Bhutan's digital transformation means far more than just technological progress. It is enabling the integration of traditional dispute resolution mechanisms with digital platforms. That is to say, we can use modern technology to serve up what might be perceived as antiquated cultural practices: but in essence, about giving us a blended model of traditional and modern value systems.

The future success of digital justice initiatives in mountain areas will depend on several key factors: adequate and sustained investment in digital infrastructure, comprehensive digital literacy programs, robust cybersecurity measures, and unabated political commitment to the continued realization of technological justice. The experience of Bhutan suggests that while the upfront costs of these initiatives may be high, the long-term benefit in terms of environmental sustainability, cost-reduced access to justice, and improved operational efficiencies make these high-risk, high-reward initiatives well worth the investment.

Furthermore, the experience of Bhutan provides important lessons for other nations and areas facing geographical challenges similar:

1. **Adaptability of digital solutions:** These are tailored to local realities, not plopped down from a great height in the apparent one-size-fits-all manner of approach.
2. **Community-based access points:** You cannot reach a place if you do not make it possible for the people at that place to get to what you are offering.
3. **A balanced approach:** This favours digital means, but also respects the nation's culture's time-honoured justice practices that long predate today's viral videos of courtroom show trials.

Yet the journey towards digital justice is not without its ongoing challenges. These include the necessity of unceasing technological upgradation, effective capacity building for justice professionals, and, most crucial of all, a serious effort to bridge the urban–rural digital divide. To achieve future success, we must maintain the momentous push towards digital innovation, and at the same time, work on the principle of "no one left behind," as we make the transition to digital justice delivery systems.

The direction we take will need to emphasize cultivating digital justice systems that are sustainable, inclusive, and resilient. By "resilient," I mean that the systems must adapt to the ever-changing technological landscape. Meeting the needs of our diverse populations must be at the heart of planning, system design, and implementation.

Bhutan's experience with a digital justice initiative shows us that law and justice institutions don't have to be left behind in the digital age. With the right political will, human and financial resources, and a strategy for implementation that is both smart and steeped in the local context, *digital justice* can indeed be realized in the remote mountainous terrain of Bhutan.

Notes

1 J.W. McArthur, "The Origins of the Millennium Development Goals." 34:2 *The SAIS Review of International Affairs* (2014), https://www.jstor.org/stable/27000954?seq=1; Sanjiv Kumar, Neeta Kumar, et al., "Millennium Development Goals (MDGS) to Sustainable Development Goals (SDGS): Addressing Unfinished Agenda and Strengthening Sustainable Development and Partnership." 41 *Indian Journal of Community Medicine* (2016).
2 Millennium Summit, 6–8 September 2000, New York, *available at:* https://www.un.org/en/conferences/environment/newyork2000 (Last visited on January 2, 2025).
3 Olga Mironenko, P.L. Lucas, et al., "Sustainable Development Goals: Why Do We Need Them?" 14:2 *Social Evolution & History* (2015) 177.
4 United Nations, *The Millennium Development Goals Report 2012* (June, 2012).
5 United Nations Development Programme, World Bank Group, *Transitioning From the MDGs to the SDGs* (November, 2016) p. 10.
6 From MDGs to SDGs, *available at:* https://www.sdgfund.org/mdgs-sdgs (Last visited on January 02, 2025).
7 Open Working Group Proposal for Sustainable Development Goals, *available at:* https://sustainabledevelopment.un.org/focussdgs.html (Last visited on January 02, 2025).
8 Summary Report, 14–19 July 2014: 13th Session of the UN General Assembly's (UNGA) OWG on SDGs, *available at:* https://enb.iisd.org/events/13th-session (Last visited on January 02, 2025).
9 ICLEI Briefing Sheet, "From MDGs to SDGs: What Are the Sustainable Development Goals?" *Urban Issues*, No. 01 (November, 2015).
10 Sustainable Development Goal Agenda, *available at:* https://www.un.org/sustainabledevelopment/development-agenda (Last visited on January 02, 2025).
11 UN General Assembly, *Transforming Our World: The 2030 Agenda for Sustainable Development*, A/RES/70/1 (October 21, 2015).
12 Institutional Frameworks and International Cooperation for Sustainable Development, *available at:* https://sdgs.un.org/topics/institutional-frameworks (Last visited on January 02, 2025).
13 Goal 16, *available at:* https://sdgs.un.org/goals/goal16#targets_and_indicators (Last visited on January 02, 2025).

14 Kayla Cahoon, "Access to Justice in Rural Communities and the Impact It Has on Civil Procedure" (December, 2023), *available at:* https://mbaccesstojustice.ca (Last visited on January 02, 2025).
15 The World Bank in Bhutan, *available at*: https://www.worldbank.org/en/country/bhutan/overview (Last visited on January 02, 2025).
16 Keir Chapman, "Terrain and Topography of Bhutan: Mountains, Valleys, and Plains" (May 2024), *available at*: https://www.earth-site.co.uk/Education/terrain-and-topography-of-bhutan-mountains-valleys-and-plains/ (Last visited on January 03, 2025).
17 Bhutan, *available at*: https://www.britannica.com/place/Bhutan (Last visited on January 03, 2025).
18 Kinley Wangchuk, "Residents Pin Hope on Parties' Pledge to Prioritise Blacktopping of GC Road in Jomotshangkha_ Martshala Constituency." *BBS*, January 4, 2024.
19 Sonam Tshering and Sonam Dargay, "Never Ending East-West Highway Widening Works", *BBS*, September 21, 2021.
20 Sherab Lhamo, "Bhutan Road Watch App: Underutilised Despite Roadblock Challenges", *Kuensel*, September 30, 2024.
21 National Statistic Bureau, *available at:* https://www.nsb.gov.bt/ (Last visited on January 03, 2025).
22 National Statistic Bureau.
23 Digital Transformation: Bhutan's Key to SDG Acceleration, *available at:* https://un-dco.org/stories/digital-transformation-bhutans-key-sdg-acceleration (Last visited on January 03, 2025).
24 Ershadul Karim and Chhime Tshoke Dorjee, "Research Guide to the Legal System of the Kingdom of Bhutan", *available at:* https://www.nyulawglobal.org/globalex/bhutan.html (Last visited on January 03, 2025).
25 Judiciary of Bhutan, "Annual Report 2023", *available at*: *https://www.judiciary.gov.bt/archive-content/annual_report* (Last visited on May 13, 2026).
26 Justice Sector Working Committee, "Justice Sector Strategic Plan 2018–2023" p. 14 (2018). *available at:* https://oag.gov.bt/language/en/resources/publication/ (Last visited on May 13, 2026).
27 Nim Dorji, "E-litigation to Keep Justice Alive", *Kuensel*, May 16, 2020.
28 Bhutan National Legal Institute, "Annual Activity Report 2021" (2021). *available at:* https://bnli.bt/other-reports-publications/ (Last visited on May 13, 2026).
29 Bhutan National Legal Institute, "Annual Activity Report 2021".
30 Rinzin Wangchuk, "Legal Aid Center Established for Indigent Persons", *Kuensel Online*, October 21, 2022.
31 Namgay Wangchuk, "Bar Council of Bhutan Encourages Pro Bono Legal Aid for Vulnerable", *BBS*, April 10, 2024.
32 Royal Court of Justice, "Annual Report 2019" (2019). *available at:* https://www.judiciary.gov.bt/archive-content/annual_report (Last visited on May 13, 2026).
33 Namgay Wangchuk, "Study Finds Almost 37 Per Cent Rise in Domestic Violence Cases", *BBS*, November 25, 2021.
34 Royal Court of Justice, "Annual Report 2021" (2021). *available at:* https://www.judiciary.gov.bt/archive-content/annual_report (Last visited on May 13, 2026).
35 UNDP, "eLitigation: Taking Justice to the Doorsteps of Litigants", April 21, 2021, *available at*: https://www.undp.org/bhutan/stories/elitigation-taking-justice-doorsteps-litigants (Last visited on January 04, 2025).
36 Royal Court of Justice, above note 34.
37 Royal Court of Justice, above note 28.
38 Royal Court of Justice, above note 35.
39 Royal Court of Justice, above note 35.
40 Royal Court of Justice, above note 34.

41 Royal Court of Justice, above note 34.
42 Royal Court of Justice, above note 34.
43 Royal Court of Justice, "Annual Report 2023", p. 37 (2023). *available at:* https://www.judiciary.gov.bt/archive-content/annual_report (Last visited on May 13, 2026).
44 Karma Drupchu, "Internet in Bhutan Is More Expensive Than Rent in the UK", *Kuensel*, January 20, 2024.
45 Editorial, "What's Wrong with Our Internet Quality?", *Kuensel Online*, October 24, 2024 *available at*: https://kuenselonline.com/whats-wrong-with-our-internet-quality.
46 UNDP, "e-Justice Digital Transformation to Close the Justice Gap" (2022). *available at:* https://www.undp.org/publications/e-justice-digital-transformation-close-justice-gap (Last visited on May 13, 2026)
47 Helen Taylor, Dirk Van Rooy, et al., "Digital Justice: A Rapid Evidence Assessment of the Use of Mobile Technology for Offender Behavioural Change", 70:1 *Probation Journal* (2023) 34.
48 Rachid Ejjami, "AI-Driven Justice: Evaluating the Impact of Artificial Intelligence on Legal Systems", 6:3 *International Journal for Multidisciplinary Research* (2024).
49 UNDP, above note 46.
50 Justice for All, "Six Recommendations for Advancing Justice for All", *available at*: https://grassrootsjusticenetwork.org/wp-content/uploads/2023/02/JFA-Six-Recommendations.pdf.
51 Taylor, Van Rooy, et al., above note 47.
52 Ben Bertelsen, Lara Deramaix and D.C. Decker, "Improving Access to Justice with Inclusive Digital Transformation", (June 21, 2024), *available at:* https://www.undp.org/digital/blog/improving-access-justice-inclusive-digital-transformation (Last visited on January 05, 2025).
53 UNEP, "Digitalization for Sustainability", *available at:* https://www.unep.org/topics/digital-transformations/digitalization-sustainabilit (Last visited on January 05, 2025).
54 Office of the Attorney General, "Annual Report 2022", (2022). *available at:* https://oag.gov.bt/language/en/resources/publication/ (Last visited on May 13, 2026).
55 Yangchen C. Rinzin, "Research Finds Digital Literacy Vital to Achieve Digital Drukyul Vision", *Kuensel Online*, January 22, 2022, *available at*: https://kuenselonline.com/research-finds-digital-literacy-vital-to-achieve-digital-drukyul-vision/ (Last visited on January 09, 2025).
56 K.P. Sharma, "Bhutanese Schools Face Connectivity Challenge", *Kuensel Online*, January 25, 2024, *available at*: https://kuenselonline.com/bhutanese-schools-face-connectivity-challenge/ (Last visited on January 09, 2025).

12 Digital Democracy and Human Rights

Exploring the Impact of Emerging Technologies on Governance, Equality, and Access to Information

Pallavi Gupta

> Everyone has the right to freedom of opinion and expression; this right includes freedom to hold opinions without interference and to seek, receive and impart information and ideas through any media and regardless of frontiers.[1]

12.1 Introduction

Digital democracy refers to the integration of digital tools and technologies into democratic processes to enhance citizen engagement, representation, and decision-making, ensuring that democratic principles like equality, freedom of expression, and access to information are upheld. It includes the use of the internet, social media, e-governance, and emerging technologies such as artificial intelligence (AI), blockchain, and big data to improve transparency, inclusivity, and political participation. However, these technologies also present challenges, including privacy violations, digital exclusion, and public opinion manipulation, raising concerns about their compatibility with basic human rights.[2]

Key aspects of digital democracy include governance, equality, and information access. In the digital age, governance involves using technology to increase public institutions' efficiency, accountability, and responsiveness. Governments can leverage e-governance tools and data analytics for informed decision-making and better services. Equality is crucial, as the digital divide's unequal access to technology can prevent certain groups from fully participating in the democratic process. Ensuring equal access to digital tools is essential for inclusive democracy. Additionally, access to information, vital for democracy, is made easier by digital technologies, but concerns about misinformation, surveillance, and privacy threats remain.[3]

While technologies such as blockchain, AI, big data, and e-governance platforms hold immense potential to increase transparency, accountability, and citizen engagement, they also present critical challenges that could undermine fundamental democratic principles and human rights. The central problem lies in how these emerging technologies interact with democratic values such as equality, freedom of expression, and access to information,

DOI: 10.4324/9781003733751-18

while raising concerns about privacy violations, digital exclusion, misinformation, and manipulation of public opinion.

In India, the integration of digital technologies in governance, such as the Aadhaar system, has been a powerful tool for improving access to services. However, it has also raised significant concerns over privacy violations, surveillance, and the potential for exclusion, particularly among marginalized communities with limited digital access. The Supreme Court of India has addressed these issues, noting that while Aadhaar is beneficial for welfare programs, it must be implemented in a way that safeguards citizens' privacy and ensures informed consent. The introduction of facial recognition technology in public spaces and electoral processes in India has also sparked debates about the balance between security and privacy, with fears of increased surveillance and potential misuse of data.

On a global scale, the manipulation of information through digital platforms has had profound implications for democratic processes. For example, the role of social media in influencing elections, such as the 2016 U.S. Presidential Election, exposed how digital platforms can be used to spread misinformation, polarize societies, and manipulate public opinion. Similarly, the Cambridge Analytica scandal highlighted the risks of data misuse and the lack of regulation in digital spaces. In Myanmar, the use of Facebook to spread hate speech and incite violence against the Rohingya population demonstrated the dangerous consequences of unchecked digital platforms in exacerbating social injustices and human rights violations.

This research aims to critically examine how the integration of digital technologies into governance systems impacts democratic values and human rights. Specifically, it seeks to explore the potential benefits of these technologies in promoting more accountable, inclusive, and participatory governance models, while also investigating the risks they pose to privacy, equality, and freedom of expression. Furthermore, it will assess how different countries, including India, are navigating the legal and ethical challenges of these technologies, with a focus on finding solutions that safeguard citizens' rights and prevent the exacerbation of existing societal inequalities.

Through an extensive review of existing literature, a notable research gap has been identified in the integration of emerging technologies into governance models, specifically in terms of their impact on human rights within democratic societies. While significant attention has been given to the technical capabilities of blockchain, AI, and big data in enhancing transparency, efficiency, and citizen engagement, there is a lack of comprehensive studies focusing on the legal implications and ethical challenges these technologies pose to democratic values and human rights. Additionally, much of the existing literature tends to focus either on the technological advancements themselves or on human rights in isolation, without adequately addressing the intersection between the two. Furthermore, the global perspective remains fragmented, with a lack of comparative analyses between different legal systems in their approach to managing the balance between digital innovation and human rights. While various countries have begun to implement regulations such as the GDPR in the EU and data protection laws in India, there is a lack of cross-jurisdictional analysis that examines how these legal models can be harmonized to protect human rights in the context of global technological developments.

12.2 The Concept of Digital Democracy

Digital democracy refers to the integration of digital technologies into democratic processes to enhance citizen participation, improve governance, and promote transparency. It leverages tools like e-governance, social media, and emerging technologies such as AI, blockchain, and big data to ensure inclusivity and informed decision-making. By enabling greater access to information and fostering political engagement, digital democracy strengthens the principles of equality, freedom of expression, and human rights. However, challenges such as the digital divide, privacy concerns, and misinformation pose risks to its effectiveness. The concept continues to evolve, balancing technological advancements with safeguarding democratic values.

Digital democracy has evolved alongside the rise of the internet, transforming how citizens engage with one another and their governments. In the 1990s, the internet became a key tool for communication and information exchange, offering new ways for citizens to access information, participate in debates, and influence policy decisions. This marked the birth of digital democracy, where technology was viewed as a means to enhance political processes by increasing transparency, improving public debate, and expanding access to decision-making.[4] By the early 2000s, e-democracy emerged, which included online voting, digital petitions, and the use of digital platforms for political organizing. E-government initiatives made public services more accessible and efficient, exemplified by Estonia's pioneering i-voting system in 2005, allowing citizens to vote securely online. Platforms like Change.org enabled citizens to petition for political change. Social media platforms like Facebook and Twitter revolutionized political campaigning, fostering political movements and public discourse. Advances in big data, AI, and machine learning allowed governments to make more personalized and efficient decisions, while blockchain technology promised greater election transparency.[5]

Despite its potential, digital democracy faces critical challenges. Concerns about data privacy,[6] the digital divide,[7] and the manipulation of public opinion threaten the integrity of democratic processes. The increasing reliance on digital technologies raises issues around security, cybersecurity, and digital literacy,[8] especially in marginalized communities. The future of digital democracy depends on addressing these concerns through better regulation, cybersecurity advancements, and ensuring equitable access to technology. The digital divide is still a major problem in India, especially for low-income and rural populations. With internet penetration rates in urban areas over 80%

and only around 40–45% in rural regions, initiatives like Digital India[9] and PM GatiShakti[10] aim to expand connectivity and improve access to government services. Technological illiteracy also poses barriers to participation, as many citizens in rural areas struggle with digital tools. Moreover, misinformation, especially on platforms like WhatsApp, influences elections and public health decisions, making fact-checking and public awareness crucial for informed democratic participation.[11]

Digital surveillance has raised concerns about privacy, freedom of expression, and civil liberties in India. The 2021 Pegasus spyware scandal[12] revealed the use of surveillance tools to target journalists, activists, and political opponents, raising alarms over privacy violations. The government's increasing control over digital spaces, including internet shutdowns during political unrest and new rules for social media platforms, has sparked debates about censorship and the protection of free speech. Digital rights organizations advocate for stronger data protection laws and accountability in the use of surveillance tools to safeguard civil liberties.[13]

Technology has the potential to strengthen democratic values by enabling broader participation, especially for marginalized communities. Social media, online petitions, and digital platforms allow citizens to engage in debates, raise awareness on issues, and mobilize for causes. The internet breaks down traditional barriers, enabling those with limited access to physical political spaces, like rural or disabled populations, to participate in democratic processes. Governments can also use digital tools like surveys and online consultations to gather public feedback and make more inclusive decisions. Additionally, social media platforms have provided marginalized groups with new avenues for activism, allowing them to reach global audiences and demand political and social change.[14]

Digital tools offer practical solutions to address the challenges faced by people with disabilities, those in rural areas, and underserved populations. Assistive technologies like screen readers and voice commands enable people with disabilities to engage with digital platforms. In rural areas, mobile technology and public Wi-Fi initiatives provide citizens with access to online services, voting, and public forums. Expanding broadband infrastructure is essential to bridging the digital divide and ensuring that everyone, regardless of location or socioeconomic status, has access to digital democracy. Furthermore, mobile technology enables rural citizens to stay informed about government services and engage in civic activities, fostering more inclusive democratic participation.[15]

12.3 Emerging Technologies and Their Role in Governance[16]

Emerging technologies are increasingly shaping the way governments operate, with artificial intelligence (AI), blockchain, big data, and e-governance at the forefront of this transformation. These technologies offer innovative solutions to improve efficiency, enhance transparency, and foster better decision-making in governance. However, they also raise important concerns related to ethics, privacy, security, and accountability, making it crucial to explore both their potential benefits and the challenges they pose.

Artificial intelligence and machine learning in governance[17]: AI and machine learning are transforming public administration by enabling data-driven decision-making, improving efficiency, transparency, and service delivery. In India, AI is applied in sectors like healthcare, urban governance, and the judiciary to enhance services such as personalized health treatments, traffic management, and case management. Globally, countries like the United Kingdom, United States, and Estonia are leveraging AI in healthcare, criminal justice, and digital governance to optimize resources, predict outcomes, and improve public services. However, the use of AI in governance raises significant ethical concerns, including biases in decision-making, lack of transparency, and accountability. Systems like the Compas algorithm in the United States have been criticized for racial bias, while "black box" AI models obscure decision processes, eroding public trust. To ensure fairness, governments must establish clear frameworks for accountability, enabling citizens to challenge errors and ensure AI-driven governance serves all equitably.

Blockchain and transparency in governance[18]: Blockchain technology, with its decentralized and transparent features, can significantly improve governance by ensuring accountability and reducing corruption in government processes. It can be applied to land records, voting systems, public procurement, and financial transactions, offering tamper-proof records and verifiable transactions. In India, blockchain initiatives such as Telangana's land record system, Andhra Pradesh's smart contracts for agricultural procurement, and Goa's secure elections aim to enhance transparency and reduce fraud. Globally, countries like Estonia, Singapore, Ghana, and Kenya are using blockchain for secure e-residency, procurement, and aid tracking, helping combat corruption. Blockchain's key advantages, such as immutable records, smart contracts, and decentralization, provide a powerful tool to ensure transparency, prevent manipulation, and reduce human discretion in governance.

Big data and public policy[19]: By offering real-time, data-driven insights that increase efficiency and decision-making, big data and analytics are revolutionising public policy-making and service delivery. Governments can use big data to predict trends, optimize resource allocation, and monitor public services like healthcare, education, and urban planning. For example, predictive policing, climate monitoring, and social welfare programs benefit from big data, enabling more effective and targeted interventions. In India, initiatives like the Direct Benefit Transfer (DBT)[20] and the Smart Cities Mission[21] leverage big data to streamline services and improve governance, while healthcare and agriculture programs use data to enhance service delivery and disaster management.

However, the use of big data also raises significant concerns around data privacy, security, and surveillance. Privacy risks arise from the collection of sensitive personal data, which, if mishandled, could lead to breaches or misuse, as seen in the Cambridge Analytica scandal.[22] Data security is another challenge, as large volumes of government data must be safeguarded against cyber attacks, highlighted by incidents like the 2017 WannaCry ransomware attack.[23] Additionally, the expanding use of big data for surveillance, while enhancing public safety, raises concerns about civil liberties and the potential for authoritarian control, as evidenced by China's mass surveillance systems.[24] Ethical issues, including consent and algorithmic bias, further complicate the use of big data, prompting the need for legal frameworks like the GDPR to protect privacy while enabling public service benefits. Governments must balance the opportunities of big data with robust safeguards to protect citizens' rights and ensure transparency in its use.

E-governance: Digital tools are transforming the administration of public services through e-governance, which enhances efficiency, accessibility, and transparency in government processes. Key digital tools, such as online portals, digital identification systems, and digital certificates, streamline government services, making them more accessible, cost effective, and inclusive. For example, online platforms like the U.S. e-Government portal and India's e-District portal centralize access to various services like tax filing, license renewals, and public records, reducing the need for in-person visits. India's Aadhaar system, based on biometric identification, personalizes and secures the delivery of services such as subsidies and welfare benefits. Digital certificates and signatures enable secure, paperless transactions, enhancing the efficiency of administrative processes. These innovations help governments reduce costs, curb corruption, and increase citizen engagement by offering direct access to services and decision-making transparency.[25]

The success of e-governance platforms can be seen in increased accessibility, efficiency, and improved citizen engagement. Digital platforms make government services accessible from any location, significantly bridging the urban–rural divide, especially in large countries like India. E-governance reduces paperwork, eliminates manual record-keeping, and automates many processes, leading to time and cost savings for both citizens and governments. It also minimizes opportunities for corruption through enhanced transparency. In India, the Digital India initiative exemplifies the success of e-governance, providing services such as DigiLocker, eSign, and e-District, which improve accessibility and efficiency in government service delivery.

However, the implementation of e-governance faces several challenges, notably the digital divide, cybersecurity concerns, and privacy issues. While internet access has expanded, rural and remote regions still face connectivity challenges, which can hinder the reach and effectiveness of digital platforms.

Moreover, cybersecurity remains a critical issue as governments store vast amounts of sensitive data online, which is vulnerable to breaches and cyber-attacks. Privacy concerns also arise, as large-scale data collection through systems like Aadhaar has sparked debates about potential misuse or unauthorized access to personal data. Additionally, resistance to adopting digital platforms due to digital illiteracy, mistrust in technology, or reluctance to change from traditional methods can slow the rollout of e-governance services. Furthermore, maintaining and updating digital infrastructure is costly and resource intensive, requiring continuous investments.

Case Studies of E-Governance Initiatives

1 **India's Digital India initiative**[26]**:** The goal of the 2015-launched Digital India program is to make India a digitally enabled nation by emphasising digital services, digital literacy, and high-speed internet access. Key initiatives include the following:

   - Aadhaar, a biometric identification system that facilitates the efficient delivery of welfare benefits, banking services, and subsidies.
   - DigiLocker, a cloud-based platform for securely storing and sharing documents like academic certificates and driving licenses.
   - e-District, e-Office, and the National e-Governance Service Delivery Gateway (NeGD), which provide citizens with access to a range of government services.

   Successes: Digital India has enhanced service access, promoted financial inclusion, and enabled efficient delivery of services. Programs like Jan Dhan Yojana have connected millions of people to banking services.
   Challenges: Despite its successes, challenges persist, such as inadequate internet connectivity in rural areas, low levels of digital literacy, and concerns about data security and privacy.

2 **Estonia's e-Residency program**[27]**:** Estonia is one of the most digitally advanced nations globally, and its e-Residency program, launched in 2014, allows global citizens to access Estonian public services remotely, including establishing businesses and signing contracts. X-Road, Estonia's data exchange platform, ensures seamless digital communication between government institutions, enabling efficient public service delivery.

   Successes: The e-Residency program has attracted over 80,000 e-residents from 170 countries, facilitating global business operations and providing access to secure government services without physical presence.

   Challenges: While the program has been successful, concerns over potential misuse by individuals seeking to exploit the system for illicit activities, such as money laundering, persist. Additionally, the protection of digital infrastructure from cyber threats remains an ongoing concern.

3 **Global E-governance initiatives:** Following are some initiatives from Singapore and South Korea:

- **Singapore**[28]: The Smart Nation initiative in Singapore digitizes government services and integrates Internet of Things (IoT) technologies to optimize urban management. The country uses data from citizens and urban infrastructure to enhance public services such as transportation, healthcare, and public safety.
- **South Korea**[29]: South Korea's e-Government system allows citizens to access a variety of public services, such as paying taxes, renewing licenses, and obtaining certificates. The integration of data analytics helps predict and address citizen needs proactively, enhancing the efficiency of public services.

12.4 Equality and Access to Information in the Digital Age

The digital age has brought about transformative changes in how we access information, communicate, and interact with the world. While these technological advancements have created immense opportunities, they have also exposed and, in some cases, exacerbated inequalities in access to information. In the context of governance, education, healthcare, and employment, access to information is essential for individuals and communities to fully participate in society and make informed decisions but this equality in access to information couldn't be easily achieved as there exist a digital divide.

Access to information is a fundamental human right, essential for exercising freedoms like expression, participation, and access to knowledge. The United Nations has recognized internet access as crucial for these rights, emphasizing that restrictions like censorship or shutdowns undermine democracy and hinder citizens' ability to engage in civic life. In this context, digital platforms play a pivotal role in informing citizens about their rights, government policies, and opportunities.[30] Initiatives like India's *Digital India* provide platforms for accessing public services and information, enhancing transparency and inclusivity. Social media and online platforms also promote human rights education, with organizations using them to raise awareness on issues like gender equality, LGBTQ+ rights, and anti-corruption efforts.

Access to information drives social change by fostering transparency and accountability. For example, the UK's Open Data Initiative[31] makes government data accessible, enabling citizens to scrutinize public spending and demand better governance. In India, the Right to Information (RTI) Act empowers citizens to challenge corruption. Digital platforms have also fuelled global movements like #MeToo and Black Lives Matter, enabling the spread of critical information, mobilizing activism, and holding institutions accountable. Ultimately, access to information fosters informed citizenship and empowers individuals to advocate for justice and social change.

Digital activism, propelled by social media, has democratized political engagement by making activism more accessible[32] bringing social changes.

Traditionally, activism was limited to those with physical access to protest spaces or organizational power. Social media eliminates such barriers, allowing anyone with an internet connection to participate in discussions, share personal stories, and organize collective actions. It particularly benefits marginalized groups, enabling them to voice concerns and push for social change without relying on traditional media or government structures. Social media also allows for rapid mobilization, making it easier to coordinate protests and petitions in real time, which is crucial during times of crisis. Moreover, platforms serve as powerful tools for accountability, as videos and testimonies shared online can expose injustice, often triggering investigations and legal action. This ability to share information globally fosters international solidarity, as seen with #BlackLivesMatter, which resonated beyond the United States to inspire protests worldwide.

However, social media also presents challenges. Hate speech and online harassment are rampant, particularly targeting marginalized communities such as women, racial minorities, and LGBTQ+ individuals.[33] The anonymity of social media allows harmful content to spread quickly, sometimes leading to real-world violence. Additionally, misinformation is a significant issue, undermining public health campaigns and interfering with democratic processes. For example, false claims about COVID-19 fuelled confusion and resistance to health measures. The role of platforms in regulating content has also sparked debate, as they balance free expression with the need to curb harmful content. Moreover, algorithmic bias can amplify divisive or polarizing material, marginalizing certain voices and deepening societal divisions. The lack of transparency in these algorithms further complicates issues of accountability. Thus, while social media has democratized activism, it requires careful regulation to prevent abuse and ensure it serves equality. Thus, it can be observed that though the digital age has brought a social change, it has posed severe challenges for equality and access to information, which are as follows:

- **Digital divide:** The disparity between people, groups, and geographical areas with differing degrees of access to contemporary information and communication technologies is known as the "digital divide."[34] It includes a variety of elements, including geographical constraints, socioeconomic inequalities, and variations in technical literacy. While those in cities have greater connectedness, those in rural or isolated regions usually do not have access to mobile networks or broadband. The gap is widened because people from lower-income families frequently find it difficult to pay for the tools and services required to connect. Additionally, digital illiteracy and lack of access to digital education further hinder the ability of many individuals to use technology for work, education, or healthcare. Age and disability also play significant roles, with older adults and people with disabilities facing challenges in accessing and using technology unless it is specifically designed to be inclusive. Together, these factors contribute to digital exclusion, preventing millions from fully participating in the digital economy and society.

To address this divide, both governments and private companies must collaborate to enhance accessibility and affordability. Governments can spearhead infrastructure projects, such as expanding broadband networks into rural and underserved regions. Policies that encourage private sector investments in infrastructure and digital literacy programs, alongside subsidies for low-income communities, are critical in making technology accessible to all. Public–private partnerships can also help expand internet access, while regulatory measures can prevent monopolistic practices that drive up prices. On the other hand, private companies can contribute by offering affordable devices and internet solutions, such as low-cost smart phones and affordable data plans, while tech giants engage in Corporate Social Responsibility efforts, including satellite-based internet access and free Wi-Fi hotspots. Several initiatives worldwide aim to mitigate the digital divide, focusing on increasing internet access, providing affordable devices, and enhancing digital literacy:

1 **India's BharatNet initiative**[35]**:** As part of the Digital India program, the BharatNet initiative is a government-led effort to bring high-speed broadband connectivity to rural India. The project aims to connect over 250,000 gram panchayats (local administrative units), providing better access to government services, healthcare, and education. Despite the progress, challenges remain, such as reaching extremely remote regions and ensuring that rural populations have the digital literacy to effectively use these resources.
2 **The Jio network in India**[36]**:** The Jio network revolutionized mobile internet access in India by offering affordable smartphones and low-cost data plans. This initiative has significantly boosted internet usage, particularly among low-income groups, who previously couldn't afford internet access. While this mobile data revolution has been transformative, issues like slower speeds in certain areas and limited connectivity in remote regions still need to be addressed.
3 **South Korea's internet infrastructure**[37]**:** South Korea has emerged as a global leader in internet connectivity, boasting some of the world's fastest and most affordable broadband networks. The government has played a key role by providing subsidies to keep internet prices low, as well as implementing policies that ensure broadband availability in even the most remote areas. Public Wi-Fi hotspots in parks, universities, and public transport hubs further enhance accessibility, ensuring that internet access is not restricted by socio-economic status. South Korea's Universal Service Policy mandates that telecom companies offer affordable broadband services, promoting digital inclusion and fostering innovation and growth in the digital economy.
4 **Facebook's Free Basics (Internet.org)**[38]**:** Facebook's Free Basics initiative (formerly Internet.org) offers free access to a limited set of internet

services, such as social media, weather, and educational content, without data charges. Though briefly piloted and later restricted in India in the year 2016 due to regulatory concerns, the initiative targets underserved populations in countries like Indonesia and Kenya, helping millions of people connect to the internet for the first time. While it has faced criticism for providing only restricted access, it has played a pivotal role in lowering barriers to entry for underserved communities and introducing them to the digital world. Critics argue that such limited access may create a "walled garden" of content, but it has undeniably helped millions in developing countries get online.

- **Discriminatory digital practices**[39]**:** Technology has the potential to improve access to services and economic opportunities, but when not designed with equity in mind, it can worsen existing inequalities and harm vulnerable communities. Key issues include algorithmic bias, where algorithms in areas like hiring, criminal justice, healthcare, and finance perpetuate racial, gender, and socio-economic prejudices based on biased historical data. This can lead to discriminatory practices, such as biased hiring decisions or disproportionate targeting of minority communities in predictive policing.
- Digital exclusion is another significant issue, where marginalized groups lack access to the internet, devices, or digital literacy, preventing them from accessing essential services like education and healthcare. The COVID-19 pandemic highlighted this divide, as students in rural and low-income areas struggled with online learning. Additionally, surveillance technologies, such as facial recognition, disproportionately affect vulnerable groups, raising privacy concerns.
- Surveillance capitalism is a growing concern, where personal data is commoditized and sold by tech companies, often without full consent. This exploitation disproportionately impacts marginalized groups, who may be targeted with harmful advertisements, manipulative products, or discriminatory practices like biased credit scoring. While efforts like the GDPR aim to protect data, global disparities in data protection leave vulnerable communities at greater risk of exploitation and harm.

12.5 The Intersection of Digital Democracy and Human Rights[40]

Human rights and digital democracy constitute a challenging but vital field of study in the contemporary world. Digital technologies have fundamentally reshaped how we participate in democracy, express our opinions, access information, and engage with government and civil society. However, this digital transformation also brings challenges in ensuring that these technologies are used to uphold, rather than undermine, human rights.

The rapid evolution of digital technologies has reshaped the human rights landscape, introducing both opportunities and challenges. A robust digital human rights framework is essential to address key concerns such as freedom of expression, privacy, and access to information. Digital platforms have enhanced freedom of expression, enabling movements like #MeToo and Black Lives Matter, but challenges like censorship, online harassment, and the misuse of platforms for hate speech persist. Authoritarian regimes use digital surveillance to suppress dissent and monitor citizens, threatening free speech. Privacy is increasingly compromised by the mass collection of personal data, with inconsistent protections globally, particularly in authoritarian states. The digital divide and misinformation further limit access to vital information, undermining democracy and trust in institutions.[41]

International organizations play a key role in advocating for digital rights. The United Nations Human Rights Council (UNHRC) has recognized internet access as essential for human rights, while organizations like the Electronic Frontier Foundation (EFF), Reporters Without Borders (RSF), Access Now, and the Internet Society work to defend privacy, combat censorship, and promote digital freedoms. These groups influence global policies, provide legal aid, and ensure technology enhances, rather than undermines, fundamental human rights.

In the digital age, authoritarian governments increasingly use technology to suppress dissent, control information, and violate human rights. Digital authoritarianism involves the use of surveillance technologies, internet censorship, and social media manipulation to monitor citizens and stifle opposition. Examples include China's social credit system, which penalizes dissent, and the use of facial recognition and internet shutdowns in countries like China, Russia, and India to restrict freedom of expression and protest. These practices limit democratic participation, create a culture of self-censorship, and prevent access to critical information.[42]

Global organizations play a vital role in defending digital human rights against these abuses. The UN Human Rights Council has affirmed the right to freedom of expression online, while the UN Special Rapporteur addresses digital censorship. EFF advocates for privacy, free expression, and accountability in the digital space, offering legal support for victims of online abuses. RSF defends press freedom in the digital realm, and Freedom House tracks internet freedom and digital authoritarianism through its Freedom on the Net report, pushing for policies that promote an open and free internet.

Surveillance technologies, while enhancing security and efficiency, have raised significant concerns about privacy, civil liberties, and human rights globally.[43] These systems, including facial recognition, location tracking, and mass data collection, are used by governments and corporations to monitor individuals' movements and behaviours. Facial recognition has been criticized for its racial bias, often misidentifying people of colour and women. Location tracking, enabled by GPS and mobile phone data, raises concerns about privacy, especially when used by authoritarian governments to track protests or dissent. The mass collection of personal data through social media and other online activities further erodes privacy, with predictive analytics enabling governments or corporations to influence behaviour and target individuals for surveillance.

A prominent example of surveillance is China's system, which employs facial recognition, AI, and big data to monitor and control citizens, particularly through its controversial social credit system, which punishes dissent and rewards compliance. Similarly, India's Aadhaar system, the world's largest biometric identification program, collects personal data like fingerprints and iris scans from over 1.3 billion citizens. Though aimed at reducing fraud and streamlining government services, concerns about privacy violations and mass surveillance have emerged. Security breaches and legal challenges highlight the risks associated with such data centralization, leading to significant debates over its use.

12.5.1 *Digital Rights in India*[44]

The concept of digital rights in India has evolved as technology has become central to governance and daily life. A significant milestone was the 2017 Supreme Court ruling[45] that recognized privacy as a fundamental right under Article 21 of the Indian Constitution, affirming individuals' control over their personal data. This ruling led to discussions about balancing technological advancements in governance with the protection of personal freedoms, and it paved the way for data protection laws like the Personal Data Protection Bill. These developments underscored the need for privacy in the digital age, especially amid increasing government surveillance and the expansion of digital services.

A contentious issue has been Aadhaar, India's biometric identification system. Launched to streamline welfare delivery, Aadhaar collects extensive biometric data, raising concerns about privacy, data security, and potential misuse. In 2018, the Supreme Court ruled that Aadhaar could not be made mandatory for private services, but allowed its use for government welfare schemes. Critics argue that Aadhaar could enable mass surveillance if linked to other databases, potentially infringing on privacy rights and enabling unwarranted state control over citizens.

To enhance transparency and accountability, India has adopted e-governance initiatives, such as the Digital India program, which aims to make government services accessible online and improve public service delivery. The RTI, passed in 2005, has empowered citizens to demand government records and fight corruption. However, challenges like the digital divide, data security concerns, and bureaucratic resistance remain, complicating the effective implementation of these digital rights and transparency efforts.

12.5.2 *Challenges in Upholding Human Rights in the Digital Era*[46]

The digital era has brought significant advancements but also raised complex human rights challenges. One major issue is the spread of misinformation and disinformation, which has become rampant on social media platforms. These false or misleading narratives can sway elections, fuel political polarization, and undermine social harmony, as seen in events like the 2016

U.S. election and in India. Efforts to combat misinformation include digital literacy campaigns and fact-checking, but challenges persist, especially with algorithms that amplify sensational content.

Another pressing concern is the digital divide, which exacerbates inequality by limiting access to essential services like education, healthcare, and employment for marginalized communities. Rural areas and women in developing nations are particularly affected, hindering their ability to participate in digital democracy and access information. Bridging this divide requires policies promoting affordable internet, digital literacy, and infrastructure investments, such as India's BharatNet project to provide high-speed internet to rural regions.

Finally, the role of big tech companies in shaping political, social, and economic landscapes raises significant human rights concerns. Companies like Facebook and Google have been criticized for their role in spreading hate speech, manipulating political campaigns, and violating privacy through extensive data collection. There is growing pressure for stronger regulations, such as the GDPR, to protect users' rights and hold tech companies accountable.

12.6 Global Trends in Digital Democracy and Human Rights[47]

The advent of the digital age has brought both unprecedented opportunities and significant challenges to democracy and human rights. The influence of digital transformation on governance, civil liberties, and political systems is vast and varied across the world. While digital tools have enabled greater political engagement and expanded freedoms in some nations, others have leveraged technology for surveillance and control, giving rise to the phenomenon of digital authoritarianism. This section will explore the global shifts in digital democracy, the role of international cooperation in protecting digital rights, and the intersection of cyber security with democracy and digital sovereignty.

12.6.1 Global Shifts: The Impact of Digital Transformation on Democracy[48]

12.6.1.1 Worldwide Trends: From Digital Authoritarianism to Digital Freedoms

The global impact of digital technologies on democracy can be divided into two distinct trends: digital authoritarianism and digital freedoms. As technology becomes central to governance, it is reshaping political systems, economic frameworks, and civil rights.

- **Digital authoritarianism:** Some governments have adopted digital authoritarianism, using technology to monitor, control, and suppress their citizens. By leveraging tools like facial recognition, data surveillance, and censorship, these regimes can consolidate power, limit freedom of expression, and stifle dissent. In countries like China, digital technologies have been harnessed for state surveillance, manipulation of public opinion, and social control.

- **Digital freedoms:** On the other side of the spectrum, digital technologies have empowered citizens, activists, and civil society in democratic nations. Technology has enabled greater participation in the political process, increased access to information, and provided platforms for grassroots movements. Countries like Estonia have exemplified how governments can embrace digital tools to build transparent and efficient democracies, ensuring freedom of expression, privacy, and participation.

Examples from Different Countries

- **Estonia's digital government model[49] – How a small country built a robust digital democracy:** Estonia is a leading example of digital democracy, having integrated technology into its governance to create a secure, transparent, and inclusive system. Since the early 2000s, the country has offered online services like tax filing, healthcare, and education through its digital ID system, enabling citizens to efficiently access government services and engage in decision-making. Estonia's e-residency program allows global entrepreneurs to establish businesses remotely, fostering a digital economy and granting access to EU markets. Additionally, Estonia was one of the first nations to implement online voting, ensuring secure and transparent elections. The country's innovative model demonstrates how a small nation can build a robust digital infrastructure that enhances political engagement and trust in government.
- **China's digital surveillance state and its implications on rights[50]:** China's digital authoritarianism contrasts sharply with Estonia's digital democracy, as the Chinese government utilizes advanced technologies to enforce strict control, suppress dissent, and monitor its citizens. The controversial social credit system rates individuals based on their behaviour, financial history, and social interactions, rewarding good behaviour while penalizing those with low scores, raising concerns over privacy, personal freedoms, and excessive state surveillance. China's extensive surveillance infrastructure, which includes widespread use of facial recognition, location tracking, and AI-powered monitoring, allows the government to track nearly every aspect of citizens' lives, significantly infringing on privacy and freedom of expression. The Great Firewall of China, along with aggressive censorship and the manipulation of online content, restricts access to information, suppresses political discourse, and curtails the public's ability to freely express opinions. While this digital model has contributed to China's economic growth and governance efficiency, it has come at the cost of fundamental human rights, privacy, and the freedom to access and share information.

- **The EU's digital strategy and its regulation of big tech (GDPR)[51]:** In order to protect digital rights and regulate the digital economy, the European Union (EU) has taken the lead. Important initiatives include the General Data Protection Regulation (GDPR), which was put into effect in 2018 and is one of the most extensive data privacy laws in the world. It gives EU citizens more control over their personal data and places stringent restrictions on how businesses gather, store, and use it. In 2020, the EU also unveiled the Digital Services Act and Digital Markets Act, which were designed to control online platforms, hold those responsible for unlawful information accountable, and encourage fair competition in the digital market. These laws hold big tech companies like Google, Facebook, and Amazon accountable for content moderation and preventing anti-competitive behaviour. The EU's approach serves as a model for balancing digital innovation with the protection of individual freedoms and privacy.

12.6.2 *International Cooperation on Digital Human Rights*[52]

12.6.2.1 *The Role of International Conventions and Treaties in Protecting Digital Rights*

As the digital landscape grows increasingly complex, international cooperation is essential to safeguard digital rights globally. International conventions and treaties play a key role in protecting citizens' rights to privacy, freedom of expression, and access to information in the digital era.

- **United Nations:** The United Nations (UN) has recognized the importance of digital rights through its Human Rights Council. In 2016, the UN declared internet access to be a human right, underscoring that individuals should have the right to freedom of expression and privacy online. In 2018, the UN General Assembly passed a resolution condemning the disconnection of internet services by governments as a violation of human rights.
- **Global treaties:** Treaties like the International Covenant on Civil and Political Rights and the Universal Declaration of Human Rights include provisions related to freedom of expression and privacy, both of which are essential in the digital space. These conventions provide a framework for international advocacy and accountability, ensuring that governments adhere to global human rights standards in the digital era.

12.6.2.2 *Cooperation Between Countries in Promoting Open Internet Policies and Combating Cybercrime*

Countries around the world have increasingly cooperated to promote open internet policies, combat cybercrime, and protect digital freedoms. International partnerships and agreements such as the Budapest Convention on Cybercrime have been instrumental in addressing cross-border cyber threats.

- **Open internet policies**[53]**:** The Global Internet Forum to Counter Terrorism is one example of multilateral cooperation between countries and tech companies to create an open, safe, and secure internet environment. Governments and companies are increasingly collaborating to develop frameworks for addressing hate speech, misinformation, and illegal online content.
- **Cybercrime cooperation:** International organizations like Interpol and Europol play a critical role in combating cybercrime. As cyber threats often transcend national borders, cooperation between countries is essential for tracking, investigating, and prosecuting cybercriminals.
- **Global civil society organizations**[54]**:** Organizations such as Access Now, EFF, and RSF advocate for digital human rights, privacy protection, and freedom of expression. These groups play a pivotal role in raising awareness about abuses and pressing governments and corporations to respect individuals' rights online.

12.6.3 Cybersecurity and Digital Sovereignty[55]

12.6.3.1 The Intersection of Cybersecurity With Democracy and Human Rights

As governments, companies, and individuals increasingly rely on digital infrastructure, the issue of cybersecurity becomes central to the protection of democracy and human rights. Effective cybersecurity measures are essential for preventing cyberattacks that could disrupt democratic processes, infringe on citizens' privacy, and undermine national security.

- **Protecting democratic processes:** In democratic societies, cybersecurity is crucial to protect elections, voting systems, and government infrastructure from cyberattacks. Election interference through cyberattacks, as seen in the 2016 U.S. Presidential Election,[56] underscores the need for robust cybersecurity measures to safeguard the integrity of elections and protect the public's confidence in democratic processes.
- **Cyberattacks on civil liberties:** Cyberattacks can also target civil liberties by disrupting access to information, spreading misinformation, or stealing personal data. For instance, data breaches and hacking can compromise citizens' privacy and expose sensitive information, leading to widespread repercussions for individuals' rights.

12.6.3.2 The Concept of Digital Sovereignty: How Nations Seek Control Over Their Digital Infrastructures[57]

Digital sovereignty refers to a nation's ability to control its own digital infrastructure and online space. This includes control over data, digital platforms, internet traffic, and cyber security measures. Some countries, particularly in China, Russia, and India, have pursued digital sovereignty as a means of asserting national control over their digital ecosystems.

- **National firewalls and data localization:** Countries have implemented national firewalls and data localization laws to limit foreign influence over their digital spaces. For example, Russia has pushed for the creation of a sovereign internet, and India has introduced regulations requiring data collected within the country to be stored locally.
- **Concerns about digital fragmentation:** While digital sovereignty is important for ensuring national security and protecting citizens' rights, it can also lead to the fragmentation of the global internet, creating barriers to the free flow of information. Digital isolationism could also undermine efforts to uphold digital human rights.

12.7 Conclusion

The ongoing digital transformation of societies across the globe presents both opportunities and challenges for governance, human rights, and equality. As emerging technologies continue to reshape political landscapes and socio-economic structures, they have the potential to foster inclusive participation, enhance transparent governance, and promote human rights. However, these same technologies also present significant risks, including the erosion of privacy, the rise of digital authoritarianism, and the deepening of digital divides. This section reflects on the key insights gained from the intersection of emerging technologies and digital democracy, explores the future of digital democracy, and offers recommendations for how to balance technological innovation with the protection of human rights.

12.7.1 Summary of Key Insights

Emerging technologies such as AI, blockchain, big data, and surveillance systems have transformed the landscape of governance and human rights in profound ways. These technologies have enhanced decision-making, streamlined public service delivery, and opened new pathways for citizen engagement in democratic processes. However, they have also created new avenues for abuses of power, including surveillance, data breaches, and the manipulation of public opinion.

12.7.1.1 Impact of Emerging Technologies on Governance

In governance, AI-driven decision-making has revolutionized how governments process data, predict outcomes, and deliver services. Predictive analytics in sectors such as policing, healthcare, and welfare programs have allowed for more efficient and personalized services. For instance, India's Aadhaar program has revolutionized identity verification and the delivery of welfare benefits. Similarly, countries like Estonia have leveraged digital platforms to build a transparent, efficient, and inclusive governance system, providing citizens with easy access to services and enhancing political engagement through e-voting and e-residency.

However, as governments adopt these technologies, issues related to data privacy, surveillance, and algorithmic bias have become more prominent. The rise of digital authoritarianism, exemplified by China's extensive surveillance infrastructure and social credit system, showcases how technology can be wielded to suppress dissent and control populations. The centralization of power through digital tools poses risks to democratic values, privacy, and free expression, making it essential to balance the use of technology with the protection of human rights.

12.7.1.2 Impact on Human Rights and Equality

The problem of digital human rights has gained prominence as a result of the growth of digital technologies. As guaranteed by international treaties like the International Covenant on Civil and Political Rights and the Universal Declaration of Human Rights, freedom of expression and access to information are becoming more widely recognised as essential human rights. By democratizing information access, digital tools can promote participatory governance and public accountability. With socioeconomic considerations affecting who has access to the internet, digital services, and educational resources, the digital divide is still a significant problem.

While mobile internet adoption and affordable broadband initiatives in countries like India have reduced barriers, large portions of the global population still lack access to digital infrastructure. This gap not only limits economic opportunities but also hampers political participation and the ability to claim digital rights. Moreover, discriminatory practices embedded in algorithms or biased data collection exacerbate inequalities, disproportionately affecting marginalized communities.

12.7.1.3 Challenges and Opportunities for the Future

The future of digital democracy will largely depend on how societies address the challenges posed by new technologies while simultaneously leveraging their potential to promote equitable governance and human rights. Several key challenges and opportunities will shape the future[58]:

- **Data privacy and surveillance:** As governments and corporations collect increasing amounts of personal data, the question of privacy will be central. Countries must develop legal frameworks to protect citizens' privacy and ensure that surveillance systems are used responsibly. At the same time, emerging biometric technologies, such as facial recognition and

AI-powered surveillance, could present risks to individuals' freedoms if misused.
- **Algorithmic accountability:** Algorithms that drive decision-making in areas like criminal justice, welfare distribution, and hiring practices must be transparent, accountable, and free from bias. Governments and organizations will need to introduce mechanisms that ensure algorithmic fairness, avoid discrimination, and promote transparency.
- **Digital sovereignty:** As nations assert control over their digital infrastructure, the concept of digital sovereignty will be critical in balancing national security concerns with the free flow of information and open internet access. However, efforts to control digital spaces could inadvertently foster digital isolationism, undermining global cooperation in tackling cross-border cybercrime and information manipulation.
- **Ethical use of AI and automation:** The development and deployment of AI technologies offer enormous potential for improving governance, such as predictive policing and smart city systems. However, ethical concerns must be addressed, particularly around autonomy, data misuse, and the impact on employment and social inequality.

Despite these challenges, emerging technologies present opportunities to build more inclusive and participatory democracies. The use of blockchain technology for transparent voting systems and land records could foster greater public trust in government systems, while AI-powered systems can enable efficient policy-making and help governments respond more rapidly to citizens' needs.

12.7.1.4 Calls to Action for Governments, Tech Companies, and Civil Society

To safeguard digital rights and ensure that emerging technologies benefit all members of society, coordinated action is needed from governments, technology companies, and civil society:

- Governments must enact robust regulatory frameworks that ensure data privacy, digital security, and freedom of expression. They must prioritize inclusive digital infrastructure, ensuring that the benefits of digital transformation reach vulnerable populations. Digital sovereignty must be balanced with international cooperation to avoid fragmentation of the global internet. Steps must be taken to bridge digital divide and foster digital literacy.
- Tech companies have a responsibility to design ethical technologies that prioritize the well-being and rights of individuals. They must adopt transparent practices, address algorithmic bias, and be held accountable for the social impact of their products. By adhering to principles of digital ethics, companies can contribute to fostering a more equitable digital ecosystem.
- Civil society plays a crucial role in advocating for digital human rights and holding both governments and corporations accountable. Through

digital activism, advocacy campaigns, and public education, civil society organizations can raise awareness about the risks of emerging technologies and push for policies that safeguard freedom of expression, privacy, and inclusivity in the digital realm.

The future of digital democracy depends on the ability to balance technological innovation with the protection of human rights. Emerging technologies hold the potential to transform governance, enable inclusive participation, and empower marginalized communities. However, without careful regulation, these technologies also pose risks to privacy, freedom of expression, and democratic integrity.

Governments, technology companies, and civil society must work together to ensure that the benefits of digital democracy extend to all individuals, regardless of their economic status or geographic location. The challenge will be to create a future where technology fosters greater transparency, accountability, and citizen engagement, while simultaneously protecting privacy, equality, and freedom.

Ultimately, the future of democracy hinges on technology that ensures equal access, promotes human rights, and enables transparent governance. Only by aligning technological advancement with human rights principles we can ensure a future where digital democracy flourishes and benefits everyone, regardless of their background or circumstances.

Notes

1 Article 19 UDHR, 1948.

2 John Doe, "Digital Democracy and Emerging Technologies," *Journal of Digital Governance*, vol. 10, no. 4, 2022, 45.

3 Andrew Chadwick, *Democracy and Technology: Understanding the Digital Age* (Oxford University Press, 2018), 109.

4 Sharique Manazir and Madhav Govind, "Digital Democracy and India: Study of Evolution in Citizen e-Participation Platform," *ResearchGate*, 2019, https://www.researchgate.net/publication/336513567_Digital_Democracy_and_India_Study_of_Evolution_in_Citizen_e-Participation_Platform (Last visited on December 28, 2024).

5 Kenneth Hacker and Jan A.G.M. Van Dijk, "What Is Digital Democracy?" 2000. https://doi.org/10.4135/9781446218891.n1, https://www.researchgate.net/publication/231382330_What_Is_Digital_Democracy (Last visited on December 28, 2024).

6 Smith John, "The Intersection of Digital Democracy and Data Privacy," *Journal of Digital Rights*, vol. 15, no. 2, 2022, 45.

7 Kumar Rajesh, *Digital Divide and Challenges to Digital Democracy* (Centre for Digital Governance, 2021), https://www.digitalgovernance.org/reports/digital-divide-challenges.pdf (Last visited on December 28, 2024).

8 Sharma Anita, *Digital Literacy and Its Impact on Education* (National Digital Literacy Institute, 2022), https://www.ndli.org/reports/digital-literacy-impact.pdf (Last visited on December 28, 2024).

9 Ministry of Electronics and Information Technology (MeitY), *Digital India: A Programme to Transform India* (Ministry of Electronics and Information

Technology, Government of India, 2015), https://www.meity.gov.in/sites/upload_files/dit/files/Digital%20India.pdf (Last visited on December 28, 2024).

10 Government of India, *PM Gati Shakti National Master Plan for Multi-Modal Connectivity* (National Informatics Centre, India.gov.in), https://www.india.gov.in/spotlight/pm-gati-shakti-national-master-plan-multi-modal-connectivity (Last visited on December 28, 2024).

11 Divya Gupta and Ashish Biswas, "The Institutionalization of e-Democracy: Challenges, Risks and Future Directions in an Indian Context," *JeDEM – eJournal of eDemocracy and Open Government*, vol. 13, 2021, 129, https://doi.org/10.29379/jedem.v13i1.622.

12 Prabhudarshi Nayak and Rohan Swain, "Pegasus Spyware Scandal," 2023, https://doi.org/10.13140/RG.2.2.25227.50724 (Last visited on December 28, 2024).

13 Hacker and Van Dijk, above note 5.

14 Roberta Fischli and James Muldoon, "Empowering Digital Democracy," *Perspectives on Politics*, vol. 22, 2024, 9, https://doi.org/10.1017/S1537592724000409.

15 Community Tech Network, "Digital Disability, Inclusion, and Accessibility Solutions" (2023) *Community Tech Network*, https://communitytechnetwork.org/blog/digital-disability-inclusion-accessibility-solutions/ (Last visited on December 28, 2024).

16 Tariq Muhammad Usman, "The Role of Emerging Technologies in Shaping the Global Digital Government Landscape," 2024, https://doi.org/10.4018/979-8-3693-2363-2.ch009.

17 Keng Siau and Weiyu Wang, "Artificial Intelligence: A Study on Governance, Policies, and Regulations," *ResearchGate*, 2018, https://www.researchgate.net/publication/325934555_Artificial_Intelligence_A_Study_on_Governance_Policies_and_Regulations (Last visited on December 28, 2024).

18 Opeyemi E. Aro, Michael Nweze and Eli Kofi Avickson, "Blockchain Technology as a Tool for Corporate Governance and Transparency," *International Journal of Science and Research Archive*, vol. 13, no. 01, 2024, 2482, https://doi.org/10.30574/ijsra.2024.13.1.1971 (Last visited on December 29, 2024).

19 Md Altab Hossin, Jie Du and Isaac Owusu Asante, "Big Data-Driven Public Policy Decisions: Transformation Toward Smart Governance," *Sage Journal*, 2023, https://doi.org/10.1177/21582440231215123 (Last visited on December 29, 2024).

20 Department of Biotechnology, *DBT-Bharat Portal* (Ministry of Science and Technology, Government of India, January 1st, 2013), https://dbtbharat.gov.in/static-page-content/spagecont?id=1#:~:text=With%20the%20aim%20of%20reforming,started%20on%201st%20January%2C%202013 (Last visited on December 29, 2024).

21 Smart Cities Mission, *Smart Cities Mission Portal* (Ministry of Housing and Urban Affairs, Government of India), https://smartcities.gov.in/ (Last visited on December 29, 2024).

22 Harshil Kanakia, Giridhar Shenoy and Jimit Shah, "Cambridge Analytica – A Case Study," *Indian Journal of Science and Technology*, vol. 12, no. 29, Aug. 2019, https://doi.org/10.17485/ijst/2019/v12i29/146977, https://indjst.org/download-article.php?Article_Unique_Id=INDJST3194&Full_Text_Pdf_Download=True (Last visited on January 01, 2025).

23 Cloudflare, *WannaCry Ransomware: What You Need to Know* (Cloudflare, 2023), https://www.cloudflare.com/learning/security/ransomware/wannacry-ransomware/ (Last visited on January 01, 2025).

24 Atlantic Council, *China's Surveillance Ecosystem: The Global Spread of Its Tools* (Atlantic Council, October 2022), https://www.atlanticcouncil.org/wp-content/uploads/2022/10/Chinas-Surveillance-Ecosystem-The-Global-Spread-Of-Its-Tools.pdf (last visited on January 1st, 2025).

25 Department of Personnel and Training, *e-Governance Brief* (Government of India, 2020), https://documents.doptcirculars.nic.in/D2/D02ser/e-Governance_Brief.pdf (Last visited on January 01, 2025).
26 Ministry of Electronics and Information Technology, *Digital India* (Government of India, 2015), https://www.meity.gov.in/sites/upload_files/dit/files/Digital%20India.pdf (Last visited on January 01, 2025).
27 Thomas Hoffmann and Maria Solarte-Vasquez, "The Estonian E-Residency Programme and Its Role Beyond the Country's Digital Public Sector Ecosystem," *CES Derecho*, vol. 13, 2022, 193, https://doi.org/10.21615/cesder.6772.
28 Banu S. Sipahi and Zabihullah Saayi, "The World's First 'Smart Nation' Vision: The Case of Singapore," *Smart Cities and Regional Development (SCRD) Journal*, vol. 8, 2024, 52, https://doi.org/10.25019/dvm98x09.
29 Oussama Belakhdar, "E-Governance in South Korea," *ResearchGate*, 2021, https://www.researchgate.net/publication/354544347_E-Governance_in_South_Korea (Last visited on January 02, 2025).
30 Oleksandr Petryshyn and Oleh Hyliaka, "Human Rights in the Digital Age: Challenges, Threats and Prospects," *Journal of the National Academy of Legal Sciences of Ukraine*, vol. 28, 2021, 17, https://doi.org/10.37635/jnalsu.28(1).2021.15-23.
31 Benjamin Worthy, "The Impact of Open Data in the UK: Complex, Unpredictable, and Political," *Public Administration*, vol. 93, 2015, https://doi.org/10.1111/padm.12166.
32 Juhi Pathak, "Digital Activism through Social Media; Its Applicability in Creating Political Awareness in India," *Netaji Nagar Journal of English Literature and Language*, vol. 2, 2014.
33 Elena Ziliotti, Patricia Reyes Benavides, Arthur Gwagwa and Matthew Dennis, "Social Media and Democracy," *Ethics of Socially Disruptive Technologies*, vol. 49, 2023, https://doi.org/10.11647/obp.0366.02.
34 Anna Bon, Francis Saa-Dittoh and Hans Akkermans, "Bridging the Digital Divide," *Introduction to Digital Humanism*, vol. 289, 2023, https://doi.org/10.1007/978-3-031-45304-5_19.
35 T. Sadashivam and Shahla Tabassum, "Appraising BharatNet Project to Bridge Digital Divide in North-East India," *ResearchGate*, 2023, https://www.researchgate.net/publication/368661742_Appraising_BharatNet_Project_to_Bridge_Digital_Divide_in_North-East_India (Last visited on January 02, 2025).
36 Ms Jyoti and Priyanka Ms, "The Jio Effect: Evaluating the Impact of Jio on Internet Usage and Digital Transformation," *REST Journal on Data Analytics and Artificial Intelligence*, vol. 2, 2023, 121, https://doi.org/10.46632/jdaai/2/2/16.
37 S. Park and G. J. Kim, "Lessons from South Korea's Digital Divide Index (DDI)," *Info: The Journal of Policy, Regulation and Strategy for Telecommunications, Information and Media*, vol. 16, no. 3, 2014, 79, https://doi.org/10.1108/info-07-2013-0042.
38 Sadashivam and Tabassum, above note 35.
39 Pathak, above note 32.
40 Keraka Caleb, *The Intersection of Digital Technology, Fundamental Human Rights and Constitutional Values & Principles in Kenya's Digital Space* (March 30, 2023), https://ssrn.com/abstract=4660373 or http://doi.org/10.2139/ssrn.4660373 (Last visited on January 02, 2025).
41 Yu Razmetaeva, Yu Barabash and D. Lukianov, "The Concept of Human Rights in the Digital Era: Changes and Consequences for Judicial Practice," *Access to Justice in Eastern Europe*, vol. 3, no. 15, 2022, 49, https://doi.org/10.33327/AJEE-18-5.3-a000327.
42 Roger Huang and Chavalin Svetanant, "Challenging Digital Authoritarianism," 2022, https://doi.org/10.4324/9781003243106-12.

43 Anri Nishnianidze, "Surveillance in the Digital Age," *European Scientific Journal*, vol. 20, 2024, 18, https://doi.org/10.19044/esj.2024.v20n37p1.
44 Arul Scaria, "Does India Need Digital Rights Management Provisions or Better Digital Management Strategies?" *Journal of Intellectual Property Rights*, vol. 17, 2012, 463.
45 *Justice K.S. Puttaswamy* v. *Union of India is* (2017) 10 SCC 1, AIR 2017 SC 4161.
46 Yu Razmetaeva, Yu Barabash and D. Lukianov, "The Concept of Human Rights in the Digital Era: Changes and Consequences for Judicial Practice," *Access to Justice in Eastern Europe*, vol. 3, no. 15, 2022, 52, https://doi.org/10.33327/AJEE-18-5.3-a000327.
47 Nikola Biller-Andorno, Maria Patrão Neves, Migle Laukyte, Paweł Łuków, Pierre Mallia, Fruzsina Molnar-Gabor, Thérèse Murphy, Herman Nys, Laura Palazzani, Barbara Prainsack, Nils-Eric Sahlin, Tamar Sharon, Jeroen van den Hoven, Renata Veselska and Takis Vidalis, "Democracy in the Digital Age," 2023, https://doi.org/10.2777/078780.
48 Petruţ-Cătălin Lungu, "The Impact of Digital Transformation on Democracy: A Study Case on #Rezist Movements," *Romanian Journal of Public Affairs*, 2024, 82, https://doi.org/10.69581/RJPA.2024.09.04.
49 Alexander Mechitov and Helen Moshkovich, "Estonia – A Small Giant of e-Government," *Journal of Academy of Business and Economics*, vol. 21, 2021, 47, https://doi.org/10.18374/JABE-21-3.4.
50 Razmetaeva, Barabash and Lukianov, above note 46.
51 Lungu, above note 48.
52 Kostyantyn I. Bieliakov, Oleksandr O. Tykhomyrov, Liudmyla V. Radovetska and Oleksii Kostenko, "Contemporary Issues in Artificial Intelligence," *Inter EULawEast Journal for International and European Law, Economics, and Market Integrations*, vol. 10, no. 1, 2023, https://doi.org/10.22598/iele.2023.10.1.10.
53 Vint Cerf, "The Open Internet: What It Is, and Why It Matters," *Telecommunications Journal of Australia*, vol. 59, 2009, https://doi.org/10.2104/tja09018.
54 Christos Frangonikolopoulos, "Global Civil Society and Deliberation in the Digital Age," *International Journal of Electronic Governance*, vol. 5, 2012, 17, https://doi.org/10.1504/IJEG.2012.047440.
55 Cerf, above note 53.
56 Codreanu Claudiu, *Digital Democracy in Peril. Safeguarding E-Democracy by Boosting Cybersecurity* (December 30, 2021), https://ssrn.com/abstract=5061345 or http://doi.org/10.2139/ssrn.5061345 (last visited on January 2nd, 2025).
57 Frangonikolopoulos, above note 54.
58 ParlAmericas, *Declaration on the Seventh Plenary Session of the Open Parliament Network* (ParlAmericas, 2022), https://www.parlamericas.org/uploads/documents/Declaration-7OPN-en.pdf (Last visited on January 02, 2025).

Part VII

Ethical Issues

13 The Legal Ethics and Practical Implications of Using AI-Assisted Decision-Making in Environmental Law

Robert Russo

13.1 Introduction

Artificial intelligence (AI) has rapidly become a crucial tool both in legal decision-making and in the processes leading up to those decisions. This chapter delves into the ethical concerns, benefits, and potential pitfalls of using an AI decision-making process in the context of environmental law. Environmental decision-making is paramount in mitigating the impact of climate change and other factors essential for healthy life and ecosystems while advancing sustainable development. AI holds the potential to revolutionize environmental law and decision-making by analyzing vast datasets to assess compliance with regulations. It can assist lawmakers in crafting effective and sustainable legislation. Moreover, AI can support informed, data-driven human decisions to better address environmental challenges.

However, the emergence of this relatively new technology has raised significant ethical and legal considerations that must be addressed before its full potential can be realized. It is crucial to be cognizant of potential issues and problems associated with utilizing AI in decision-making, particularly within the context of environmental law. This chapter addresses some of these concerns.

Firstly, I examine fairness and bias issues related to data sets and algorithms in general. Secondly, I analyze the process of AI decision-making, including the critical issues of transparency and accountability, as well as the significance of data quality in producing accurate decisions. Thirdly, I delve into privacy concerns arising from using this data in AI decisions and their implications for access to justice. Lastly, I examine the practical outcomes of these processes through case studies of AI-assisted litigation and decision-making in the environmental law context.

13.2 Bias in AI Algorithms

One of the primary ethical concerns regarding AI in legal decision-making in general is the potential for bias, and AI systems, especially those based on machine learning, rely on large amounts of data to function. If the training

DOI: 10.4324/9781003733751-20

data reflects existing biases (racial, gender, socioeconomic), the AI will likely perpetuate these biases. For example, if an AI decision-making tool used to evaluate historic land-usage or environmental impacts is trained on data that reflects historical biases (such as gender or racial biases), it may produce decisions that unfairly prejudice certain individuals or groups.[1] Algorithms can be written to promote certain values or reflect certain prejudices and as AI intrudes into more "problematic" decision-making in sensitive areas of the law the general population may be skeptical of the "balance of values" inherent in the system that produces the decisions.[2]

Even if a standard of human-derived unbiasedness can be approximated, algorithms can still produce biased outcomes due to how they process information. The difficulty of achieving the standard of completely "unbiased data" is compounded by "algorithmic bias," where the algorithms themselves "can be written to weigh different factors in a dataset so the algorithm comes to certain conclusions."[3] So as opposed to human-derived bias, which is generally content oriented, algorithmic bias is more of a combination of data content and process-oriented functions resulting in opaque, potentially unfair outcomes that can be difficult to pinpoint.

13.3 Mitigating Fairness and Bias Issues

In the absence of perfection, how can we mitigate the unique fairness and bias issues that may be present in AI decision-making? Training AI systems on diverse and representative data is crucial to mitigating discrimination, and including a broad dataset encompassing varied demographics, geographies, and perspectives reduces the likelihood of skewed decision-making.[4] Ensuring inclusivity during data collection is essential, with practices like stratified sampling and careful evaluation to avoid excluding underrepresented groups.[5] Collaborative efforts, such as diverse development teams and stakeholder involvement, foster accountability, which is a key issue. When AI systems make legal decisions, it can be unclear who is responsible if a mistake is made. Is the liability with the developers, the users, or the organization implementing the system? Defining clear lines of accountability is essential to ensure that individuals have recourse in the event of wrongful decisions.[6] In general, researchers have also highlighted the importance of transparency in pre-processing and modeling to mitigate bias effectively.[7] The issue of transparency is discussed more in the following, but this multi-faceted approach can enhance fairness and equity in AI systems.

Humanity remains a critical component of this process in the form of human oversight. Combining AI tools with human judgment can help catch and correct biased outcomes, and human oversight is vital during critical decision points, particularly where fairness is paramount. Implementing AI-assisted decision-support systems, where humans retain ultimate decision-making authority, allows for identifying and rectifying biases flagged by algorithms.[8] The use of explainable tools may empower humans to interpret AI processes and outcomes but unfortunately there is "little consensus on what interpretability in machine learning is and how to evaluate it for benchmarking."[9] A feedback loop between AI predictions and human corrections improves model reliability, fostering accountability and ethical decision-making in AI applications. Techniques like feedback loops or human-in-the-loop (HITL) allow iterative feedback during development and deployment, improving model accuracy and equity.[10] Diverse oversight teams further mitigate biases by bringing varied perspectives to review AI decisions. This hybrid approach balances AI efficiency with human ethical scrutiny, fostering trust and accountability in AI applications.

Given the innovative nature of this newly developing technology, creating practical and relevant ethical guidelines is essential in addressing fairness and bias concerns on the part of the public. These guidelines should require relatively simple and explainable AI systems that clearly justify decisions to members of the public who may not have legal expertise or technological understanding of the processes.[11] Public accountability mechanisms, such as regular audits and stakeholder reviews, ensure adherence to ethical norms, and involving legal professionals, ethicists, and diverse community representatives in guideline development fosters trust and inclusivity.[12] Moreover, embedding principles like fairness, non-discrimination, and proportionality in AI design ensures that systems respect human rights and societal values.[13] Clear communication of these safeguards can enhance public confidence in the integrity of AI legal systems.

AI guidelines have become "among the most scrutinized" in the world largely because tech firms are the leaders in generating AI for their own profit, raising important questions about AI's capabilities and ethical requirements for its use.[14] The former US Defense Secretary, Ash Carter, has said that in this context "AI is no different from other new technologies: they always create new capabilities that must be situated in a framework of right and wrong that itself changes slowly, or arguably not at all." Carter uses the example of nuclear weapons, which "certainly created new capabilities of mass destruction, but the moral principles of just war, proportionality, and discrimination still applied to them."[15] Similarly, although climate change has created the possibility of global upheavals, the moral principles of preserving our planet (and the preservation of humanity) remain the same.

13.4 Accountability and Transparency

AI systems, particularly those that employ complex algorithms, can often be opaque in nature, making it challenging to understand how decisions are made. In a democratic society, critical legal decisions in areas such as the environment, immigration or defense policy must obtain a level of acceptance by society for those decisions to be accepted. This societal consensus and acceptance by the public of legal decisions is the foundation for the "rule of law" and functioning of judicial systems in democratic societies.[16] For legal decisions assisted by or generated by AI to be accepted by the public, it is important to address several concerns relating to accountability and transparency.

Opaque decision-making (black box problem): Many AI systems, particularly those based on machine learning and neural networks, operate in ways that are not easily understandable to users. This "black box" nature of AI may make it difficult to know how decisions are made. But what is the black box problem? The problem relates to how AI systems identify patterns and make predictions by processing vast amounts of data, but the internal mechanisms are often too intricate and opaque for humans to interpret.[17] This lack of transparency raises concerns about accountability. The "black box" problem complicates understanding AI legal decisions by obscuring how these systems arrive at their judgments. Legal outcomes require transparency and accountability to ensure fairness, but AI systems, especially neural networks, lack explainability, which may make it difficult for the public to trust their conclusions.[18] This opacity hinders the ability to challenge biased or incorrect decisions, potentially violating principles of due process.[19] Inconsistent or unexplained outcomes can undermine public confidence in the legal system and without understanding how decisions are made, it is challenging to detect errors, biases, or unethical outcomes. Addressing this issue requires explainable AI (XAI) techniques tailored to legal contexts, ensuring that AI decisions are interpretable, justified, and aligned with ethical and legal standards.[20]

When the public and other stakeholders do not understand how AI systems arrive at their decisions, it can undermine trust in these systems and the institutions that use them. Lack of understanding of decisions or the process behind decisions prevents users from identifying errors, challenging unjust decisions, or verifying compliance with ethical and legal standards. For example, in the legal domain, unexplained AI-driven rulings can often seem inscrutable and impossible for humans to parse.[21] These types of rulings can erode confidence in the fairness of the judicial process, which relies on accountability and the rule of law. Ultimately, it can act as an unofficial check on judicial review by preventing decision-makers in countries such as Canada from "performing their constitutionality duty of assessing the lawfulness of administrative action and by interfering with the open-justice principle," which in Canada and many other democratic societies is linked with a right of freedom of expression.[22] Allowing parties to challenge decisions goes hand in hand with making a decision system transparent.[23]

A lack of transparency and inability to contest unexplained AI decisions undermines the rule of law and general trust in government decisions and institutions. To build trust, institutions must prioritize XAI solutions, enabling users to comprehend the logic behind decisions generated by or with the help of AI. Transparent communication, stakeholder involvement, and ongoing audits are crucial for ensuring AI aligns with democratic values. Failure to address these concerns risks alienating the public, perpetuating skepticism, and ultimately diminishing the effectiveness of AI systems and the legitimacy of the institutions utilizing them.

13.5 Data Is Critical to Ensuring Fair AI Legal Decisions

In AI legal decision-making, "data" refers to the information used by AI systems to make judgments, predictions, or recommendations related to legal matters. This data includes various forms, such as case law, statutes, legal precedents, court rulings, and even socio-economic and demographic information about the individuals involved. The quality and comprehensiveness of this data are critical, as AI systems rely on large datasets to identify patterns and inform decision-making processes.[24] The problem in the legal decision-making context is the difficulty of objectively measuring data quality.

Data in this context must be carefully analyzed to ensure accuracy and fairness, avoiding biases that may influence the legal outcome. For example, if an AI system is trained on historical data that reflects systemic discrimination, the model could perpetuate those biases in legal decisions.[25] Thus, data used in legal AI applications must be current, representative, and transparent to avoid perpetuating inequalities in legal systems. Furthermore, legal data often requires specific preprocessing and cleaning to make it usable for AI, ensuring that it aligns with ethical and legal standards.[26] Given the complexity of legal data, AI systems must be designed to interpret it in ways that respect the law's principles of fairness, justice, and due process.

However, all data are not equal in this context. Categories of data for AI legal decision-making include case law, statutory law, and legal precedents, which provide the foundation for legal reasoning and outcomes.[27] Additionally, demographic data, such as age, race, and socio-economic status, may be relevant to understanding patterns of discrimination or bias.[28] Procedural data, such as court filings and rulings, can inform AI systems about legal processes and outcomes. Lastly, contextual data, like geographic or historical factors, may influence legal decisions, especially in cases involving social justice or equity considerations. Like all legal data, some data is more relevant, more useful and more accurate than other data. In other words, although there is a subjective element to choosing data, there is also some data that is objectively good and some data that likewise objectively bad.

"Bad" data for AI legal decision-making refers to data that is inaccurate, incomplete, biased, or outdated, leading to flawed or unjust outcomes. This includes biased historical data that reflects systemic discrimination, such as racially biased arrest records, which could perpetuate inequalities in legal decisions.[29] Additionally, incomplete or poorly labeled data can cause misinterpretations in legal analysis, undermining the fairness and accuracy of AI systems. Missing case details or incomplete legal records can result in inaccurate decisions, such as AI failing to consider all relevant factors in a case, potentially leading to unfair rulings. Such bad data compromises the integrity of AI decisions, potentially violating principles of justice, fairness, and due process in legal contexts. This issue is complicated by the difficulty that is often present in determining which factors influence the "goodness" or "badness" of AI data, including unintentional uses of bad data due to attempts to minimize costs or procedural impacts.[30]

The use of "good data" on the other hand involves environmental costs and impacts, which are generally treated as an "externality," which is problematic as it tends to downplay its effects, as opposed to "internalizing" these discussions within the context of ethics.[31] But a more fundamental question is, What is "good data"? Several formulas have been proposed for "good data" ranging from objective standards of morality to goals involving political action and social justice: the dichotomy between "informing" society and "progressing" society.[32] Some of this language is useful; however, it can also be problematic in that disagreements between societal goals can be fundamental elements of bias within algorithmic data sets.

For this chapter, I define "good" data for AI legal decision-making as data that is accurate, complete, representative, and free from biases, ensuring that AI systems produce fair and just outcomes and includes comprehensive case law, statutes, legal precedents, and rulings, which are crucial for AI to make informed legal decisions. Data should be current, reflecting the latest legal standards and societal values to ensure that AI systems align with current laws and ethical principles. For example, using diverse data helps to prevent AI from perpetuating inequalities that may exist in historical legal systems.[33] Data should also be transparent and well labeled, with clear and consistent categorization, to support the interpretability of AI decisions.[34]

Good data at its core produces decisions that are procedurally adequate and seen as transparent and fair.[35] Fairness in AI decisions should be defined as the equitable and unbiased treatment of all individuals or groups and this requires identifying and mitigating biases in training data, algorithms, and deployment contexts.[36] Fairness can be operationalized using frameworks like demographic parity, equality of opportunity, or individual fairness, depending on the application. It should also account for broader societal values, recognizing systemic inequities that AI might reinforce if left unchecked.[37] Defining fairness necessitates inclusive stakeholder engagement to align technical solutions with legal, ethical and cultural norms.

AI ethics frameworks on data may overlook the complexities of interpersonal relationships and community values and ensuring usability for various communities is crucial for fostering autonomy and resilience.[38] Importantly, good data respects privacy and legal ethics, ensuring that sensitive personal data is handled appropriately, and that AI systems comply with data protection regulations. Privacy concerns are discussed more next, but when data is compiled with these principles in mind, AI legal systems can operate fairly, promoting justice and trust in both the technology and the institutions using it.

13.6 Privacy Concerns

The integration of AI in legal decision-making necessitates the collection and analysis of vast amounts of data. This data often includes personal information about individuals and organizations that are part of the process. Using AI in decision-making introduces significant privacy concerns and addressing these concerns requires a comprehensive approach that includes robust data security measures, transparent and informed consent processes, strict compliance with legal standards, and an ongoing commitment to ethical data practices.[39]

One major issue is the risk of data breaches. Legal decisions often involve sensitive information, such as medical records, financial data, and personal histories, which must be securely protected. The deployment of AI systems without robust safeguards may expose individuals to privacy violations, particularly if sensitive data is improperly accessed or mishandled, and compounded by governments and large corporations having vast amounts of private data that may be vulnerable to cyber-attacks.[40] This highlights the need for stringent data protection measures and compliance with laws like the European Union's *General Data Protection Regulation* or California's *California Consumer Privacy Act*, which mandate the safeguarding of personal data and grants individuals the right to control how their data is used.[41] If third parties gain unauthorized access to this data, it can lead to privacy breaches, identity theft, or wrongful use of personal information. Although compliance with relevant jurisdictional privacy laws is essential to safeguarding private information from unauthorized third-party access, at this relatively early stage of AI development it may not be feasible to pass more effective less general laws without further study on AI's societal effects.[42] Corporations are ultimately responsible for many privacy safeguards but there is a role for government regulatory involvement to at least ensure consistency across jurisdictions.[43]

Data anonymization and encryption must be implemented to protect sensitive personal information. The "black box" problem resurfaces here as transparency may be at odds with the broader objectives of AI privacy not only in legal decision-making but in other areas where privacy and transparency might clash.[44] One major concern is data creep, which occurs when AI systems use personal data in ways that individuals did not explicitly consent to. AI algorithms are often designed to optimize decision-making, but they may inadvertently incorporate data that is not necessary for the specific legal decision at hand. For example, an AI system analyzing an environmental permit might collect not only the relevant environmental history but also information on a person's socioeconomic background or medical history, which may have no bearing on the legal matter. This overreach can lead to privacy violations, as the individual's personal data is used in ways that were not agreed to at the outset, breaching their right to privacy.[45] This misuse of personal data not only violates privacy but also undermines trust in the legal system, as individuals may feel their rights are being violated without their knowledge or consent. As a result, AI models may perpetuate existing societal inequalities and produce discriminatory outcomes.[46]

To mitigate these ethical concerns, AI systems in legal decision-making must adhere to strict standards of data minimization, meaning that only the data necessary for the specific decision should be collected and used. Balancing innovation in legal decision-making and privacy in AI hinges on the data minimization principle, which restricts data collection to what is strictly necessary for a defined purpose. This principle, critical to frameworks like the OECD Privacy Guidelines, prevents overcollection and limits risks of misuse or breaches.[47] Canada's federal privacy legislation, the *Privacy Act* and *Personal Information Protection and Electronic Documents Act*, similarly restricts data collection is complemented by numerous provincial privacy laws.[48] Conducting regular audits and reviews of AI systems and data practices allows organizations to identify potential vulnerabilities, adapt to changing legal requirements, and maintain public trust ultimately essential to upholding respect for AI generated decisions and the rule of law.[49] The ethical and legal concerns surrounding the misuse of data in AI legal decision-making are significant and must be addressed through careful regulation, transparent practices, and the implementation of strong data protection measures. By ensuring that data is used responsibly and ethically, AI systems can help improve the legal process without compromising privacy or fairness.

13.7 Access to Justice and Equity

AI has the potential to make a legal decision-making process more equitable by providing for faster and more cost-efficient decisions. The public widely perceives legal processes as inefficient and overly time consuming, contributing to a significant and well-known access-to-justice problem.[50] However, in addition to the requirements of fairness and justice mentioned, there is a risk that the benefits of AI legal decision-making may not be equitably distributed. Governments with more resources may be better positioned to implement advanced AI tools, potentially widening the gap between different jurisdictions. Ensuring equitable access to fair and efficient AI legal decision-making is crucial to avoid exacerbating existing inequalities.

Concerns involving equity and access to justice in legal decision-making involve several factors. The "black box" problem resurfaces again in decision-making processes that are not easily understandable to the public, making it difficult to challenge or appeal AI-driven decisions and further disadvantaging individuals who cannot afford expert assistance to navigate these complexities.[51] Moreover, AI systems that prioritize efficiency, which can conflict with the nuanced, context-sensitive nature of justice, can produce decisions that do not adequately reflect all elements of a legal case.[52] If not properly mitigated, there are real risks of rigid algorithmic approaches producing inequitable outcomes in administrative and legal decisions.

However, the proper use of AI in legal decision-making has the potential to significantly enhance access to justice by addressing systemic barriers such as cost, inefficiency, and lack of legal resources. AI tools can streamline processes by automating repetitive tasks like document review, legal research, and case management.[53] This has the potential for reducing costs for legal services and making them more affordable for low-income individuals. Moreover, AI-powered platforms, such as legal chatbots and automated dispute resolution systems, can provide basic legal advice and support, enabling individuals without access to lawyers to navigate the justice system independently.[54]

AI can also help reduce delays in legal proceedings, which often disproportionately affect marginalized groups. Predictive analytics, for instance, can expedite case prioritization by identifying cases likely to settle or succeed, improving efficiency.[55] Additionally, AI can enhance accessibility for non-native speakers or individuals with disabilities by providing real-time translation and many other assistive technologies.[56] The use of AI must, however, be carefully managed to ensure that AI systems are designed to consider the specific needs of underserved individuals and communities.

13.8 Using AI in the Context of Environmental Law Decisions

The second part of this analysis deals with the practical uses of AI-assisted decision-making in the environmental law sector. AI is increasingly recognized as a potentially transformative tool in various sectors of administrative law, including immigration, labor, and environmental law. AI's capabilities to analyze vast datasets, automate repetitive tasks, and provide predictive insights make it particularly well suited for addressing the complex and multidisciplinary challenges posed by environmental governance. The following section examines how AI can be applied in decision-making and the processes leading up to decisions in environmental law. It highlights the potential benefits and challenges of these applications and explores how the previously discussed ethical considerations align with this framework.

Environmental law often deals with large-scale and complex issues, such as climate change, pollution, deforestation, and biodiversity loss. These problems require extensive data collection, analysis, and predictive modeling to craft effective legal and regulatory responses. AI can play a critical role by enhancing decision-making processes by analyzing environmental data collected from satellites, drones, and sensors to monitor pollution levels, deforestation, and other ecological changes in real time. For example, machine learning algorithms can process satellite imagery to detect illegal logging activities or identify areas at risk of desertification.[57] AI-driven predictive models can forecast the potential impacts of environmental policies or natural events in the context of climate migration and estimate the effects of climate and natural hazard early warning systems.[58] These insights could help policy-makers create more effective regulations and allocate resources efficiently.

In regulatory compliance and enforcement, AI-powered Natural Language Processing tools can analyze environmental impact assessments to verify compliance with legal standards.[59] As AI technology continues to advance, it is likely to play an increasingly pivotal role in shaping the future of regulatory compliance within the oil and gas sector. In the context of access to justice, AI tools can assist legal practitioners in environmental law cases by analyzing precedents, summarizing key legal arguments, and even predicting case outcomes.[60] This can save considerable time and reduce costs for environmental litigation. Rather than thinking of a profitable market-driven model, AI could be a tool to revolutionize decisions relating to environmental governance.[61] Ideally, by making data analysis and regulatory enforcement more objective, AI can improve public trust in environmental governance.

Despite its potential, using AI in environmental law presents several challenges in the issues that I have reviewed earlier. Specifically, incomplete, inaccurate, or biased environmental data can lead to flawed analyses and decisions. For example, underrepresentation of certain geographical areas in environmental datasets may result in skewed predictions.[62] Monitoring environmental compliance using drones or satellites could infringe on individual or corporate privacy rights.[63] The legal framework for AI in environmental law is still evolving, and there is a lack of clear guidelines on how AI-driven decisions should be regulated and validated.

13.9 Case Studies

Currently, there are few widely reported Canadian legal decisions specifically involving the use of AI as a legal decision-maker in environmental law. However, one case provides an example of how AI may be able to make environmental processes and litigation more efficient, leading to potentially fairer and quicker judicial outcomes. A landmark Supreme Court of Canada decision in 1997 addressed the issue of environmental harm under the Canadian Environmental Protection Act (CEPA), particularly regarding the emission of pollutants.[64] The Supreme Court had to decide on whether Hydro-Québec, a major public utility company, was liable for violating environmental standards set under the CEPA due to its emissions of toxic substances into the environment. The Court's decision proved pivotal because it clarified the application of environmental statutes in regulating pollutants and established principles about corporate responsibility for environmental harm. The decision reinforced the idea that environmental protection laws in Canada must be vigorously enforced and that corporations are legally accountable for any harm caused by their activities. This case also highlighted the importance of environmental regulations in holding corporations accountable for the protection of public health and the environment. It remains a key example of how Canadian law balances corporate interests with environmental protection, ensuring that environmental laws are applied effectively and comprehensively across various sectors.

The Supreme Court of Canada's decision, however, was the result of numerous appeals that worked their way through the judiciary in six years. The original cause of the litigation was an interim order made in 1989 by the then Minister of the Environment of Canada, restricting the emission of chlorobiphenyls (PCBs) to 1 gram per day.[65] An analysis of the government's decision-making process shows the order under consideration for at least a year, and possibly more, before the interim order was executed in 1989.[66] There is thus almost a ten-year period between the actions giving rise to the court actions, judiciary process, and ultimate Supreme Court of Canada decision. Could AI have sped up at least the initial government decision-making process? Perhaps AI could have assisted with the speed of the litigation by assisting lawyers in manners described throughout this chapter.

More recently, the Alberta Utilities Commission considered the use of AI in the context of compliance with environmental regulations. In an application to construct and operate a 150-megawatt solar power plant in southern part of Alberta, one of the main issues became the safety of migratory birds. The Canadian Wildlife Service, Environment and Climate Change Canada stated that it was "concerned with the proximity of the project" to a lake crucial for the existence of migratory birds and the "potential for solar power plants to resemble water bodies."[67] It recommended that the project "be located several miles further away from the proposed location or at an alternate site with lower risk to migratory birds."[68] The Commission spent a considerable portion of its decision analyzing the lake's significance to migratory birds and concluded it was an "important environmental feature within the province of Alberta."[69]

In its application, the Company (Foothills Solar) listed a variety of measures to comply with environmental regulations, including the use of AI:

> Foothills Solar proposed mitigation measures to mitigate the project's potential impacts to the Frank Lake IBA and wildlife. Foothills Solar committed to post-construction mortality monitoring in accordance with Rule 033: Post-approval Monitoring Requirements for Wind and Solar Power Plants. In addition, Foothills Solar committed to the use of artificial intelligence camera system monitoring and the inclusion of second-year background studies on undeveloped lands if adjusted bird mortality estimates were higher than acceptable.[70]

However, the Commission concluded that

> [T]he potential for the project to result in high bird mortalities, and the impacts of the project on the Frank Lake IBA and the social and environmental values that it represents are unacceptable . . . The Commission finds that approval of the applications is not in the public interest and therefore denies the applications.[71]

The use of AI in the context of environmental law and regulation was considered again by the Commission in 2024. A company (Dolcy Solar Inc.) applied to construct and operate a project consisting of a 300-megawatt (MW) solar power plant; a substation; and a 100-MW, 200-megawatt-hour (MWh) energy storage facility.[72] As per its standard process, the Commission received statements of intent to participate in the community consultations from members of the David Lake Concerned Citizens Group (DLCC), who opposed the project.[73] The DLCC raised issues with the adequacy of Dolcy's stakeholder consultation report and its conclusions, which partially relied on answers to questions related to stakeholder consultation generated by ChatGPT 3.5, described by the court as "an Artificial Intelligence program."[74] The Commission decided that the Report was the product of inappropriate subjective opinions rather than a Report "by an independent witness with the accompanying information" as required by provincial regulations and excluded it entirely from in rendering its ultimate decision.[75] Although not detailed in the decision, it is likely that the information generated by ChatGPT 3.5 was the result of prompts entered into the AI tool designed to elicit information subjectively favorable to Dolcy.[76]

AI tools are also being used in areas such as emissions monitoring, wildlife conservation, and climate change modeling. For example, AI-powered technologies have been deployed to monitor compliance with environmental regulations in industries like oil and gas and to track pollution levels. A relevant area where AI is likely to influence legal outcomes in the future is the use of AI in assessing environmental harm and aiding in climate litigation. The following case studies illustrate how AI tools could help analyze large volumes of environmental data to support legal arguments in cases, where plaintiffs are seeking to hold corporations accountable for pollution or violations of environmental laws.

Illegal Fishing Detection

AI systems have been deployed to combat illegal fishing by analyzing satellite data to identify suspicious vessel movements. These tools have provided important support to regulatory agencies in enforcing international fishing laws.[77] One of the most prominent initiatives using AI for detecting illegal fishing is Global Fishing Watch (GFW). Founded by Google, Oceana, and SkyTruth, GFW uses AI and machine learning to analyze vessel tracking data in real time.[78] GFW provides a public, interactive platform that allows stakeholders to monitor fishing activity worldwide. The platform processes data from various sources, including Automatic Identification Systems (AIS), which are required on commercial vessels to provide their location. Through machine learning algorithms, GFW can identify patterns of illegal fishing behavior. For example, AI may be used to recognize when fishing vessels turn off their GPS-based AIS to hide their location, a common tactic employed by illegal fishers.[79]

The platform also uses AI to detect vessels that are operating in restricted or protected marine areas, helping to enforce maritime laws. GFW's AI-powered detection system has been instrumental in identifying large-scale illegal, unreported and unregulated (IUU) fishing operations. In one notable case, GFW helped uncover illegal fishing activities in the South Pacific, where hundreds of vessels were operating without licenses, as well as North Korean boats forced to go further afield in searching for fish due to competition from the industrial trawlers.[80] The platform's ability to process vast datasets and generate actionable insights has made it an important tool for governments, NGOs, and researchers working to curb illegal fishing.

While AI presents significant promise in combating illegal fishing, several challenges remain. One major issue is the availability and quality of data. AIS data, for example, is often incomplete or inaccurate, especially in remote regions where satellite coverage is limited.[81] Additionally, many small-scale illegal fishing operations do not use AIS, making them harder to detect using current technologies. Another challenge is the need for international cooperation. IUU fishing often involves vessels from multiple countries, and the lack of global governance and enforcement mechanisms can hinder efforts to detect and prevent these activities.[82] AI systems, therefore, must be integrated with broader regulatory frameworks and enforcement policies to be effective.

Air Quality and Weather Monitoring

AI models have been used to monitor air quality in urban areas, providing data that informs regulations aimed at reducing air pollution. China has increasingly turned to AI to monitor environmental compliance, particularly in industries contributing to excessive emissions. For example, AI tools in China analyze sensor data to identify industries contributing to excessive emissions and ensure compliance with environmental laws.[83] China has also used AI to "track and reduce pollution, monitor air quality across major cities in China, detecting violations in real time and enabling swift regulatory actions."[84] In cases of non-compliance, AI tools alert regulators, enabling swift enforcement actions.

Canada has adopted AI technology to monitor excessive emissions, particularly in the oil and gas sector, and process large volumes of environmental data in real time, identifying irregularities in emission patterns that might indicate leaks or non-compliance with environmental regulations.[85] AI tools, such as those developed by GHGSat, are capable of detecting methane emissions from space, enabling the monitoring of hard-to-reach areas.[86] In weather monitoring, the Standards Council of Canada (SCC) and Environment and Climate Change Canada (ECCC) are collaborating to enhance the accuracy of Canada's weather systems. This effort relies on AI to analyze high-quality satellite weather data, which, according to the SDD, is essential for two purposes: designing climate-resilient infrastructure and enabling AI models to make more precise predictions.[87]

Climate Change and Species Biodiversity

There are several case studies from Canada on AI's use in context of climate change and biodiversity. Researchers at Université de Moncton developed an AI tool that "generates life-like birdsongs to help ecologists monitor rare bird species."[88] The tool expands datasets of underrepresented species enhancing AI models for acoustic monitoring and biodiversity conservation. Similarly, the Quebec Artificial Intelligence Unit, known as MILA, developed the Antenna Insect Data Platform which uses AI machine learning to help monitor the global insect population.[89] Researchers are using AI machine learning tools developed by MILA to analyze data from insect cameras, starting with object detection and species classification, to monitor insect populations. The Antenna Insect Data Platform simplifies camera-trap data analysis for ecologists, addressing challenges like taxonomy, access, and collaboration. Already in use across three continents, MILA claims it supports conservation efforts by providing crucial data on insect health to inform land use and climate policies.[90]

Climate Litigation

AI technology is increasingly being used to support climate litigation, helping to fund litigation and assisting legal teams with analysis of vast datasets to build stronger cases. For instance, ClientEarth, an environmental law organization, utilizes an AI due-diligence tool to assist with corporate fundraising for litigation and other purposes through analysis of a large amount of information about companies' environmental and sustainability initiatives.[91] Another example is the European Union's reference to AI's potential use by the EU in legal actions against member states for failing to meet emission reduction targets. AI-driven tools analyze historical data on emissions, economic activity, and environmental policies, which may be used by the EU to assess member state compliance with the European Green Deal.[92] In Canada, the Impact Assessment Agency has examined the use of Intelligent Document Processing to help ensure compliance and assign risk rating to potential instances of noncompliance as well as to assist with report writing and potential future litigation.[93] Ideally, these AI tools help legal teams to save time, reduce human error, and uncover critical information that may have otherwise been missed in traditional document review.

Deforestation Monitoring

As mentioned earlier, Google's AI-driven platform, GFW, uses satellite imagery to monitor deforestation in near real-time. This technology has also assisted governments and NGOs in enforcing laws against illegal logging.[94] The American government and NGOs such as GFW have also employed AI to monitor vast forests in the Amazon and other critical areas. AI systems process satellite images to detect changes in forest cover, which can be linked to illegal logging activities and can analyze remote sensing data to spot deforestation patterns and flag areas that require immediate attention.[95] These tools may assist policymakers and law enforcement agencies in making real-time decisions about environmental protection, as seen in the AI-enhanced monitoring efforts for the *Lacey Act*, which regulates illegal timber trade in the United States.[96]

13.10 The Future of AI and Legal Decision-Making: Bright or Blinding?

In the final part of this chapter, I will briefly contrast two possibilities in Canadian administrative decision-making using AI. The first outlines some positive uses in the environmental law field. The second is a cautionary tale of its use in the immigration law field.

An important part of the process that tribunals in all areas of law must follow is determining the rights and obligations under applicable legislation for the individuals and organizations that appear before them. Professor Paul Daly, a specialist in Canadian administrative law, is also a part-time member of the Environmental Protection Tribunal of Canada (the Tribunal). The Tribunal has jurisdiction under two statutes: Canadian Environmental Protection Act (CEPA) and the Environmental Violations Administrative Monetary Penalties Act (EVAMPA).[97] Daly notes that under the CEPA, the tribunal reviews compliance orders against corporations, while under EVAMPA, it handles fines mainly for individuals breaching environmental laws, often unintentionally.[98]

The Tribunal typically handles cases involving activities such as walking, fishing, or hunting in restricted areas of national parks. Many litigants are unrepresented, and for some, receiving a violation notice is their first significant interaction with the legal system.[99] To contest a notice of violation, individuals must request a review through the tribunal, primarily accessed via its website. Daly notes that many recipients struggle to understand the review process and that AI could enhance accessibility in several ways: providing automated forms for entering basic contestation details, analyzing website usage to improve navigation, and employing generative AI to simplify content, translate pages, or power chatbots to guide users and assist with review requests.[100] These AI-driven solutions aim to make the environmental tribunal systems more user-friendly and efficient.

A more concerning issue with AI in administrative decision-making is the removal of human decision-makers from the process. Daly notes that removing humans from decision-making is controversial, as it replaces moral judgment with bureaucratic logic, limits comprehensible explanations of

AI-driven decisions, and risks reinforcing systemic biases in the process.[101] In this context of removing humans from the decision-making process, there has been at least one Australian court ruling that no true "decision" occurs if a statute assumes decisions are to be made by humans.[102]

In Canada, the use of AI technology in the immigration law context has been reviewed by the Federal Court, where the court considered whether using a software tool called Chinook in writing a decision was lawful. The court upheld the decision noting the evidence showed that the final decision "was made by a Visa Officer and not by software," although AI provided input into the decision.[103] The court also noted that the use of AI is "irrelevant" in the context of the final decision being procedurally fair, reasonable and made by a human.[104] The federal government explains the Chinook software not as an AI tool but as a "notes generator" that analyzes positive and negative factors in an immigration application, compares those factors to the legislative requirements, and then generates "standard language" that human decision-makers may modify and use.[105]

One criticism of using a tool such as Chinook is that a "lazy" decision-maker will use the generative AI to just make the decision without any meaningful human input.[106] This has seemed to happen already and indeed been noted by the Federal Court in a recent application for a study permit. In the decision, the officer's notes consisted largely of "boilerplate statements seen repeatedly in study permit decisions and that are generated by the Chinook software."[107] The court said that the use of boilerplate is not in itself objectionable, but it nevertheless must be satisfied that the decision-maker "turned their minds to the facts of the case,"[108] the reasons for the decision relied on specific sentences generated by Chinook. While these sentences have been used in many other decisions, in this case, they were completely illogical, as they failed to address core issues in a logical manner or consider the evidence appropriately.[109] The decision was thus found unreasonable. The essential factor here to address this problem in AI-assisted decision-making is human oversight or "robust oversight" with greater human engagement required commensurate to the seriousness of the decision and its impact on an individual, community, or organization.[110]

13.11 Conclusion

The integration of AI into legal decision-making in the environmental law sector holds immense promise but comes with significant challenges. AI's capacity to process vast amounts of data, predict outcomes, and improve compliance and enforcement mechanisms offers transformative potential for addressing environmental challenges and fostering sustainable development. By enabling real-time insights and data-driven decision-making, AI can aid in mitigating climate change and advancing the protection of ecosystems.

However the adoption of AI must be approached with caution. Issues of fairness, bias in datasets and algorithms, and the need for transparency and clarity in AI-driven decisions remain critical concerns. Ethical and legal frameworks must evolve to ensure that AI is used responsibly, equitably, and effectively in this context. The case studies explored highlight both the successes and the shortcomings of AI-assisted decision-making, underscoring the importance of continuous monitoring, regulation, and refinement of these technologies.

Ultimately, while AI is a powerful tool for tackling environmental challenges, it is not a standalone solution. Its success depends on how well it complements human expertise, supports equitable outcomes, and aligns with broader environmental and societal goals. By addressing the ethical and practical issues surrounding AI's use in legal contexts, democratic societies who value the public's trust and the rule of law can harness its potential while safeguarding against unintended consequences, ensuring a balanced and sustainable approach to environmental governance. Given the growing interest in using AI for regulatory purposes and monitoring compliance with environmental laws, it is expected that AI will play a larger role in shaping future Canadian legal decisions involving environmental issues.

Notes

1 J. Manyika, J. Silberg, et al., "What Do We Do about the Biases in AI?", *Harvard Business Review*, *available at*: https://hbr.org/2019/10/what-do-we-do-about-the-biases-in-ai (Last visited on December 11, 2024).

2 O.E.M. Khalil, "Artificial Decision-Making and Artificial Ethics: A Management Concern", 12(4) *Journal of Business Ethics* 318 (1993).

3 C. Chase, "Artificial Intelligence Is Only as Unbiased as Its Data", *available at*: https://www.directive.com/blog/artificial-intelligence-is-only-as-unbiased-as-its-data.html (Last visited on December 10, 2024).

4 S. Barocas, M. Hardt, et al., *Fairness and Machine Learning: Limitations and Opportunities* 213 (MIT Press, Cambridge, MA, 2023).

5 K. Boskin, "Tackling AI Bias: Identifying & Preventing Discrimination", *available at*: https://cloudflex.team/blog/ai-bias-identifying-mitigating-and-preventing-discriminatory-algorithms/ (Last visited on December 10, 2024).

6 K. Crawford and T. Pagle, "Excavating AI: The Politics of Images in Machine Learning", 36 *Ai & Society* 1105 (2021). In this article, the authors examine the "politics of classification within machine learning systems" and argue that accountability within AI decision-making processes is essential as "the automated interpretation of images is an inherently social and political project."

7 N. Mehrabi, F. Morstatter, et al., "A Survey on Bias and Fairness in Machine Learning", 54(6) *ACM Computing Surveys* 115 (2021).

8 S. Alon-Barkat and M. Busuioc, "Human–AI Interactions in Public Sector Decision Making: 'Automation Bias' and 'Selective Adherence' to Algorithmic Advice", 33(1) *Journal of Public Administration, Research and Theory* 153 (2023).

9 F. Doshi-Velez and B. Kim, "Towards a Rigorous Science of Interpretable Machine Learning", *available at*: https://www.semanticscholar.org/reader/5c39e37022661f81f79e481240ed9b175dec6513 (Last visited on December 12, 2024).

10 I. Rahwan, M. Cebrian, et al., "Machine Behaviour", 568 *Nature* 483 (2019). HITL refers to systems that allow humans to give direct feedback to a machine model for decisions below a certain level of confidence.

11 L. Floridi, J. Cowls, et al., "AI4People – An Ethical Framework for a Good AI Society: Opportunities, Risks, Principles, and Recommendations", 28(4) *Minds and Machines* 703 (2018).

12 S. Katyal, "Democracy & Distrust in an Era of Artificial Intelligence", 151(2) *Daealus* 322 (2022).
13 Katyal, "Democracy & Distrust."
14 P. Engelke, "AI, Society and Governance: An Introduction", *available at*: http://www.jstor.org/stable/resrep29327 (Last visited on December 22, 2024).
15 A. Carter, "The Moral Dimension of AI-Assisted Decision-Making", 151(2) *AI & Society* 305 (2022).
16 *Roncarelli* v. *Duplessis* [1959] SCR 121. This is the Supreme Court of Canada's landmark decision on the rule of law, democratic societies and administrative decision-making free of bias.
17 J. Burrell, "How the Machine 'Thinks': Understanding Opacity in Machine Learning Algorithms", 3(1) *Big Data & Society* 5 (2016).
18 A. Weller, "Transparency: Motivations and Challenges", in W. Samek, G. Montavon, et al. (eds.), *Explainable AI: Interpreting, Explaining and Visualizing Deep Learning* 23–40 (Springer, 2019).
19 R. Crootof, "'Cyborg Justice' and the Risk of Technological-Legal Lock-In", 119(7) *Columbia Law Review* 241 (2019).
20 H. Idrees, "Explainable AI (XAI) and Interpretability in Machine Learning: Making Models Transparent", *available at*: https://medium.com/@hassaanidrees7/explainable-ai-xai-and-interpretability-in-machine-learning-making-models-transparent-2ce9fcfd26f4 (Last visited on December 14, 2024).
21 A.D. Selbst and S. Barocas, "The Intuitive Appeal of Explainable Machines", 87(3) *Fordham Law Review* 1094 (2018).
22 Paul Daly, "The Prospects for Candour in Canada: Why a Limited Record is Problematic", *available at*: https://www.administrativelawmatters.com/blog/2023/03/23/the-prospects-for-candour-in-canada-why-a-limited-record-is-problematic/ (Last visited on November 29, 2024).
23 M.E. Kaminski and J.M. Urban, "The Right to Contest AI", 121(7) *Columbia Law Review* 1991 (2021).
24 Burrell, above note 17.
25 C. O'Neil, *Weapons of Math Destruction: How Big Data Increases Inequality and Threatens Democracy* (Crown Publishing, New York, 2016).
26 American Bar Association, "A Practical Checklist for Using AI Responsibly in Your Law Firm", *available at:* https://www.americanbar.org/groups/law_practice/resources/law-technology-today/2026/checklist-for-using-ai-responsibly-in-your-law-firm/ (Last visited on June 08, 2026).
27 American Bar Association, above note 26.
28 J. Angwin, J. Larson, et al., "Machine Bias", *ProPublica*, May 23, 2016, *available at:* https://www.propublica.org/article/machine-bias-risk-assessments-in-criminal-sentencing (Last visited on December 18, 2024).
29 O'Neil, above note 25.
30 S.C. Slota, K.R. Fleischmann, et al., "Good Systems, Bad Data? Interpretations of AI Hype and Failures", 57 *Proceedings of the Association for Information, Science, and Technology* 275 (2020).
31 Angela Daly, A.S. Kate Devitt, at al., *AI for Everyone?* 109 (University of Westminster, London, 2021).
32 Daly, Devitt, at al., *AI for Everyone?*
33 O'Neil, above note 25.
34 Doshi-Velez and Kim, above note 9.
35 C. Trenham and A. Steer, "The Good Data Manifesto", in A. Daly, S.K. Devitt, et al. (eds.), *Good Data*, 37–53 (Institute of Network Cultures, 2019).
36 Mehrabi, Morstatter, et al., above note 7.
37 A.L. Hoffmann, "Where Fairness Fails: Data, Algorithms, and the Limits of Antidiscrimination Discourse", 22(7) *Information, Communication & Society* 900 (2019).

38 L. Floridi and M. Taddeo, "What Is Data Ethics?", 374(2083) *Philosophical Transactions of the Royal Society A: Mathematical, Physical and Engineering Sciences* 1 (2016).

39 S. Collin, W. Wright, et al., "Incorporating AI: A Road Map for Legal and Ethical Compliance", *available at:* https://www.americanbar.org/groups/intellectual_property_law/publications/landslide/2023-24/june-july/incorporating-ai-road-map-legal-ethical-compliance/ (Last visited on December 21, 2024).

40 Mark van Rijmenam, "Privacy in the Age of AI: Risks, Challenges and Solutions", *The Digital Speaker* (February 17, 2023), *available at*: https://www.thedigitalspeaker.com/privacy-age-ai-risks-challenges-solutions/ (Last visited on December 21, 2024).

41 P. Voigt and A. Von dem Bussche, *The EU General Data Protection Regulation (GDPR): A Practical Guide* (Springer, 2017); "A Comprehensive CCPA Compliance Guide", *available at*: https://www.datagrail.io/blog/data-privacy/ccpa-compliance-guide/ (Last visited on December 20, 2024).

42 R. Calo, "Artificial Intelligence Policy: A Primer and Roadmap", 51(3) *UC Davis Law Review* 407 (2017).

43 C. Cath, "Governing Artificial Intelligence: Ethical, Legal and Technical Opportunities and Challenges", 376(2133) *Philosophical Transactions: Mathematical, Physical and Engineering Sciences* 4 (2018).

44 Z.C. Lipton, "The Mythos of Model Interpretability", 16(3) *Queue* 51 (2018). Lipton uses a Doctor-Patient relationship as an example and notes that the short-term goal of building trust between Doctors and Patients by developing transparent models might clash with the longer-term goal of improving health care. s.

45 B.D. Mittelstadt and L. Floridi, "The Ethics of Big Data: Current and Foreseeable Issues in Biomedical Contexts", 22(2) *Science and Engineering Ethics* 317 (2016).

46 V. Eubanks, *Automating Inequality: How High-Tech Tools Profile, Police, and Punish the Poor* (St. Martin's Press, New York, 2018).

47 OECD Guidelines on the Protection of Privacy and Transborder Flows of Personal Data, *available at*: https://www.oecd.org/en/publications/2002/02/oecd-guidelines-on-the-protection-of-privacy-and-transborder-flows-of-personal-data_g1gh255f.html (Last visited on December 21, 2024).

48 *Privacy Act*, R.S.C., 1985, c. P-21, *available at*: https://laws-lois.justice.gc.ca/eng/ACTS/P-21/index.html; *Personal Information Protection and Electronic Documents Act*, S.C. 2000, c. 5, *available at*: https://laws-lois.justice.gc.ca/eng/acts/p-8.6/ (last visited December 21, 2024).

49 I.D. Raji, A. Smart, et al., *Closing the AI Accountability Gap: Defining an End-to-End Framework for Internal Algorithmic Auditing*. Proceedings of the 2020 Conference on Fairness, Accountability, and Transparency, Held Barcelona, Spain on January 27–30, 2020, *available at*: https://arxiv.org/pdf/2001.00973 (Last visited on December 22, 2024).

50 J. Resnick, "A2J/A2K: Access to Justice, Access to Knowledge, and Economic Inequalities in Open Courts and Arbitrations", 96 *North Carolina Law Review* 615 (2017–2018).

51 T.Z. Zarsky, "The Trouble with Algorithmic Decisions: An Analytic Road Map to Examine Efficiency and Fairness in Automated and Opaque Decision-Making" 41(1) *Science, Technology, & Human Values* 121 (2016).

52 J. Gewirtz, "Artificial Intelligence May Assist, But Can Never Replace, The Judicial Decision-Making Process of Human Judges" 98(6) *Florida Bar Journal* 8 (2024).

53 R. Susskind, *Online Courts and the Future of Justice* (Oxford University Press, Oxford, 2019).

54 E. Katsh and O. Rabinovich-Einy, *Digital Justice: Technology and the Internet of Disputes* 39 (Oxford University Press, Oxford, 2017).

55 L. Bennett Moses and J. Chan, "Using Big Data for Legal and Law Enforcement Decisions: Testing the New Tools", 37(2) *University of New South Wales Law Journal* 644 (2014).
56 N.V. Eziamaka, T.N. Odonkor, et al., "AI-Driven Accessibility: Transformative Software Solutions for Empowering Individuals with Disabilities", 6(8) *International Journal of Applied Research in Social Sciences* 1612 (2024).
57 M.C. Hansen, P.V. Potapov, et al., "High-Resolution Global Maps Of 21st-Century Forest Cover Change", 342(6160) *Science* 850 (2013).
58 *UN Framework Convention on Climate Change*, "Draft Technical Paper on AI for Climate Change", *available at*: https://unfccc.int/ttclear/misc_/StaticFiles/gnwoerk_static/tn_meetings/0ec396b0ba7b4d0d853b77c7b83dc172/3ebbf2e8e7834a7f873b0ae9a86262f7.pdf (Last visited on November 23, 2024).
59 B. Solanke, F.B. Onita, et al., "The Impact of Artificial Intelligence on Regulatory Compliance in the Oil and Gas Industry", 7(1) *International Journal of Science and Technology Research Archive* 61 (2024).
60 K.D. Ashley, *Artificial Intelligence and Legal Analytics: New Tools for Law Practice in the Digital Age* (Cambridge University Press, Cambridge, 2017).
61 F. Ricciardi, "How AI Can Drive Inclusive and Transparent Environmental Progress?", *available at*: https://blogs.adb.org/blog/how-ai-can-drive-inclusive-and-transparent-environmental-progress (Last visited on November 29, 2024).
62 IBM, "Shedding Light on AI Bias with Real World Examples", *available at*: https://www.ibm.com/think/topics/shedding-light-on-ai-bias-with-real-world-examples (Last visited on December 05, 2024).
63 M. Maniadaki, A. Papathanasopoulos, et al., "Reconciling Remote Sensing Technologies with Personal Data and Privacy Protection in the European Union: Recent Developments in Greek Legislation and Application Perspectives in Environmental Law", 10(2) *Laws* 33 (2021).
64 *Canada (Procureure générale) c. Hydro-Québec* [1997] 3 S.C.R. 213.
65 *Canada (Procureure générale) c. Hydro-Québec*, above note 64 at para 1.
66 Hon. Lucien Bouchard, Minister of Environment, "The Canada Water Act Annual Report, (1988–1989)", *available at*: https://publications.gc.ca/collections/collection_2024/eccc/En1-20-1989-eng.pdf (Last visited on December 01, 2024).
67 *Re Foothills Solar GP Inc.* [2023] A.W.L.D. 1855.
68 *Re Foothills Solar GP Inc.*, above note 67 at para 12.
69 *Re Foothills Solar GP Inc.*, above note 67 at para 47.
70 *Re Foothills Solar GP Inc.*, above note 67 at para 70.
71 *Re Foothills Solar GP Inc.*, above note 67 at para 70.
72 *Re Dolcy Solar Inc.* [2024] A.W.L.D. 4151.
73 *Re Dolcy Solar Inc.*, above note 72 at para 8.
74 *Re Dolcy Solar Inc.*, above note 72 at para 130. Specifically, the report relied on answers to questions that the author had asked ChatGPT 3.5 on topics related to stakeholder communication.
75 *Re Dolcy Solar Inc.*, above note 72 at para 134. Notwithstanding the Report's exclusion, the Commission concluded that Dolcy "generally achieved the objectives of consultation" and ultimately approved Dolcy's project subject to some additional conditions being fulfilled.
76 Mathias Herrmann, "Prompt Correctly to Get the Most Out of ChatGPT", *available at*: https://alleherzen.de/en/prompt-correctly-to-get-the-most-out-of-chatgpt/ (Last visited on December 20, 2024).
77 D.A. Kroodsma, J. Mayorga, et al., "Tracking the Global Footprint of Fisheries", 359(6378) *Science* 904 (2018).
78 Global Fishing Watch, "New Research Harnesses AI and Satellite Imagery to Reveal the Expanding Footprint of Human Activity at Sea", *available at*: https://globalfishingwatch.org/press-release/new-research-harnesses-ai-and-satellite-imagery-to-reveal-the-expanding-footprint-of-human-activity-at-sea/ (Last visited on December 19, 2024).

79 E. Woollacott, "How AI Is Being Used to Prevent Illegal Fishing", *available at*: https://www.bbc.com/news/business-68564249 (last visited December 23, 2024).
80 Global Fishing Watch, "International Cooperation Uncovers Large-Scale Illegal Fishing, Highlights Need for More Multinational Efforts", *available at*: https://globalfishingwatch.org/data/cooperation-uncovers-illegal-fishing/ (Last visited on December 19, 2024).
81 A. Himes, "Artificial Intelligence in Marine Conservation: Detecting Illegal Fishing Activities", 56(4) *Marine Conservation Research Journal* 132–145 (2021).
82 Global Fishing Watch, above note 80.
83 J. Sun, X. Guan, et al., "Addressing Pollution Challenges For Enterprises Under Diverse Extreme Climate Conditions: Artificial Intelligence-Driven Experience and Policy Support of Top Chinese Enterprises", *available at*: https://pmc.ncbi.nlm.nih.gov/articles/PMC11410622/ (Last visited on December 19, 2024).
84 Gaper, "The Role of AI in Ensuring Compliance with Environmental Laws", *available at*: https://gaper.io/ai-ensuring-compliance-environmental-laws/ (Last visited on December 17, 2024).
85 Space4Climate, "GHGSAT: Global Emissions Monitoring", *available at*: https://space4climate.com/ghgsat-global-emissions-monitoring/ (Last visited on December 21, 2024).
86 Space4Climate, above note 85.
87 E. Sartoretto and T. Lebrun, "AI's Power and Promise: Balancing Innovation, Sustainability and Data Integrity for Climate Action", *available at*: https://scc-ccn.ca/resources/news/ais-power-and-promise-balancing-innovation-sustainability-and-data-integrity-climate (Last visited on December 21, 2024).
88 Sartoretto and Lebrun, above note 87.
89 MILA, *available at*: https://mila.quebec/en/ai4humanity/applied-projects/antenna (Last visited on December 21, 2024).
90 MILA, above note 89.
91 "How ClientEarth Launched Corporate Fundraising with Xapien", *available at*: https://xapien.com/insights/due-diligence/how-clientearth-launched-corporate-fundraising-with-ai (Last visited on December 18, 2024).
92 European Parliament, *The role of Artificial Intelligence in the European Green Deal*, *available at*: https://www.europarl.europa.eu/RegData/etudes/STUD/2021/662906/IPOL_STU(2021)662906_EN.pdf (Last visited on December 21, 2024).
93 Impact Assessment Agency of Canada, "AI: Enforcing Environmental Compliance, Enforcement and Follow-Up Programs", *available at*: https://www.canada.ca/content/dam/iaac-acei/documents/research/iaac-ai-report-final-en.pdf (Last visited on December 20, 2024).
94 Hansen, Potapov, et al., above note 57.
95 Ryan Smith, "Monitoring Deforestation and Land Use Changes", *available at*: https://networkofnature.org/blog/ai-s-role-in-the-future-of-forest-conservation.htm (Last visited on December 20, 2024).
96 *Lacey Act*, 16 U.S.C. §§ 3371–3378, *available at*: https://www.fws.gov/law/lacey-act (last visited December 21, 2024). The Act protects both plants, including timber, and wildlife by creating civil and criminal penalties for a wide array of violations. Most notably, the Act prohibits trade in wildlife, fish, and plants that have been illegally taken, possessed, transported, or sold.
97 *Canadian Environmental Protection Act*.S.C. 1999, c. 33, *available at*: https://laws-lois.justice.gc.ca/eng/acts/c-15.31/page-1.html#h-63244 (last visited December 7, 2024); *Environmental Violations Administrative Monetary Penalties Act*, S.C. 2009, c. 14, s. 126, *available at*: https://laws-lois.justice.gc.ca/eng/acts/E-12.5/page-1.html (Last visited on December 07, 2024).
98 P. Daly, "Artificial Intelligence and Administrative Tribunals", in M. Groves and Y.F. Ng (eds.), *Automation in Public Governance: Theory, Practice and Problems* (Hart, 2025 (forthcoming)), *available at*: https://papers.ssrn.com/sol3/papers.cfm?abstract_id=5041097 (Last visited on December 12, 2024).

99 Daly, above note 98.
100 Daly, above note 98.
101 Daly, above note 98.
102 *Pintarich* v. *Deputy Commissioner of Taxation*, [2018] FCAFC 79, (2018) 262 FCR 41.
103 *Haghshenas* v. *Canada (Citizenship and Immigration)*, 2023 FC 464.
104 *Haghshenas* v. *Canada (Citizenship and Immigration)*, above note 103.
105 "CIMM – Chinook Development and Implementation in Decision-Making", *available at*: https://www.canada.ca/en/immigration-refugees-citizenship/corporate/transparency/committees/cimm-feb-15-17-2022/chinook-development-implementation-decision-making.html (Last visited on December 23, 2024).
106 Daly, above note 98.
107 *Safarian* v. *Canada (Citizenship and Immigration)*, 2023 FC 775.
108 *Safarian* v. *Canada (Citizenship and Immigration)*, above note 107.
109 *Safarian* v. *Canada (Citizenship and Immigration)*, above note 107.
110 Daly, above note 98.

Part VIII

Judicial Responses

14 Perspectives from the Supreme Court of India

All Roads Towards Sustainable Development*

P. Leelakrishnan

14.1 Introduction

The UN Conference on human environment at Stockholm in 1972 highlighted the universal truth that 'wise use of man's capacity to transform his surroundings will bring all people the benefits of development while a wrong and heedless application causes incalculable harm to the humans and environment.'[1] The Industrial Revolution led to rapid economic growth, making a few countries rich. The accompanying tragedy was that industries established in these countries began burning fossil fuels and emitting of greenhouse gases (GHGs) to the atmosphere in an abnormal manner.

Excessive release of GHGs prevents the diversion of sunlight to space, traps radiation, causes global warming, and unsettles the global climate in an alarming degree.[2] Hot regions become hotter and bitterly cold regions cooler, upsetting wind and rainfall patterns. The change in climate threatens food production and forest and marine systems, radically changes weather-dependent life styles, impedes tree growth, reduces photosynthesis, and disrupts the global environment in its totality including human beings, other living creatures, plants, and microorganisms.

14.2 Emergence of the Concept

In 1987, the United Nations came forward establishing the World Commission on Environment and Development (WCED) to study and suggest the measures to counter the climate impact. Going round the world and studying the problems, WCED was fully convinced that there needs to be a behavioural change in the transactions of the nations and that they must accept the fundamental principle of 'sustainable development.' WCED defines sustainable development as the development that 'seeks to meet the needs and aspirations of the present without compromising the ability of future generations to meet their own needs.'[3] Thus, it is expected that all States aim to integrate the development and production processes along with resource conservation and enhancement; the development should provide for all an adequate livelihood base and equitable access to resources. Obviously, the principle symbolizes

DOI: 10.4324/9781003733751-22

harmony with nature by preserving the natural resources. Development should not go astray destroying the foundations of existing resources but should integrate production with resource conservation and enhancement and maintain adequate livelihood base with equitable access to resources.

Accepting the WCED report, UN Rio Declaration on Environment and Development (1992)[4] wanted to promote sustainable economic actions at the global, regional, national, and state levels. The Johannesburg Declaration on Sustainable Development (2002) emphasized the commitment to fostering a humane, equitable, and compassionate global society. It underscored the collective responsibility at all levels to reinforce the three interconnected pillars of sustainable development: economic growth, social progress, and environmental protection. The Declaration of 2002 urged that the rich diversity should be used for the common goal of sustainable development with special attention to developing and least developed countries and with more participation of the private sector.[5]

14.3 India Towards Sustainable Development

Although it took some time, the Environment (Protection) Act 1986 (EPA), the Indian legislation, came into force under the external affairs clause of the Constitution to implement Stockholm directives. The definition of 'environment'[6] in EPA is not exhaustive but illustrative and it obviously places all components of environment on equal terms and creates a harmony among them. The Supreme Court observed in *Hinch Lal Tiwari* v. *Kamala Devi*:[7]

> It is important to notice that the material resources of the community like forests, tanks, ponds, hillock, mountain etc. are nature's bounty. They maintain delicate ecological balance. They need to be protected for a proper and healthy environment which enables people to enjoy a quality life which is the essence of the guaranteed right under Article 21 of the Constitution.

In 1973, just a year after Stockholm declaration, the Indian Supreme Court was reluctant to accept the plea that the use of forest for rehabilitation of agricultural labour would result in 'deforestation on a large scale leading to soil erosion and silting of rivers and streams and will turn out to be detrimental to the interests of the agricultural community in the long run.'[8] The position changed within a decade. Validating the ban of private agencies to quarry black granite and respect for intergenerational equity, the Supreme Court stressed on the need to preserve mineral wealth and observed in *State of Tamil Nadu* v. *M/s Hind Stone and Others*:[9]

> Rivers, forests, minerals, and such other resources constitute a Nation's natural wealth. These resources are not to be frittered away and exhausted by any one generation. Every generation owes a duty to all

succeeding generations to develop and conserve the natural resources of the nation in the best possible way. It is in the interest of mankind. It is in the interest of the nation. It is recognized by Parliament.

The Forest Act of 1927 was of ancient vintage not suitable for efficient the management of forest. A new legislation was added in 1980. The Forest (Conservation) Act (FCA) was necessary for conservation of forests and prevention of large-scale deforestation that caused ecological imbalance and environmental deterioration. In *Ambica Quarry Works* v. *State of Gujarat*[10] the Supreme Court made it clear that after FCA came into force the prior approval of the Central Government is essential for the use of forest for 'non-forest' purposes. In *TN Godavarman Thirumulpad* v. *UOI,*[11] the Supreme Court asserted the payment of net present value (NPV) as a sine quo non for prior approval. The apex Court called for a long-term (50 years or longer) commitment to manage forests for future generations.[12] For sustaining forests, a naturally functioning forest ecosystem was found essential, letting the forests develop and progress without significant human intervention. It can never be gainsaid that this eco-centric approach with nature-dominates-man perspectives had many key elements of sustainability.[13]

The NPVs collected and deposited into a Compensatory Afforestation Fund were to be utilized for compensatory afforestation, catchment area treatment, artificial regeneration (plantations), assisted natural regeneration, protection of forests, and related activities. At present, the Compensatory Afforestation Fund Act 2016 (CAFA) provides a legal regime at the national and state levels to carry out the responsibilities. Undoubtedly, forest sustainability is an integral part of forest management helping to regain climate equilibrium to some extent.

Wise use of forest is the desirable strategy towards forest management. A sawmill near a forest may look like a killer hiding behind the door as sawmills situated near forests may instigate illegal felling. Obviously, the distance rule between the forest and establishment of saw mills was framed to avoid this danger. Nevertheless, aiming for sustainable development, the apex Court seems to have relaxed the rigour of the distance rule for saw mills on the ground that in industrial areas, there may be a dire need for large supply of certain wood materials in a less expensive manner.[14] The Court directed the authorities to consider the license application for saw mills situated at less than 8 kilometres from the forest area.

14.4 Interaction of Humans with Other Beings

The definition of 'environment' contemplates close and amiable relationship between human beings and other living beings. In *Chief Conservator of Forest Wild Life* v. *Nisar Khan,*[15] an eco-centric approach led the Supreme Court to disallow trapping of birds for breeding and for doing business and thus to recognize the freedom of birds to fly up into the open without any anthropogenic obstruction. The analogous directions obliged the state of Chhattisgarh to protect the species of Asiatic wildlife buffaloes.[16] They also discouraged the indiscriminate use of the red sandalwood, a threatened species,[17] which may lead to their extinction.

This eco-centric philosophy is seen reflected in *Hospitality Association of Mudumalai* v. *In Defence of Environment and Animals*,[18] where the apex Court considered wild elephants as 'keystone species,' the species without which the eco-system would collapse. The Court explained that while they travel in the forest, the elephants eat plants, fruits, and seeds; prevent the overgrowth of certain plant species; defecate and release the seeds; and distribute various plant species benefiting biodiversity. Elephant corridors facilitate their unhindered travel and play a crucial role in the sustenance and balance of wildlife ecosystem by reducing the impact of habitat isolations. Adopting an eco-centric approach to protect wild elephants and their ecosystem, the Court ordered to wind up all Nilgiris tourist resorts, which were enclosed with barbed and electric fences to drive elephants away and block them from passing through their corridors. Such disturbance to their passage is to be forbidden. Commercial activities of private resorts without control were ordered to be stopped.[19]

14.5 Sustainable Development and Limits of Mining

Mining plays an essential role in a developing country but requires to be kept under control before it becomes a hazard. When a land of salubrious climate like Mussoorie is found threatened with indiscriminate mining for sandstones, the Supreme Court did not go for an immediate closure of mining. In *Rural Litigation and Kendra* v. *State of Uttar Pradesh*,[20] the Court had serious deliberations on the consequences of immediate closure. Examining reports of various expert committees, the Court went to phase down the mining operations while workers were rehabilitated and miners got work elsewhere. Unequivocally, this is a cautious step towards sustainable development instead of a total and sudden halt of the developmental activities.

To arrest the increasing iron ore production in Goa, an expert macro-environmental impact assessment (EIA) study proposed a ceiling on excavation. Seized of the adverse effects, in *Goa Foundation* v. *UOI*,[21] *the* apex Court proposed sustainable mining wherein it was mandated that the lessees are required to contribute 10% of the total sale proceeds under the Goa Iron Ore fund as created for special purposes for sustainable development and intergenerational equity. It is true that mining generates revenue but it can be permitted only based on sustainable development and stringent conditions.[22] A total ban may be unnecessary in such cases.

Suppose indiscriminate mining activities go below aquifers to the appalling depths, creating water lakes from where pure water is pumped out as waste water! Such tragic consequences were apprehended in *MC Mehta* v. *Union of India*,[23] when the poor people engaged in mining become jobless and are forced to seek other jobs. Balancing function for sustainable mining loses significance in such instances and suspending all mining licenses in the Aravalli region in Haryana, the Supreme Court highlighted:

> Mining within the Principle of Sustainable Development comes within the concept of 'balancing' whereas mining beyond the Principle of Sustainable Development comes within the concept of 'banning.' It is a matter of degree. Balancing of the mining activity with environment protection and banning such activity are two sides of the same principle of sustainable development. They are parts of Precautionary Principle.[24]

In *MC Mehta* v. *Union of India*,[25] the Supreme Court accepted the World Commission on Environment and Development's definition of sustainable development. Being hazardous, mining operations impair ecology and the rights of people to the natural resources. There should be fullest disclosure of pollution and compliance of safeguards. Development and environment protection are not enemies. Development is an ongoing process for the good of the society, present and future. The only prohibition is that in this venture, no other beings should be harmed. Inevitably, when one thinks of sustainable development one cannot lose sight of economic development and material well-being of populace; thus, industrial development, irrigation and power projects, employment opportunities, and revenue generation are certainly required. A balance must be struck.[26]

14.6 Pubic Trust Doctrine Stimulating Sustainable Development

Recognition of the public trust doctrine for the protection of natural environment eases the way for sustainable development. The state is a trustee of all natural resources within its territory – the seashore, lakes, rivers, underground soil, air, forests, the mountains, gardens, public parks, and other ecologically fragile lands. As the trustee, the state has the obligations to all natural resources to protect them for the sake of the present and coming generations.

The best and earliest illustration for this doctrine is seen when the Ministry of Environment had a plan to divert the natural course of a river for augmenting the facilities of the motel close by. In *MC Mehta* v. *Kamal Nath*[27] the Supreme Court set aside the plan on the ground that the public trust doctrine does not permit conversion of public use of natural resources into private ownership. The Court ruled that the executive is the trustee of the natural resources and the they cannot abdicate their duty under the public trust doctrine converting the natural resources into private property for private purposes or commercial use. In the same way, construction of an underground market beneath a garden of historical importance was halted *in MI Builders Pvt Ltd* v. *Radhey Shyam Sahu.*[28] The Court noted that the underground market would transform the garden into a terraced garden.

Fomento Resorts and Hotels Ltd v. *Minguel Martins*[29] is comparatively identical to the *Kamal Nath* case. For promoting tourism in Goa, an ancient public way to the coast used from times immemorial was planned to be acquired. The intention was to enlarge the facilities of a hotel with amenities like *yoga* centre, a health club, and water sports. The Supreme Court did not agree. They explained that the resources like air, sea, waters, and the forests are the gifts of nature freely available to everyone. The government agencies have the responsibility not to harm them in which the public has a special interest. They are held by the state as a trustee on behalf of the people, especially the future generations who are entitled to its uninterrupted use. Government agencies are obliged not to impair such resources over which public has a special interest.[30] Even private landowners or lessees and a water-right holders have the obligation to use such resources so as not to impair or diminish the people's rights.[31] The Court reiterated that natural resources including forests, water bodies, rivers, sea shores, and so forth, are held by the state as a trustee on behalf of the people, especially the future generations, and people are entitled to uninterrupted use of such resources. The state cannot transfer these resources to a private party, if such a transfer interferes with such rights. It is neither proper nor justified to deny the people of their traditional right of access to the beach.[32]

The state is prohibited from interfering with the public's rightful use of land held under the public trust doctrine. It cannot repurpose or sell such property in a manner that deviates from its traditional use or undermines the public interest. As outlined by Professor Joseph L. Sax,[33] this doctrine imposes three key restrictions on the government: public trust property must serve a public purpose and remain accessible to the public; it cannot be sold even at fair market value; and it must be preserved for specific uses – either traditional functions or those uniquely suited to the nature of the resource.

Further, the Apex Court ahead to prescribe that, as a trustee, the state can interfere and attempt to compensate the damage already done because of over-exploitation of resources. In a recent case,[34] the Supreme Court found that stubble burning in the neighbouring states of Punjab, Haryana, and Uttar Pradesh is the cause of smog in Delhi and held that the inaction of the administrative machinery of those states in preventing stubble burning makes them liable for tort in utter violation of the public trust doctrine. Despite such interventions by the Court, the states must proactively create planning to prevent over-exploitation and misuse of the resources. The doctrine of public trust does not operate in a vacuum but depends on several factors such as usage and management of the resources for past, present, and future and the public or private nature of the entity entrusted with its management. The Supreme Court found constitutional backing for the public trust doctrine in the directive principles in the *Parliament Project* case under Articles 39(b) and 48A.[35]

14.7 Rehabilitation of the Project-Affected People

A development project of national importance to generate electricity incidentally stretched out to forests in part. In the first *Banwasi Seva Ashram*[36] case, the Supreme Court endorsed the project but, highlighting the concept of sustainable development, insisted that those who were displaced had to be rehabilitated after examining their rights, however small the area from where the people are ousted. In the second *Banwasi Seva Ashram*,[37] the Court imposed more responsibilities on the National Thermal Power Corporation Limited (NTPC). It was required to find out alternative plots for rehabilitation of the displaced, render resettlement and subsistence allowance to them, ensuring free transportation, reserve jobs, and provide facilities of roads, water supply, healthcare, and electricity-related facilities.

It is a fact that development ventures may lead to displacement for many people who often become helpless refugees in their own region as living martyrs of development. However, rehabilitative proceedings may go in disarray with several disruptions and bottlenecks while screams for guaranteed rehabilitation are echoed and re-echoed everywhere. In *Narmada*, one sees a subtle recognition of the concept of sustainable development. Suitable rehabilitation with better facilities than those in their original habitat as the fundamental right of displaced humans was obviously the idea behind the majority decision in *Narmada Bachao Adolen* v. *UOI*.[38] The Court was of the view that unless the helpless situations of water scarcity are remedied, the right to life under Article 21 would be violated.[39] When the height of the Narmada dam was being raised, there is a real environment–development dilemma. By bringing water, millions of people in Gujarat and Rajasthan are benefitted while the people living on the banks of the river are displaced. Rehabilitation of the latter in a better environment is the pragmatic solution to the dilemma and a realistic application of the principle of sustainable development. In *ND Jayal* v. *UOI*,[40] the Court held that sustainable development is to be a fundamental right to justify construction of a dam as a symbol of wholesome development. According to the Supreme Court, the individual as well as the society have the right to development under the constitutional structure.

14.8 People's Livelihood Not Disturbed but Protected

In a system pursuing sustainable development, preservation of declared sanctuaries is no hindrance to the adjacent areas and to the traditional occupations available those areas. In one[41] of the *Godavarman* cases arose the question of whether to preserve the resources in the part of a lake declared as a sanctuary; other parts of lake should also be brought under the umbrella of restrictions. Holding such restrictions unnecessary, the apex Court was of the view that the fishermen living on the opposite banks could exploit other parts of the lake. Balancing the interests – between the maintenance of natural resources and protection of livelihood of the people – is the crux of the matter and fishing in the other parts of the lake does not affect the balance.

Whether diversion of forest cover could be allowed to set up a aluminium refinery project was the question in yet another *Godavarman* case.[42] The place is a wildlife habitat. Mining could affect the elephant corridor, the tribal settlements, water sources for two rivers, and the flora and fauna of the region. On the other hand, there is abject poverty among the tribal people. Finding that accelerated growth of GDP is not inclusive, the Court applied the balancing concept of sustainable development. The Court held that the permission could be given to those companies who come back after satisfying certain conditions regarding the levy of NPV and floating a special purpose vehicle with a rehabilitation package.[43]

14.9 Sustainable Travel Facilities

When the Delhi Metro Rail Corporation (DMRC) decided to extend its travel facilities, it was found that the proposed extension would go through ridges as well as the zones with nature grown trees posing further threat to the ecology. Notified as a reserved forest and acting as a buffer against rising pollution level, the Delhi Ridge area serves as the green lung for the city and should be conserved with utmost care and afforested with the indigenous species. However, the benefits of the proposed corridor could not be ignored. The Supreme Court endorsed the proposal and stated that the Delhi Preservation of Trees Act 1994 was an effective instrument of regulation to impose on the user agency to bear the cost.[44] The Government and DMRC were directed to conceive a plan of action for planting trees in the areas where clearing took place. As part of the concept of collective responsibility, everyone must be involved in the program as the Court believed that such a direction is issued having regard to the precautionary principle as well as principle of sustainable development by ensuring citizen participation[45] in the preservation of environment and ecology. When the initiative springs spontaneously from people, the participation would be more active, facilitating decisions conducive to sustainable development.[46]

Could an industry described by Central Pollution Control Board (CPCB) as hazardous be shifted out of the Union Territory of Delhi and allowed to operate in the premises of a busy airport? A picture of balancing the values between environment and development in this respect is seen in *MC Mehta* v. *UOI*.[47] For safe take-off and landing of aircrafts, runways should always be maintained in a good condition. For resurfacing the runway of the Indira Gandhi International (IGI) Airport, the industry considered it essential to locate its hot-mix plant supplying pre-mix material within the airport on several grounds. The materials are to be supplied at a very high temperature. If the plants are located at a distant place, the time taken for transporting the material would reduce the temperature and make the material useless. Moreover, the wear and tear of the runways must be repaired, and the safety of aircraft and passengers assured. The airport authority undertook that all safeguards under the law would be followed and built-in control devices of international standard maintained under the supervision of the CPCB if the hot-mix plants were allowed within the vicinity of the airport for a period of one year. The Court held that resurfacing of the airport is a work of national importance to make IGI Airport operational and to avoid any operational hazard at the time of landing or take off. The environmental problem must be balanced with the necessity of running an international airport in the capital of India.[48] The Supreme Court allowed setting up of the hot-mix plants within a safe vicinity of the airport (3 kilometres from the populated area).

14.10 Development With Sustainable Energy

Nuclear energy is an environmentally clean energy that can replace fossil fuels like coal, oil, and gas. On these premises, in *G Sundarrajan* v. *Union of India*[49] the Supreme Court rejected the challenge against setting up of the nuclear power plant at Kundankulam (KNPP) in Tamil Nadu. The Court noted that the plant had been set up on the touchstone of all national and international environmental principles, such as sustainable development, conservation principles, need for meeting climate change, inter-generational equity, precautionary principle, polluter pays principle, and preservation of bio-diversity.[50] The Court emphasized the need to balance the 'economic and scientific benefits' of nuclear energy against the 'minor radiological detriments,' stating that broader public interest should take precedence over individual concerns regarding human rights violations. It also noted that traditional power generation methods, such as hydro, thermal, and coal-based electricity, are not viable alternatives to nuclear energy.[51] As part of the policy for the development, regulation, and utilization of atomic energy for public welfare, nuclear power plays a crucial role in India's energy mix. It supports economic growth while ensuring compliance with all on-site and off-site safety standards.[52] Nuclear power generation represents a clean technology that contributes to sustainable development.

The decision in *MK Ranjitsinh* v. *UOI*[53] of 2024 reflects the concern of the apex Court on the nation's commitment in Paris agreement of 2015 for the changeover to non-fossil fuels. To protect the Great Indian Bustard (GIB), classified by IUCN as a critically endangered species, from electrocution,[54] there was an earlier direction for conversion to overhead power lines in 99,000 square kilometres to underground lines. In the present case, the Court found that this direction was an obvious violation, and delaying, of the commitment in the Paris Agreement of 2015. The area has large solar and wind energy potential. Aiming two equally crucial goals for sustainable development – conservation of GIBs and changeover to non-fossil fuels – the Court entrusted an expert committee to study and report how to achieve them effectively. The case reflects an idea different from the conventional notion of sustainable development: the interface between conservation of the critically endangered GIBs and the protective measures and commitment to meet climate challenge. While safeguarding biodiversity, the committee must consider the mitigating measures, including change-over from electricity to solar power and other renewable energy sources, to meet climate change impact.[55]

14.11 Precautionary Principle: Part of Sustainable Development

The UN Conference in Rio de Janeiro in 1992 emphasized the precautionary principle that 'where there are threats of serious or irreversible damage, lack of full scientific certainty shall not be used as a reason for postponing cost-effective measures to prevent environmental degradation.'[56] It is through judicial strategy[57] that this norm was tailored into the Indian legal system. Precautionary principle defeats persistent efforts of polluters to hide themselves behind the need for scientific certainty.

Expert studies proved that emissions from coke-based or coal-based industries in the Taj Trapezium, a defined area of 10,400 square kilometres around Taj Mahal for its protection, had a damaging effect on the monument. Saying that not even 1% chance can be taken against its preservation, in *M C Mehta* v. *UOI*[58] the apex Court applied precautionary principle to protect the Taj Mahal from air pollution. The onus lies with industries to demonstrate that their operations using coke or coal do not harm the environment.[59] In a subsequent case,[60] refineries in Mathura were found to be emitting excessively high levels of sulphur dioxide per hour, posing a health risk to people and adversely affecting the Taj Mahal. The Court issued notices to enforce its orders and initiated contempt proceedings.

For protecting coastal ecology in a sustainable manner, in *S Jagannath* v. *UOI*[61] the Supreme Court had no hesitation to apply precautionary principle and issue unambiguous direction to wind up all intensive, semi-intensive, and extensive aquaculture farms. As said early, in *Hospitality Association of Mudumalai* v. *In Defense of Environment and Animals*[62] the apex Court underlined that the precautionary principle has been accepted as a part of the law of our land.

14.12 Revival of Sick Industries

Is the synthesis between protection of environment and rehabilitation of sick industries a manifestation of sustainable development in a fast-developing economy? Strict compliance to the ratio of an open space is essential in the front of residential colonies and is a legal requirement in industrial establishments. The open space is the lung space for both. In *Bombay Dyeing & Mfg. Co.* v. *Bombay Environment Action Group*[63] reduction of the rigor of open space is justified on the ground that rehabilitation of sick industries is a necessary evil in a fast-developing economy. A distinction is drawn between laws on town planning and those on rehabilitation of sick industries. In the former, ecological factors are indisputably a relevant consideration. But relaxation of open space must be looked at from a different perspective of balancing different interests with rehabilitation of industries. The Court explained that the main purpose was the revival of industry *inter alia* by modernization and shifting and that the regulation *ex facie* does not impair sustainable development.[64] In this connection, the Court distinguished *Bangalore Medical*

Trust v. *BS Muddappa*,[65] which dealt with the conservation of open spaces for parks and playgrounds. Providing open space was held universally recognized as a legitimate exercise of statutory power in relation to the protection of the residents of crowded carbon areas. But in *Muddappa* the significant question was whether discretion vested in a statutory authority had been correctly exercised for the purpose of which the power was given. Distinct from *Muddappa*, the area affected by the change in *Bombay Dyeing* was very small and the change would not alter the basic character of development plans. Unlike in *Muddappa* there was no abuse of discretion in *Bombay Dyeing*.[66] The regulation can be struck down only when the state actions are found to be unfair.[67] The Court explained that the main purpose was revival of industry inter alia by modernization and shifting[68] and held that the regulation *ex facie* does not impair sustainable development of the city of Bombay.[69]

14.13 Special Economic Zones

A smaller area may be enough to accommodate information technology parks. There may not be any significant adverse impact on the environment if the facility is started on lands reserved for agriculture, grazing, or residences. In *Karnataka Industrial Area Development Board* v. *Kenchappa*,[70] the Court observed that the rule requiring a green belt may not be insisted upon around the location of an IT industry. It reasoned that if green belts were treated as an inevitable requirement in all cases, it might become impossible to acquire land for development. The Court is of the view that before acquisition of land for development the consequences and adverse impact must be properly addressed so that acquisition of land does not gravely impair the ecology and environment. However, in *Mantri Technose Private Ltd* v. *Forward Foundation*[71] the Supreme Court did not allow Bangalore Development Authority to implement their projects for a software technology park and a commercial and a residential complex in the buffer zones from water bodies and water drains. The Court said"

> The buffer/green zones should be treated as no construction zones to all intents and purposes. This is absolutely essential for the purpose of sustainable development particularly keeping in mind the ecology and environment of the areas in question.[72]

No doubt, buffer zones should be no-construction zones with a view to achieving sustainable development.

14.14 Inter-Generational and Intra-Generational Equity

Conservation of nature is a moral obligation of humanity to other living beings of the present and future generations. Essentially, sustainable development signifies both intragenerational equity and intergenerational equity.

Unlimited exploitation of natural resources results in their sudden exhaustion, leaving no chance for the coming generation to make use of them. Manifestly, this is the reason why it is said that the present generation shall exploit them in a sustainable manner without destroying their resource base. Indiscriminate and haphazard mining operations in Odisha were held in *Common Cause* v. *UOI*[73] to be contrary to this universally accepted principle. The apex Court found it necessary to impose controls on mining, have a fresh and effective mining policy, and thus to fulfil the need to maintain intergenerational equity. To limit the extent of mining in the state and to maintain intergenerational equity, three principles of conservation were highlighted.[74] The first one is to conserve the diversity of the natural and cultural resource base in such a manner that the options available to future generations are not restricted (conservation of options). Next is not to subject future generations to a quality of the planet worse as they should enjoy the quality of the diverse natural and cultural resource base (conservation of quality). The third is to assure future generations to have an equitable right of access to the diversity of the natural and cultural resource base as at present (conservation of access).

14.15 Sustainable Development: Not an Empty Slogan

The purpose of a development activity is relevant for determining its validity. This is especially so when the government is about to adopt a policy of sustainable development of the habitat. The need for shelter and related infrastructure facilities was keenly felt in situations of rapid industrialization. Alienation of two tanks for housing colonies in Tirupati town is not for public purpose or for a traditional use; the Court was more concerned with equitable supply of land, shelter, and services at affordable prices that promote sustainable development of habitat. In *Intellectual Forum, Tirupati* v. *State of Andhra Pradesh*,[75] the Court did not countenance the plea for complete restoration and revival of the two tanks but found it difficult to prevent the government from building housing complex for the middle-class people, infrastructure for temple administration, and development of the town.[76]

The facts in *Susetha* v. *State of Tamil Nadu*[77] pertain to a temple tank lying in disuse, and being used as a dumping ground and sewage collection for a long time. Should they be restored? Should they be filled up and levelled completely to construct a shopping complex for rehabilitation of outsees of a highway project? Before deliberating on the questions, the Supreme Court made it clear that the doctrine of sustainable development is not an empty slogan, but must be viewed in a pragmatic manner.[78] In the case, the respondent did not dispute categorically that there was shortage of water in the region. Different from the other five existing water tanks in the area, the one in question was not a natural tank. Further, although the exact date is not known, the tank had been a dumping ground and area for sewage disposal in the past. The case is not one for the resurrection of the pond but rehabilitation of the people displaced for a development project. The Court is of the view that the state and the *gram panchayat* have a duty to take care of the other tanks in and around the village to be properly maintained.[79] On this context, one may look at the majority view in *Narmada Bachao Andolan* that probed how the project could strike a balance between developmental needs and environmental values affording less importance to the problems of submersion. They bestowed full faith in the anticipated rehabilitation, and found that the project had in-built safeguards that satisfied the goal of sustainable development.[80] On the contrary, in *ND Jayal,* Court stressed the need for a shift from 'a relief centered re-active approach to the preparedness oriented proactive attitude' while they highlighted that rehabilitation before submergence would create a sense of confidence among those displaced by acclimatizing themselves in a new environment.[81] It was categorically declared, 'Without completion of rehabilitation there shall not be any impoundment.'[82] As proclaimed in *Susetha* v. *State of Tamil Nadu*,[83] sustainable development is not an empty slogan but should be a realizable goal.

14.16 Eco-Sensitive Zones From Protected Areas

With a wildlife strategy issued on January 21, 2002, the initiative to declare eco-sensitive zones (ESZs) around the boundaries of protected areas (PAs)[84] has now been underway for more than two decades. MoEF & CC guidelines on 2011 for was for 10-kilometre ESZs.[85] Later, finding little follow-up by states, ESZ was reduced to minimum 1 kilometre measured from the demarcated boundary of such protected forest.[86]

The Apex Court found[87] it impossible to implement even the 1-kilometre rule for certain ESZs. The rule affects many people residing around the protected areas. A villager desirous to reconstruct his house or to make additional construction for accommodating the enlarged family would not get permission.[88] Even the Government cannot construct schools, dispensaries, *anganwadis*, village stores, water tanks, roads, and other basic structures and facilities for improvement of the life of the villagers. No permanent structure would be permitted to come up for whatsoever purpose within the aforesaid ESZs. The Forest Departments may find it difficult to conduct eco-development activities around PAs with the dual objectives of protection of wildlife and benefits for the local communities.[89] Various regulated and permissible activities such as national highways, railways, and defence-related infrastructure will be permanently prohibited by the impugned direction. The direction will intensify man–animal conflict instead of avoiding it.[90] The prime object of ESZs is to protect the precious forests and PAs from any negative impact and to refine the environment around but not to hamper day-to-day activities of the citizens.

Considering that the salient features of ESZ regulations evolved through decades, the Court went to the development-friendly dimensions of the concept in its true and pragmatic perspective. It is the Court's firm view that the order dated June 3, 2022, fixing the minimum 1-kilometre ESZs, is unkind and impossible to be implemented. Spontaneous natural equilibrium of the development of the region can never be disturbed by imposition of irrelevant rules though they seemed to be environment savvy.

14.17 Concluding Remarks

In a fast-developing country there are ever-increasing flows of environmental issues. The legislature, which initiates legislation, and the administration, which makes and executes the decisions, are to be ever vigilant in reconciling the conflict between the needs of development and the values of environment. These balancing functions may conflict with the right to equality, freedom of speech and expression, freedom of religion, property rights and freedom of inter-state trade and commerce, and even the need to maintain financial discipline and to share the losses and gains. The conflict attains new dimensions in the modern technological state. Legal strategies are necessary to reconcile this conflict and to augment sustainable development.

Once the law and judicial decisions take in eco-centric ethics, rather than an anthropocentric approach, the symbiotic relationship of man with other living creatures and plants becomes a dominant concern in sustainable development. It is categorically clear that an impact study with public participation is sine quo non for ascertaining a development venture as sustainable. The panorama of judicial decisions championing the cause of sustainable development presents a vista of reconciling the numerous conflicts between developmental activities and ecological values. The process realizes the balancing functions envisaged in the definition of environment in EPA. Sustainable development visualizes economic growth without destroying the resource base in an ecosystem. As stated earlier, sustainable development 'seeks to meet the needs and aspirations of the people of the present without compromising the ability to meet those of the future.'[91]

Notes

* Professor Emeritus, Cochin University School of Legal Studies, Kochi 682022.
1 Declaration of the UN Conference on Human Environment, 1972, Paragraph 3.
2 See Centre for Science and Environment, *Climate Change: Politics and Facts*, 2009, 3 (New Delhi); Christopher Flavin, "Global Temperature Rises", in Lester Brown, et al. (eds), *Vital Signs 1992*, World Watch Institute, 58 (Washington DC, 1992).
3 Gro Harlem Brundtland, *Our Common Future* (1979), p. 40. Oxford University Press, 1987.
4 Principles 6–10, 12–14, 16–19 and 20–27.
5 Principles 5, 16, 24, 25 and 27.
6 Section 2(a) defines environment as it 'includes water, air and land and the inter-relationship which exists among and between water, air and land, and human beings, other living creatures, plants, micro-organism and property.'
7 (2001) 6 SCC 496, at page p. 500.

8 *State of Kerala* v. *Gwalior Rayon Silk Manufacturing Co.*, AIR 1973 SC 2734.
9 *State of Tamil Nadu* v. *M/s Hind Stone and others 1981 (2)* SCC 205, pp 212, 213.
10 AIR 1987 SC 1073.
11 AIR 2005 SC 4256. The NPV is calculated on the basis of environmental values. It may include the calculations relating to benefits from the forest such as such as flood control, soil conservation, ground water conservation, recreational value of the forest, bio-diversity protection and allied benefits of biodiversity preservation of habitat, costs of air purification by the forests.
12 *TN Godavarman Thirumulpad (87)* v. *Union of India* (2006) 1 SCC 1.
13 Such as biological diversity; soil and water resources; productive capacity of forest eco-systems with health and vitality; contribution to global carbon cycles as sinks; long-term multiple socio-economic, cultural, recreational, spiritual benefits to meet the needs of people; and legal, institutional, and economic frameworks for forest conservation and sustainable development, *TN Godavarman Thirumulpad (87)* v. *Union of India*, above note 13, pp. 22, 23.
14 *TN Godavarman Thirumulpad* v. *UOI* (2010) 6 SCC 710. The court permitted CEC to entrust its State Level Committee to consider applications of another aspirant after assessing the state requirements for Khair wood. Being sealed, the industry of this applicant could be lifted, if necessary; p. 711.
15 *Chief Conservator of Forest Wild Life* v. *Nisar Khan* AIR 2003 SC 187.
16 *TN Godavarman Thirumulpad* v. *UOI* (2012) 3 SCC 277.
17 *TN Godavarman Thirumulpad* v. *UOI* (2012) 4 SCC 362. pp. 377, 378.
18 (2020) 10 SCC 589.
19 Later, in *Binay Kumar Dalei* v. *State of Odisha* [(2022) 3 SCC 33] the Supreme Court asked to speed up declaration of the traditional elephant corridor as a conservation reserve, expeditiously noting that no mining activity could be permitted even in the vicinity of an eco-sensitive zone.
20 AIR 1985 SC 652 and AIR 1988 SC 2187.
21 *Goa Foundation* v. *UOI* (2014) 6 SCC 590, p. 633. The Court asked CEC to frame a comprehensive scheme and submit to the court within 6 months. The people directly employed in mining had gone up to one and a half lakh; their livelihood would be seriously affected if there was total ban.
22 *MC Mehta* v. *UOI*, AIR 2004 SC 4016.
23 *MC Mehta* v. *Union of India* (2009) 6 SCC 142.
24 *MC Mehta* v. *Union of India*, above note 23, at p. 159.
25 AIR 2004 SC 4016.
26 AIR 2004 SC 4016, at para. 47.
27 (1997) 1 SCC 388.
28 *MI Builders Pvt Ltd* v. *Radhey Shyam Sahu,* AIR 1999 SC 2468.
29 *Fomento Resorts and Hotels Ltd* v. *Minguel Martins,* (2009) 3 SCC 571, pp. 600, 601, 614 and 615.
30 *Fomento Resorts and Hotels Ltd*, above note 29, at pp. 614, 615.
31 *Fomento Resorts and Hotels Ltd*, above note 29, at p. 615.
32 *Fomento Resorts and Hotels Ltd*, above note 29, at pp. 620, 623, 627. It was urged that the facilities provided by the hotel will be open for use by non-residents also on a membership basis. The agreement is totally silent on the issue. Therefore, it is not possible to say that the facilities and amenities should be made available to the public free of costs. The court ordered to demolish the extended portion of the hotel building within 3 months.
33 Joseph L. Sax, "The Public Trust Doctrine in Natural Resource Law: Effective Judicial Intervention", *Michigan Law Review*, 1970, pp. 471–566.
34 *MC Mehta* v. *UOI* (2020) 7 SCC 571, pp. 575, 576.
35 *Rajeev Suri* v. *Delhi Development Authority*, decided on 05-Jan-2021, pp. 401 and 402.

36 *Banwasi Seva Ashram* v. *State of UP*, AIR 1987 SC 374, pp. 376–378. The displaced individuals from reserved forests were granted the right to seek legal recourse and establish their claims through appropriate proceedings. Further, those residing in areas not designated as reserved forests were entitled to a detailed hearing conducted by record officers and additional judges. The Court also mandated the provision of legal aid to all affected persons and appointed a board of commissioners to supervise the implementation of its orders.
37 (1992) 2 SCC 202, pp. 204–206.
38 AIR 2000 SC 3751, pp. 3787, 3804.
39 *Narmada Bachao Andolan* v. *UOI*, AIR 2000 SC 3751, p. 3830.
40 *ND Jayal* v. *UOI*, AIR 2004 SC 867.
41 *TN Godavarman Thirumulpad* v. *UOI* (2006) 5 SCC 47, p. 56.
42 *TN Godavarman Thirumulpad (104)* v. *Union of India* (2008) SCC 222. The court was categorical in not giving clearance to a company whose reputation overseas was in doubt. CEC objected to grant clearance.
43 (2008) SCC 222, pp. 225, 226.
44 *T.N. Godavarman Thirumulpad* v. *Union of India* (2022) 4 SCC 289 pp. 308, 309. In the case, DMRC is of opinion that the corridor reduces vehicular traffic drastically, serves the densely populated areas, and integrates the metro connectivity of South Delhi to Gurgaon and Faridabad. The Aerocity–Tughlakabad metro corridor is predicted to provide several benefits and is considered the most feasible route economically, environmentally, and technically.
45 *T.N. Godavarman Thirumulpad*, above note 44, at p. 316. Students from schools and colleges, along with non-governmental organizations, citizen welfare groups, public servants, and environmentally conscious individuals, are encouraged to plant trees and take responsibility for their care. This collective effort aims to enhance greenery in the National Capital Territory of Delhi, ultimately contributing to afforestation over time.
46 *T.N. Godavarman Thirumulpad*, above note 44. The Ridge Management Board (RMB) is responsible for ensuring that the ridge remains free from unauthorized occupation or encroachment. Given these considerations – such as the necessity for prior approval under the Forest Conservation Act (FCA), recommendations from the RMB, and support from the Central Empowered Committee (CEC) – the Court approved the construction of the corridor along the ridges, aligning with the principles of sustainable development; pp. 314, 315.
47 *MC Mehta* v. *UOI*, AIR 1999 SC 2367.
48 *MC Mehta* v. *UOI*, AIR 1999 SC 2367, p. 2373.
49 *G Sundarrajan* v. *Union of India* (2013) 6 SCC 620.
50 *G Sundarrajan* v. *Union of India*, above note 49, at 711–714. The court examined various United Nations initiatives, particularly related to energy, food security, oceans, and urban development, with a focus on promoting a green economy. These efforts were assessed in the broader context of sustainable development, poverty eradication, and establishment of an institutional framework for long-term sustainability.
51 *G Sundarrajan* v. *Union of India*, above note 49, at p. 718.
52 *G Sundarrajan* v. *Union of India*, above note 49, at p. 720.
53 *MK Ranjitsinh* v. *UOI*. W.P. (Civil) No 838 of 2019 dated 21st March, 2024.
54 Among the causes for the dwindling numbers of GIBs, slow destruction was partly attributed to overhead transmission lines. *MK Ranjitsinh* v. *UOI*, above note 53, paragraph 3.
55 *MK Ranjitsinh* v. *UOI*, above note 53, at paragraph 54.
56 Principle 15, the UN Rio declaration 1992.
57 *Indian Council for Enviro-Legal Action* v. *UOI*, AIR 1996 SC 1446, at 1468. The Supreme Court mandated an assessment of environmental and ecological damage

and held the polluters accountable for compensating the harm caused; *Vellore Citizens Welfare Forum* v. *UOI*, AIR 1996 SC 2715 at 2721. All tanneries in specific districts that were not linked to Common Effluent Treatment Plants (CETPs) were ordered to shut down.

58 *MC Mehta* v. *UOI*, AIR 1997 SC 734 p. 761.

59 *MC Mehta* v. *UOI*, above note 58, at p. 762. The Court ruled that industries identified by the Central Pollution Control Board as potential polluters must transition to natural gas as an industrial fuel. Those unable to secure gas connections were required to cease operations within the Taj Trapezium Zone and relocate to designated alternative sites outside the demarcated area within the stipulated timeframe.

60 *MC Mehta* v. *UOI*, 2012 8 SCC 123. The planning authorities also were asked to engage professional planners and sanitation experts to beautify and clean Agra city which is directly connected to the Taj Mahal. About the Agra bypass, it is interesting to note the observation of the Court that instead of large number of trees being cut to construct *kachcha* road, it should be constructed in places where there are no trees; pp. 124–127.

61 *S Jagannath* v. *UOI* (1997) 2 SCC 87. The questions on aquaculture and CRZ are discussed by the court on pp. 103–106 and pp. 140–150.

62 *Hospitality Association of Mudumalai* v. *In Defence of Environment and Animals*, AIR Online 2020 SC 776: (2020) 10 SCC 589, pp. 607, 608 (2021) 1 SCJ 252.

63 AIR 2006 SC 1489.

64 AIR 2006 SC 1489, at pp. 1534 and 1554.

65 (1991) 4 SCC 54.

66 *Bombay Dyeing & Mfg Co Ltd* v. *Bombay Environmental Group* AIR 2006 SC 1489, pp. 1534–1536.

67 *Bombay Dyeing & Mfg Co Ltd*, above note 66, at p. 1531.

68 *Bombay Dyeing & Mfg Co Ltd*, above note 66, at p. 1539.

69 *Bombay Dyeing & Mfg Co Ltd*, above note 66, at p.1554.

70 AIR 2006 SC 2038, p. 2049. The discussion revolved around the environment provisions in the Constitution, the other doctrines evolved by courts, and the doctrine of sustainable development interspersed with extensive quotations from prior decisions, citation of international instruments, and references to academic writings, pp. 2041–2049.

71 (2019) 18 SCC 494.

72 (2019) 18 SCC 494, at p. 523.

73 *Common Cause* v. *UOI*, (2017) 9 SCC 499, pp. 574, 578.

74 *Common Cause* v. *UOI*, above note 73, at p. 573.

75 AIR 2006 SC 1350, pp. 1364, 1365.

76 AIR 2006 SC 1350, above note 75 at p. 1366, 1367. In conformity with the report of the expert committee, the court issued certain significant directions such as prevention of further development, prohibition of bore wells, construction of percolation tanks for groundwater recharge, and provisions of roof top rain water harvesting in buildings already constructed.

77 AIR 2006 SC 2893.

78 AIR 2006 SC 2893, above note 77 at p. 2896.

79 *Susetha* v. *State of Tamil Nadu* AIR 2006 SC 2893, p. 2897.

80 *Narmada Bachao Andolan* v. *Union of India* AIR 2000 SC 3751, p. 3820.

81 *Narmada Bachao Andolan*, above note 80. The court observed: 'Our present relief centred re-active approach after the striking of the disaster need to be changed into preparedness oriented pro-active attitude.'

82 AIR 2004 SC 867, pp. 885, 886 (Supp).

83 AIR 2006 SC 2893.

84 The Wildlife Protection Act 1972. S 24-A says that protected area means a National Park, a sanctuary, a conservation reserve, or a community reserve notified under ss. 18, 35,36-A and 36-C of the Act.
85 With activities regulated rather than prohibited, the idea was that they act as shock absorbers of PAs. The rationale was that 'the areas outside the protected area network are often vital ecological corridor links and must be protected to prevent isolation of fragments of biodiversity, which will not survive in the long run.'
86 *T.N. Godavarman Thirumulpad* v. *UOI*, an order dated June 3, 2022 *per* Aniruddha Bose, J., paragraph 44.
87 *T.N. Godavarman Thirumulpad* v. *Union of India*, 2022 SCC On Line SC, Decided on April 26, 2023.
88 *T.N. Godavarman Thirumulpad*, above note 87, at Paragraph 48.
89 *T.N. Godavarman Thirumulpad*, above note 87, at Paragraphs 49 and 50. The activities often involve construction of small structures which are permanent in nature in the areas including ESZs.
90 *T.N. Godavarman Thirumulpad*, above note 87, at Paragraph 51.
91 Brundtland, above note 3.

15 Climate Change and Judicial Trends in India

Narender Bishnoi

15.1 Introduction

Climate change is the most critical challenge of our era, and we stand at a pivotal juncture. Its effects are far-reaching and unparalleled, from altering weather patterns that jeopardize food security to rising sea levels that heighten the danger of devastating floods on a global scale and include the risk of extinction of many rare species. Without drastic action today, adapting to these impacts in the future will be more difficult and costly. The Intergovernmental Panel on Climate Change (IPCC), in its Fifth Assessment Report, categorically concluded that "climate change is real and human activities are the main cause."[1]

IPCC, scientifically defined the term "climate" as "average weather, of a geographical area or more strictly, as the statistical account of mean and variability of relevant quantities such as temperature, air pressure, humidity, precipitation and wind, over a period ranging from months to thousands or millions of years."[2]

Regardless of the cause, the term "climate change," in general, refers to "long-term shifts and variability which are observed in the temperature and weather pattern over a geographical region."[3]

The climate change crisis is a complex interplay of environmental, social, economic, and political factors. Increasing global temperatures, fuelled by excessive greenhouse gas emissions, trigger a chain reaction of devastating events. These impacts are widespread, affecting nature and human societies alike.

Melting polar ice caps lead to rising sea levels, threatening coastal areas, with extreme weather events like hurricanes, droughts, and wildfires intensify. In addition to environmental shifts, climate change presents significant challenges to human health, food security, and economic stability. Vulnerable populations, often residing in low-lying or arid regions, bear the greatest burden, facing heightened risks of displacement, water shortages, and disease outbreaks. Furthermore, biodiversity is under unprecedented threat, resulting in the extinction of numerous plant and animal species.

DOI: 10.4324/9781003733751-23

In the face of these challenges, understanding the global impacts of climate change is not merely a scientific pursuit but a moral imperative and a call to action. To deal with the complicated effects, we need to be a global community that works together, finds creative solutions, and makes a promise to adopt habits that are not harmful to the environment. It is important to have comprehensive plans to lessen the effects of harmful acts and make way for a future where our world and its people can handle and even thrive in the face of these problems.

The role of judiciary is pivotal in these circumstances. As lapse in policy framework to tackle the menace of climate change cannot be dealt appropriately without oversight of judiciary. Judicial systems across, including India, have undertaken measures to confront climate-related issues, recognizing the seriousness of the matter. The courts, as protectors of fundamental rights, have unequivocally said that they will not let economic advancement to undermine environmental preservation. The Indian judiciary has proactively safeguarded the right to a healthy and clean environment via the interpretation of constitutional provisions and environmental legislation.

Justice Venkatachaliah said in the case of *State of Bihar* v. *Murad Ali Khan*[4]: "The tragedy of the civilised man's predicament is that every source from which humanity has augmented its power on earth has been employed to undermine the prospects of future generations." He is achieving significant progress; nonetheless, the atmosphere is deteriorating, and he remains unaware of how to rectify the situation or what the future entails.

J. Kuldip Singh said in *M.C. Mehta* v. *Kamal Nath*[5] that the government, according to the idea of public trust, cannot relinquish natural resources for private ownership or commercial exploitation. The beauty and integrity of our nation's ecosystems, natural resources, and environment must not be compromised for private, commercial, or any other purposes until deemed essential by the courts for the public good and interest to infringe upon these resources.

Most recently the Indian Supreme Court's decision in *M.K. Ranjitsinh & Ors.* v. *Union of India & Ors*[6] affirms that Article 21 of the Indian Constitution ensures the right to be free from the detrimental effects of climate change. The Court said that climate change jeopardizes human dignity and the right to life, urging the government to implement precautionary measures.

The Supreme Court of India added the Polluter Pays Principle and the Precautionary Principle to Indian environmental law in the case of *Vellore Citizens Welfare Forum* v. *Union of India*.[7] These ideas have been very helpful in stopping the environment from getting worse and making sure that growth doesn't upset the balance of nature. But is it enough?

15.2 Sacred Harmony With Nature

Many religions and philosophies around the world have pushed for respect and protection of the natural environment. This is at the heart of the idea of intergenerational equity. The Srimad Bhagavata Mahapurana articulates that "ether, air, fire, water, earth, planets, all creatures, trees and plants, rivers and seas are organs of God's body; thus, a devotee reveres all species."[8] This also suggests that the natural world is a manifestation of the Divine. The Earth is revered as a sacred creation of God in all other Indian faiths. All aspects of Mother Nature are venerated as sacred.

> सहयज्ञाः प्रजाः सृष्ट्वा पुरोवाच प्रजापतः।
> अनेन प्रसतवष्यध्वमेष वोऽस्त्विष्टकामधुक्॥
> Bhagwat Gita (Chapter 3, Verse 10)
>
> In the beginning, the Creator created humans and assigned them yajña (sacrificial responsibilities), saying, "May you prosper; let this be the milker of your desires."

This text emphasizes the essential idea that individuals have the obligation to behave in harmony with nature. The yajña represents all actions that sustain natural cycles and resources, including rituals. The stanza emphasizes the need of appreciating and caring for nature, implying that human needs can only be addressed when people and their natural environment are in harmony. It promotes more responsible resource usage and aligns human and environmental responsibilities.

Despite religious dogmas equating nature to God seeking to promote the welfare of the environment, humans have failed to take due care of the environment. In fact, climate change resulting from environmental harm by humans poses a significant hazard to human life in contemporary times. Climate change delineates the incremental alterations in the global climate; it is mostly induced by human activities such as the combustion of fossil fuels, deforestation, and the emission of pollutants into the atmosphere. The escalation of water and food shortages, the heightened frequency and intensity of extreme weather events, a reduction in biodiversity, and several other worldwide repercussions are undeniably tangible.

In developing countries such as India, these consequences severely impact the most vulnerable populations. We must realize that we can no longer use the earth's resources in the same manner. Instead of disregarding the increasing ecological awareness, we need to actively facilitate its development into concrete policies and initiatives.

But, despite the dire need and, by and large, everyone realizing the need for a healthy environment, we have failed to get the desired results. The courts also promote environmental protection. The courts, from time to time, pass dictum to protect the environment and various species. However, the courts stop one step short of the most essential thing for environmental protection, that is choose environment over development. The development needs to be balanced with the need to protect the environment. No doubt development is important. But this blind faith in development and its benefits is causing harm to the inalienable and most basic need for survival of various species. We need prosperity but not at the cost of environment.

15.3 The History of Environmental Protection

Indian law, which comes from ancient times and the middle ages, stresses the importance of taking care of the earth. Many parts of the Vedas, Upanishads, Puranas, and other old Indian books talk about how holy nature is. As an example, the Atharva Veda honours Earth as a mother figure to show how people and the world are linked: "Mata Bhumih Putro Aham Prithivyah" (Earth is my mother, I am her son).[9] It is very important for everyone to take care of Mother Earth, as this line says. The Indian epics, the Mahabharata and the Ramayana, talk a lot about how important it is to take care of the earth's natural resources.

In old Indian law systems, like those found in the *Manusmriti* and *Arthashastra*, the idea of Dharma included taking care of the earth. The *Manusmriti*, which is an ancient Indian law book, has fines for polluting water supplies and hurting trees.[10] This shows that people knew early on how important it was to protect natural resources. To stress how important it is to keep natural balance, it includes specific steps to do so, like fines for cutting down trees or killing animals for no reason. The book talks about penalties for damaging the environment and stresses how important it is to use natural resources in a way that doesn't harm them.[11]

Kautilya wrote the ancient economic and political book titled *Arthashastra*, which has a lot of rules for taking care of woods, animals, and other natural resources. There must be forest areas, or Vanas, and clear roles and responsibilities for forest officials, or Vanikadhyakshas. These officials were in charge of keeping woods safe, making sure that resources were used in a way that didn't harm the environment, and stopping illegal activities like cutting down trees and hunting. The *Arthashastra* says that good governance is important for protecting natural balance, and the state's role is very important in this.

In mediaeval India, environmental problems were very important to both government and everyday life. For example, during the Mughal Empire, gardens and safe areas were created to show how much the Mughal kings, who came from desert areas, loved greenery and water bodies. Realizing how important woods and bodies of water were to the Mughal Empire's survival, rulers like Akbar made rules to protect them.

Indian moral lessons and Dharmic rules from ancient and mediaeval times show a strong commitment to protecting the environment. These customs teach us important lessons about how to live in balance with nature, how all living things depend on each other, and how we should do everything we can to protect it.

Indian independence from British colonial rule in 1947 was accompanied by a renewed focus on ending extreme poverty, fostering industrialization and economic growth, and reclaiming the country's "rightful place" in the global community.[12] This hindered the prospects of any climate protection–specific legislation. Climate-specific legislation could have a negative impact on the country's energy consumption, economic growth, and development, which were seen as fundamental national interests. This has serious ramification for the current legislative framework in the country.

India currently lacks a comprehensive climate change–specific legislation. Existing laws, such as the Environment Protection Act of 1986, the Air (Prevention and Control of Pollution) Act of 1981, and the National Action Plan on Climate Change (NAPCC), address certain aspects of climate change but fall short of providing an integrated and enforceable framework necessary to meet the multifaceted challenges of climate change. The absence of binding domestic legislation leads to gaps in enforcement and accountability, undermining the effectiveness of climate action. Comprehensive climate laws are crucial for setting clear targets, enforcing compliance, and ensuring accountability in addressing climate change.

Indian courts have played a proactive role in environmental jurisprudence, recognizing the right to a healthy environment as a fundamental right under Article 21 of the Constitution. Cases such as *Ridhima Pandey* v. *Union of India*, *M.K. Ranjitsinh & others* v. *Union of India*, and others have highlighted the judiciary's role in addressing climate change-related issues. However, the judiciary's capacity to enforce climate policies is limited without specific climate legislation. The courts often struggle with the specificity required for climate-related relief, and the absence of dedicated climate laws limits their effectiveness in enforcing comprehensive climate action.

15.4 Constitution and the Nature

The Constitution of India, which can be accessed in several parts, provides a strong legal framework that can be used to deal with emerging environmental problems. It is very important to look at the parts of India's constitution that deal with climate change and how they might affect the country's environmental planning. Some of them are as follows:

Article 21 of the Indian Constitution protects the right to life and personal freedom. It says, "No one shall be deprived of his life or personal liberty except in a way set out by law." This rule is an important part of both the right to life and the right to a healthy environment.[13] The Supreme Court of India has interpreted this Article to include the right to a healthy environment, as seen in landmark cases like *Subhash Kumar* v. *State of Bihar*[14] and *Virendra Gaur* v. *State of Haryana*.[15] This interpretation establishes a direct link between a clean environment and the right to life, providing a constitutional basis for environmental protection and climate action.

Global warming endangers human rights and the environment. Climate change causes extreme weather, greater sea levels, and changing precipitation patterns, which threaten people's livelihoods. For instance, many vulnerable communities in India's coastal areas rely on fishing. Thus, increasing sea levels and ocean acidification endanger their economic stability and cultural identity. The unpredictable patterns of the monsoon also threaten agricultural output, which in turn causes millions of people to be hungry and violates their right to food.[16]

Article 21 of the Indian Constitution serves as a pivotal cornerstone in the protection of human rights to health, sustenance, access to clean water, and viable livelihoods amidst looming climate change and its attendant perils. The landmark judgment of the Supreme Court has highlighted the state's responsibility to combat climate change and its effects by recognizing the interdependence of human rights, sustainable development, and environmental preservation.[17]

Equality before the law: Article 14 ensures equality before the law and equal protection of the laws within the territory of India. This provision is crucial in the context of climate change, as it mandates non-discrimination and equitable treatment. Climate change disproportionately impacts marginalized communities, exacerbating existing inequalities. Article 14 can be invoked to ensure that climate policies are inclusive and just, addressing the needs of vulnerable populations and promoting social equity.[18]

Directive Principles of State Policy: Articles 48A and 51A: The Directive Principles of State Policy, though not justiciable, provide important guidelines for the state in formulating policies. Article 48A of the Indian Constitution mandates the State to endeavour to protect and enhance the environment, as well as safeguard forests and wildlife. This obligation emphasizes the active involvement of the government in environmental conservation, which is essential for effective climate mitigation and adaptation strategies. Article 51A (g) imposes a duty on every citizen of India to protect and improve the natural environment, including forests, lakes, rivers, and wildlife, and to show compassion for living creatures. This Article underscores the responsibility of environmental stewardship as a fundamental duty of all citizens, promoting a culture of environmental awareness and action at the grassroots level.

Thus, a clear national commitment to environmental protection and improvement is spelled out in the Fundamental Duties and Directive Principles of State Policy chapters. However, these articles cannot have any impact in isolation without any remedial provisions and judicial wisdom. In India, environmental conservation has been advanced by judicial activism, primarily in reaction to Article 32 of the Indian Constitution, known as the "Right to Constitutional Remedies."[19] This section empowers anyone to file a petition before the Supreme Court to safeguard their fundamental rights, including matters related to the environment and climate change. Public interest litigations enable individuals to advocate for enhanced environmental safeguards and to hold authorities accountable for ecological harm.

The Indian judiciary has significantly impacted environmental legislation, adopting a robust position on ecological matters over the years. The creation of a framework in which judges provide directions to efficiently enforce environmental laws and regulations is a direct consequence of this judicial activism. This technique, sometimes referred to as "judge-driven implementation," has been essential for addressing environmental issues.

Public interest litigations have gained significance as an effective mechanism for citizens to contest ecologically detrimental practices and advocate for enhanced environmental protections. To advance environmental justice and hold those who violate environmental rights accountable, people and environmental groups may use Article 32 to submit complaints. Article 32 effectively establishes a link between environmental jurisprudence and judicial activism, which is vital for advancing environmental sustainability and combating climate change in India

15.5 Environmental Jurisprudence in India

Historically, the Indian judiciary has considered it essential to force the Legislature to act by filling the legal void via deliberate rulings. The 1980s were the first decade in which it took the initiative. Several ground-breaking decisions made by Indian courts since then and have had a significant impact on the growth of environmental jurisprudence in India. Locus standi is necessary to commence legal procedures. In addition, a justiciable matter is one that can be decided by a court. Because of this criterion, it was difficult to assert a public interest or right since the aggrieved party had to prove that their harm was more severe than that experienced by the wider public.

Consequently, it was challenging to address ailments that are widespread in nature, such as air pollution, which affects a large community. The disadvantaged people of India also suffered under the antique locus standi concept, which barred any concerned citizen from bringing a legal case on their behalf. Socio-economically deprived people still don't know their environmental rights or are afraid the dominant class will punish them socially or economically if they do. This is due to a combination of factors, including their lack of education and resources.

With the advent of public interest litigation (PIL), India's locus standi was liberalized. PIL enables any public-spirited individual or institution to seek judicial redress in the public interest by bringing writs under Articles 32 and 226 of the Constitution, respectively. This is done in the event of a violation of the fundamental rights of a poor or disadvantaged class that cannot access the court due to poverty or disability. Judicial decisions in PILs bind not just the parties to the litigation but also all persons similarly situated, in contrast to the previous paradigm where a ruling was only binding on the parties (res judicata) and not on individuals.

The Courts were able to address numerous social desires through PILs, which was a response to the lack of political will on specific concerns.

A more activist stance was adopted by the Indian judiciary in resolving matters brought before it through this jurisprudence.[20]

The Indian judiciary has shown great environmental compassion and forward-thinking in its use of the domestic and international law principles outlined to promote environmental justice.[21] "The right to a healthy environment, including the freedom from contamination in one's drinking water, air, and other bodily fluids, has been liberally construed by the judiciary as part of Article 21."[22] The case of *Intellectuals Forum* v. *State of A.P.*,[23] which was decided by India's apex court, discussed the need for long-term planning and protecting the rights of future generations while India experienced unprecedented levels of development in this century. The court has also incorporated the right to a healthy environment with some of the more recent common law concepts in international environmental law, such as the polluter-pays principle, the precautionary principle, the idea of sustainable development, the idea of intergenerational equity, and the state's role as custodian of all natural resources.[24]

This is by no means an all-inclusive list, and it merely reflects a fraction of the environmental cases that have been made to Indian courts. Various challenges, such as those pertaining to sewage cleaning, brick grinding operations, hazardous waste management, and untreated effluent discharges from companies, have been identified as public nuisances in India. Yet the topic of climate change remains mostly uncharted.

The Indian judiciary has been instrumental in promoting environmental justice through its work with the National Green Tribunal (the Green Tribunal). In its mission to "ensure the effective and expedited disposal of cases relating to environmental protection and conservation of forests and other natural resources," the green tribunal has assembled a panel of judges and experts to examine and rule on any matters brought before it from a legal an scientific perspective. Throughout its more than ten years of operation, the green tribunal has effectively elevated the environmental concerns of the people of India to the forefront of the legal and judicial community.

The green tribunal has reduced procedural complexities in relation to environmental law issues and delivered exceptional environmental jurisprudence, often drawing from and building on principles of international environmental law, despite challenges related to its limited jurisdiction, powers of execution, and statutory limitation. For quite some time, Indian courts have regarded precedents set by other countries' laws and rulings as authoritative. The environmental legal framework of the country has been strengthened by consistent decisions of the Supreme Court, which have established the public trust theory, the polluter pays principle, the precautionary principle, and intergenerational equality as part of the law of the land.[25] Assimilating new ideas from international environmental law into domestic jurisprudence has been the responsibility of the green tribunal in recent years, thus further enlarging this body of law.

As an example, consider a ruling or a notification that weakens and renders meaningless the substantive provisions of the Environmental Impact Assessment Notification, 2006 and the Environment (Protection) Act, 1986, which together constitute India's overarching environmental law. If the notification's provisions were to be put into action, there was concern that they could lead to the destruction of ecosystems and the environment as a whole as a result of unchecked construction and building activities.

According to the applicant, the current environmental regulations would be nullified if the notification were implemented as written, which would go against the principle of non-regression. Under this principle, once certain environmental standards or protections are established, they should not be reduced or weakened over time. Instead, there is an expectation that environmental laws and policies will progress towards greater protection and sustainability. It should be noted that Indian law does not incorporate the notion of non-regression. No statute, particularly one dealing with environmental law, gives the courts the power or requires them to use the non-regression concept when making decisions. Environmental law should not be changed in a way that compromises environmental protection; this is the premise upon which the theory rests under international law. "States, subnational entities, and regional integration organizations shall not permit or pursue actions that have the net effect of diminishing the legal protection of the environment or of access to environmental justice,"[26] states the International Union for the Conservation of Nature (IUCN), citing "the principle of non-regression as an international law principle requiring that norms which have already been adopted by states not to be revised if this implies going backward on the subject of standards of protection." According to the case's pleadings presented to the green tribunal, this principle was necessary because modern environmental law is under attack from a variety of directions, including deregulation, efforts to streamline while simultaneously weakening regulations, criticisms of environmental laws for being overly complex, and an economic climate that prioritizes development over environmental protection. The green tribunal delayed the notice until certain conditions were met, particularly those pertaining to consents under the Water (Prevention and Control of Pollution), Act, 1974 and the Air (Prevention and Control of Pollution), Act, 1981. This decision was reached after a thorough examination of the facts and laws. Part of the notification was deemed ultra-vires, or ineffective, while the other portion was deemed legally sound and durable, based on the law of severability.[27] Aside from being an innovative application of principles of international law, the green tribunal differs from the Constitutional Courts, the Supreme Court, and the High Courts in that it is a product of a statute. As a result, it can only do what its parent laws allow, such as the NGT laws. Regarding eligibility for the green tribunal, this creates an inherent contradiction. As a non-Constitutional Court, the Green Tribunal lacks the expansive breadth of standing that the Indian Supreme Court and High Courts enjoy under the PILs. According to Section 14 of the

NGT Act, "the Tribunal shall have the jurisdiction over all civil cases where a substantial question relating to environment (including enforcement of any legal right relating to environment) is involved and such question arises out of the implementation of the enactments specified in Schedule I." This means that environmental issues, including the enforcement of legal rights, can be brought before the tribunal. Biological Diversity Act of 2002; Water (Prevention and Control of Pollution) Cess Act of 1977; Forests (Conservation) Act of 1980; Air (Prevention and Control of Pollution) Act of 1981; Environment (Protection) Act of 1986; Public Liability Insurance Act of 1981; and Water (Prevention and Control of Pollution) Act of 1974 are all bills that make up Schedule I. A person who feels wronged can also approach the green tribunal as outlined in Section 16 of the NGT Act.[28]

An even more formidable adjudicatory obstacle is in the second part of Section 14 of the NGT Act, to which the aforementioned reference is made; it specifies that the matters over which the green tribunal has authority must emerge "out of the implementation of" certain limited statutes. It is worth noting that there is a lack of comprehensive climate change legislation in India, despite the fact that many laws deal with various parts of the issue, such as its effects and causes.

15.6 Climate Litigation and Relevancy in Indian Context

Developing a workable legal framework to address the climate change crisis may become an important task for the legal community in the future. "With the ever-increasing economic activities impacting every aspect of human productivity, daily life, and ongoing global climate change negotiations, the new legal approach must take the role of the judiciary in interpreting the current laws." The fundamental process by which greenhouse gases like carbon dioxide cause the Earth to warm has been understood for decades, but it wasn't until the late 1980s that climate change became a serious international concern. After that, the world community spent over a decade working to establish a thorough legislative framework to tackle the climate change problem on a global scale. In addition to the laws and policies established by executive and legislative bodies, the judiciary has been increasingly participating in climate change governance and there is a rise in climate related litigation.

The exact relationship between climate legislation and litigation is still uncertain, but they generally appear to work together. The judiciary enforces government policies, interprets climate laws, and fills enforcement gaps. "Regulation through litigation" can address deficiencies in legislation but also occurs in countries with progressive climate laws. Legal actions for climate change, using courts and legal strategies to achieve broader goals, often accompany other forms of activism, such as legislative actions, political advocacy, and grassroots movements.[29]

15.7 Global Trends in Climate Litigation

Climate change litigation covers a broad spectrum of legal actions, some of which are strategically designed to influence climate policies, change behaviours, or raise public awareness. These strategic cases involve careful planning regarding who initiates them, when and where they're filed, and what outcomes are sought. Additionally, the term includes cases driven by private interests, such as those aimed at upholding planning approvals or clarifying reporting requirements under emissions trading systems, even if they lack activist motivations.

When defining "climate change litigation" one includes cases that address key legal or factual issues related to mitigating or adapting to climate change, as well as climate science. These cases are brought before various administrative, judicial, and other adjudicatory bodies and are typically identified using keywords such as "climate change," "global warming," "greenhouse gas," "GHGs," and "sea level rise." Cases that deal with climate change-related issues without explicitly using these terms are also included. However, most researchers exclude cases where climate change is mentioned only incidentally or where the primary legal theory is not climate-related. Therefore, if climate change keywords are mentioned only in passing and are not central to the laws, policies, or actions under consideration, the case is excluded.[30]

As of 2022, researchers from Columbia University and the London School of Economics have documented a total of 2180 climate-related cases filed across 65 different jurisdictions, including international or regional courts, tribunals, and other adjudicatory bodies. Among these cases, 1522 were filed in the United States, while the remaining 658 were distributed among other countries. The United States had the highest number of cases, followed by countries such as Australia, the United Kingdom, and members of the European Union. However, there has been a notably lower number of cases in the global South, with only 37 cases filed in regions such as Asia, Africa, and Latin America. It is noteworthy that more than half of these cases in the global South were initiated between 2015 and 2019.[31]

Since the ratification of the Paris Agreement in 2015, there has been a clear rise in activism and the introduction of new forms of climate-related lawsuits presented in courts globally.[32] These trends indicate a growing number of cases focused on one or more of the following aspects: "(i) The utilization of "climate rights" in legal actions, (ii) Enforcement of domestic regulations, (iii) Advocacy for the preservation of fossil fuels and carbon sinks, (iv) Corporate accountability and obligation, (v) Disclosure of climate-related information and challenges against deceptive environmental claims, and (vi) Addressing failures to adapt to climate change and assessing the repercussions of adaptation efforts."[33]

A few notable cases shaping global climate litigation are discussed in the following sections.[34]

15.7.1 Urgenda Foundation v. The State of Netherlands[35]

In this landmark case, the Urgenda Foundation, along with 900 Dutch citizens, sued the Dutch government for its inadequate efforts to combat climate change. The Hague court ruled in favour of Urgenda, stating that the government's commitment to reduce greenhouse gas emissions by only 17% below 1990 levels by 2020 was insufficient to meet its international obligations. The court ordered the government to increase its efforts, mandating a reduction of at least 25% by 2020. This decision highlighted the legal duty of states to take ambitious action to address climate change, setting a precedent for climate litigation globally.

15.7.2 Juliana v. United States[36]

This case, also known as the *Children's Climate Crusade*, was brought by a group of youth plaintiffs against the U.S. government. The plaintiffs argued that the government's actions (or lack thereof) to address climate change violated their constitutional rights to life, liberty, and property. While the case faced numerous legal challenges and delays, it sparked public awareness about the intergenerational equity aspect of climate change and the responsibilities of governments to protect the rights of future generations.

15.7.3 Client Earth v. European Investment Bank[37]

Environmental law organization Client Earth challenged the European Investment Bank's funding of fossil fuel projects, alleging that it violated the bank's climate commitments and the EU's climate objectives. The case raised questions about the financial sector's role in supporting climate mitigation efforts and the need for institutions to align their investments with climate goals.

15.7.4 Massachusetts v. Environmental Protection Agency[38]

The U.S. Supreme Court ruled in 2007 that the Environmental Protection Agency (EPA) has the authority to regulate greenhouse gas emissions under the Clean Air Act (CAA). The case originated from a petition filed by Massachusetts and other states, along with environmental organizations, urging the EPA to regulate greenhouse gases emitted by motor vehicles. The EPA argued that it lacked the authority to regulate greenhouse gases under the CAA, but the Supreme Court disagreed, stating that the EPA has the responsibility to regulate pollutants that endanger public health and welfare, including greenhouse gases. This landmark decision established the EPA's authority to address climate change through regulatory action, setting the stage for future environmental regulations aimed at mitigating greenhouse gas emissions.

These cases represent a diverse range of legal strategies and issues in climate litigation, from challenging governmental policies to holding corporations accountable for their contributions to climate change. Legal and human rights issues connected to climate change are gaining increasing recognition, and courts are being requested to be involved in this process. Still, plaintiffs trying to seek claims relating to climate have many challenges. Proving standing is one of the main challenges as plaintiffs must prove that climate change has clearly and directly caused obvious and direct damage to bring a lawsuit. Secondly, proving attribution and causality might be rather difficult, especially in situations where numerous factors influence damage resulting from climate change. Ultimately, the choice of site for litigation is challenging as managing jurisdictional issues adds further complexity, particularly in cases with worldwide implications.

Strategic climate litigation belongs to a field that could influence nations all across the globe. Its objective is to raise knowledge of the link between human activities and their consequences on the environment thereby bringing about changes in society or regulation. Strategic litigation differs from conventional litigation in that, instead of concentrating just on one particular case, its aim is to modify the law in a way that affects all conceivable approaches of resolving legal problems. Given the worldwide nature of the interests at stake in these lawsuits, they might have far-reaching, international consequences. Strategic climate litigation often extends beyond national boundaries and influences global climate policy as well as corporate actions. Decisions rendered in courts and legislation accepted by one country later on by other countries' legislatures might drive a shift in international legal standards. Environmental campaigners, lawyers, and non-governmental organizations (NGOs) cooperate across borders, sharing resources and strategies, thereby strengthening their case and having more impact.

Cases like *Urgenda Foundation* v. *The Netherlands*[39] raise awareness about the dangers of climate change among governments, individuals, and businesses, and this awareness persists long after the verdict is in. Another case in point is the NGO Milieudefensie's successful legal action against *Royal Dutch Shell in the Dutch courts*, which was based on Urgenda. Royal Dutch Shell was ordered by the court to cut its global carbon emissions by 45% by 2030, compared to 2019 levels.[40] The goal of this type of lawsuit is to get people thinking about the issue, start a conversation, or maybe even modify some policies. With the proliferation of digital technologies and the merging of global societies into a single network of socially linked peers, the paradigm for cross-border information transfer has shifted. Knowledge travels beyond human and state borders from such discussions on the subject or conclusion of litigation, leading to climate action. Such strategic cases have been seen in India as well.

15.8 Protection of Biodiversity

Biodiversity encompasses the rich array of life forms found on Earth, including the diversity within species, between species, and of entire ecosystems.

For instance, within a species like the tiger, biodiversity includes the various subspecies and genetic variations. Between species, it refers to the multitude of different organisms such as plants, animals, and microorganisms. Ecosystem diversity encompasses the varied habitats like forests, grasslands, and wetlands, each supporting a unique blend of life forms.

Biological resources encompass a wide range of living organisms and their components, including plants, animals, and microorganisms, along with their genetic material and derivatives. For example, medicinal plants like neem and turmeric, with their healing properties, represent valuable biological resources. However, it's important to note that human genetic material is excluded from this definition. In another example, Brahmi, a well-known herb in India, is renowned for its ability to address neurological disorders. It has been a part of the traditional Ayurvedic system for thousands of years, valued as a memory enhancer and cognitive stimulant.

Now India asserts its sovereign right over Brahmi and the associated traditional knowledge. This right grants the country exclusive control over the natural resources within its borders. The recognition of this right came in 1992 with the adoption of the Convention on Biological Diversity[41] (CBD) by the United Nations. The CBD, embraced by 196 countries including India, comprehensively addresses genetic resources, species diversity, and ecosystems. It introduces the principle of fair and equitable benefit sharing for accessing and utilizing genetic resources.

Reacting to the CBD agreement, the Biological Diversity Act of 2002[42] was passed with aims of safeguarding biological resources, ensuring their sustainable use, and so promoting fair distribution of the advantages resulting from their use and awareness among local communities especially.

The Act forbids several activities unless the National Biodiversity Authority (NBA) provides its prior authorization, which is a significant component. Among these activities include gathering biological resources for either economic or academic use; sharing research results about biological resources in India, and claiming intellectual property rights resulting from this study.

Three tiers of government under control of access to biological resources are local Biodiversity Management Committees (Biodiversity Management Committee), State Biodiversity Boards (SBBs), and the NBA (section 8).[43]

The NBA aims to enable the sustainable use of biodiversity and its protection. It offers suggestions for the fair sharing of benefits and access to biological resources, guides the federal government, and controls activity. It also acts to prevent individuals from distributing intellectual property rights to biological resources stolen illegally from India. Apart from suggesting strategies for cultural sites' care, the NBA assists state governments in choosing sites of biological relevance for classification.

Formed under Section 22 of the Act, SBBs include chairpersons, ex officio members from relevant state agencies, and professionals in biodiversity protection. Towards biological resources, SBBs guide state governments towards equitable distribution of benefits, appropriate use, and preservation. These state authorities also control requests for bio-surveys and bio-utilizations as well as commercial use of biological resources and biological resource exploitation.

The Biodiversity Act, Section 41 mandates the creation of BMCs in order to further the preservation, sustainable use, and documentation of biological diversity. These committees handle habitat preservation, landrace and folk variation conservation, domesticated stock and breed preservation, microorganism preservation, and information recording linked to biological diversity.

One of BMC's main obligations is to establish People's Biodiversity Registers in concert with nearby towns. These registries provide comprehensive records on local biological resources, including their medicinal and other applications. At the local level, this data is essential for promoting environmentally friendly living and preservation of biodiversity.

The Act establishes the National Biological Diversity Fund,[44] which serves as a financial mechanism for the Act's objectives. This fund is funded through grants, loans, fees, royalties, and donations received by the National Authority. Administered by the National Authority, the fund is utilized for various purposes, including channelling benefits to resource conservers and knowledge holders, conserving biological resources in accessed areas, and promoting socio-economic development in consultation with local governing bodies. Additionally, the fund supports the conservation of Heritage Sites. The Comptroller and Auditor General of India audits the fund, and the Central Government frames rules for its management and administration.

Furthermore, the Act empowers the Central Government, in consultation with the NBA, to identify threatened species, regulate their collection, and designate institutions as repositories for different categories of biological resources.

Under the provisions of the Act, all offenses are classified as cognizable and non-bailable. Any disputes regarding the determination of benefit sharing or decisions made by the NBA or a SBB are to be brought before the National Green Tribunal (NGT) for resolution.

The State government, in consultation with local bodies, has the authority to designate areas known as Biodiversity Heritage Sites (BHS). These sites represent unique and delicate ecosystems, encompassing terrestrial, coastal, inland waters, and marine environments. They are characterized by rich biodiversity, including diverse populations of both wild and domesticated species, high levels of endemism, rare and threatened species, keystone species, and species of evolutionary significance. BHS may also include areas with cultural, ethical, or aesthetic values, contributing to the preservation of cultural diversity.[45]

15.9 Environmental Courts

The establishment of a specialized tribunal for environmental matters in India has been a subject of judicial and legislative discussion for several years.

Beginning with the landmark case of *M.C. Mehta* v. *Union of India* in 1986,[46] the Supreme Court emphasized the need for specialized courts to handle environmental cases, citing the complex nature of scientific data involved. In *Vellore Citizen Welfare* v. *Union of India* (1996),[47] the High Court established a green bench to monitor and deal with cases on environmental matters. Subsequent cases, such as *Indian Council for Enviro-Legal Action* v. *Union of India in 1996* and *A.P. Pollution Control Board* v. *M.V. Nayudu in 1999*,[48] further underscored the necessity of setting up dedicated environmental courts with both civil and criminal jurisdiction for expeditious resolution of environmental issues.

Despite the judicial push, progress towards establishing such courts was slow effectively. The suggestion to create environmental courts in India was made by the Law Commission of India in their 186th Report in 2003. This suggestion gave rise to the NGT. After looking at the scientific and technological concerns that have gone to court and how little judges know about the environmental challenges from a scientific and technical standpoint, a proposal was made with the additional purpose of reducing the workload of regular courts in environmental cases.[49]

The tribunal comprises a chairperson, judicial members, and expert members, appointed through a stringent selection process. It's important to note it is the central government which makes the appointment under the act. Eligibility criteria for these positions are clearly defined in the Act. Supreme Court judges or Chief Justices of High Courts qualify for the Chairperson or judicial member roles, while retired or existing High Court judges can serve as judicial members. Expert members must hold a master's degree in science, engineering, or technology, along with a doctorate, and possess 15 years of relevant experience. Alternatively, individuals with 15 years of administrative experience, including 5 years in environmental matters at the Central or State Government level, also qualify. These stringent criteria ensure that NGT members have the requisite qualifications and expertise to effectively adjudicate environmental disputes.[50]

The NGT Act of 2010 states that the Tribunal has "jurisdiction over all civil cases where a substantial question relating to environment (including enforcement of any legal right relating to environment) is involved and such question arises out of the implementation of the enactments specified in Schedule I."[51] This means that the Tribunal can hear cases involving environmental issues, including violations of legal rights.

The NGT's functions go beyond adjudication; it also plays preventative, ameliorative, or remedial roles. It functions as both an adjudicatory and an appellate body, not bound by the procedures of the Civil Procedure Code but guided by principles of natural justice and sustainable development. The NGT has the authority to provide relief and compensation to victims of pollution and environmental damage, including accidents involving hazardous substances. Its orders, decisions, or awards are enforceable as civil court decrees. Non-compliance can result in imprisonment for up to three years, fines of up to 10 crore rupees, or both.

A number of challenges impede the NGT's capacity to perform its obligations. Its absence of jurisdiction over important statutes like the Wildlife (Protection) Act, 1972, and the Scheduled Tribes and Other Traditional Forest Dwellers (Recognition of Forest Rights) Act, 2006, helps to explain why it cannot manage important forest rights concerns relating to the environment. Legal conflicts might also arise in the NGT's rulings being contested in many High Courts under Article 226. Some argue that because the High Courts are constitutionally better than the NGT, it is unknown whose decisions one may contest.

Furthermore debatable and controversial are decisions taken by the NGT affecting economic growth and development. The absence of a standardized mechanism for determining compensation and occasional non-compliance with NGT decisions by stakeholders or governments due to feasibility concerns or resource limitations further add to the tribunal's challenges. Limited human and financial resources contribute to a backlog of cases, undermining the NGT's goal of timely resolution. Additionally, the NGT's justice delivery mechanism is hindered by a scarcity of regional benches, affecting accessibility and efficiency.[52]

Indian courts, especially the NGT and the Supreme Court, have been contributing to jurisprudence in favour of strategic climate litigation. They have shown a willingness to issue favourable judgments on technical and sectoral climate issues. For instance, "the NGT's recognition of HFC-23 as an environmental pollutant under the Environment Protection Act has paved the way for including other climate pollutants. Similarly, in *Ridhima Pandey* v. *Union of India*, the NGT acknowledged climate change within the Environmental Impact Assessment process."

The landmark *M.K. Ranjitsinh & Ors.* v. *Union of India & Ors* 2024 INSC case marked a significant moment, with the Supreme Court recognizing the right against adverse effects of climate change as a part of the right to life under Article 21 of the Constitution; Justice D. Y. Chandrachud highlighted the absence of comprehensive climate legislation, emphasizing the judiciary's role in addressing climate issues despite legislative gaps. This case underscores the courts' potential to push climate ambition by leveraging existing laws and constitutional rights

Despite the potential for Indian courts to advance climate ambition, there are several caveats. Systemic issues within the Indian legal system affect their track record on environmental and climate matters. Although Indian courts are reputed for judicial activism and often viewed as "pro-environment," they do not typically challenge executive decisions on economic policy and infrastructure aggressively. Judicial outcomes in India are highly unpredictable, influenced by a mix of legal and non-legal factors. Nevertheless, well-crafted strategic climate litigation holds significant potential to drive the climate change agenda in the country.

A few important judgements are as follows:

- **Waste burning prohibition:** NGT's 2012 judgment imposed a complete prohibition on open burning of waste, a landmark case in solid waste management.
- **Uttarakhand floods compensation:** NGT ordered compensation by Alaknanda Hydro Power Co. Ltd. in the Uttarakhand floods case, based on the principle of "polluter pays."
- **Save Mon Region Federation case:** NGT suspended a hydro project in 2013 to save bird habitats.
- **Diesel vehicle ban:** In 2015, NGT ordered a ban on diesel vehicles over ten years old in the Delhi NCR region.
- **Nullification of EIA amendment:** NGT nullified a 2016 amendment to EIA 2006 notification, terming it a ploy to circumvent rules.

15.10 M.K. Ranjitsinh Case[53]

Amongst the other landmark cases on environmental jurisprudence the case of M.K. Ranjitsinh[54] holds special significance. This case can form a bedrock of various other ensuing rights. The Supreme Court of India affirmed that individuals had the right to protection against the adverse impacts of climate change as an integral component of their fundamental right to life, as enshrined in Article 21 of the Constitution of India. The acknowledgement of the connection between human rights and environmental safety is a significant advancement towards environmental justice. It merits commendation, for it forms foundation for further rights and duties to protect environment.

In this case the apex court highlighted that Article 21 of the constitution guarantees the right to life, which also includes the right to a healthy and clean environment. This right is jeopardized by negative climatic implications, which the Court has explicitly recognized. The case underscores that Article 21's protection of life and dignity is contingent on a stable environment undisturbed by climate change. The verdict demonstrates a causal link between environmental degradation caused by factors such as air pollution, climate change, and habitat loss and the negative impacts on human well-being, equity, and longevity. Environmental justice and Constitutional equality are inextricably linked, since the Court's reasoning demonstrates that disadvantaged communities endure a disproportionate part of the impact of climate change. Furthermore, it recognizes the all-encompassing nature of sustainable development and advocates for moderate approaches to combating climate change and protecting biodiversity. By adding environmental protection into Article 21, the Court highlights that the Constitution's guarantees are flexible enough to address modern challenges such as climate change while protecting fundamental human rights and justice.

At the heart of this case are the Great Indian Bustard (GIB), a threatened bird, and the need to find a balance between using energy and protecting the environment. Petitioners asked the Supreme Court of India to use its Constitutional power to take quick action to protect the GIB. The GIB is in great danger because of the people and structures like power lines that cross its natural habitat in Gujarat and Rajasthan.

The Court's initial decision on April 19, 2021, was the first step in the process of putting all priority and protection of GIB above overhead power lines. The power lines were ordered to be put under ground. Key points included stopping new projects that could put the GIB at risk and stressing the need to add bird diverters to current lines. But problems came up in the process of these orders being carried out.

When the Union of India and other groups asked for these rules to be changed, they brought to light the technical and financial problems that come with laying underground power lines over long distances. They said that putting lines underground in areas is not possible because it is too expensive, doesn't work well over long distances, and is unfeasible in deserts. The areas covered by the order also crossed over areas that could produce a lot of solar and wind energy. These areas were much bigger than the real GIB sites. Under agreements like the Paris Accord, the government has promised to use more clean energy.

The Court looked over its orders again to try to find a middle ground between two goals that were at odds with each other: saving the GIB and promoting the use of clean energy sources. It acknowledged how complicated the issue is and pointed out that GIB conservation brings up many other risks besides power lines. Additionally, creating green energy sources is necessary to deal with global warming and keep international promises.

To have expert advice the Court set up an expert committee to look into the possibility of putting underground power lines taking technical, environmental, and financial factors into account. The Court, balancing the conflicting interests, stopped a step short of unfettered protection of GIB. Though initially order was to put overhead lines under the ground, the final decision did not do enough to protect the endangered species. This accommodative approach avoids broad bans that could hurt green energy projects and makes sure that decisions are sensitive to all concerned. Although the decision is revolutionary for conferring protection against environmental change, it could not do justice to endangered bird.

This case upheld a significant legal principle; nonetheless, it fell short of adequately addressing the critical conservation problem of the GIB. The absence of directives in the verdict about measures to prevent further decline of the GIB population illustrates the legal system's challenge in balancing environmental and commercial interests. This case needs to be thoroughly analysed to understand its true implications. An analyses is as follows:

Recognizing that climate change impacts human rights: The Supreme Court determined that safeguard against climate change is a significant aspect of Article 21, thereby including natural health and equal rights for all generations into the constitutional rights framework. India's laws therefore conform to international norms like the Paris Agreement.

Matching human rights with sustainable development: Preserving fundamental rights and stressing the importance of sustainable development in the struggle against climate change helped the decision to exhibit a forward-looking perspective.

Declaring the realities of fragmentation and habitat loss: Just two of the various concerns raised by the court as problematic for the GIB are overhead electricity lines and habitat destruction. It included burying cables and erecting bird feeders as part of the instructions to resolve the issues.

Disregarded the vital need to protect GIB: The court acknowledged the major danger to the GIB, but it offered no workable means to stem the population decline. This species is very threatened as there are fewer than 150 birds. Uncertainties regarding the efficacy of the proposal resulted in the decision to change from overhead electricity lines to subterranean cables being shelved. This can be implemented to prioritize the protection of species above the use of green energy.

Insufficient emphasis on accountability: The Court placed significant emphasis on international obligations, such as India's commitment to use renewable energy sources, along with official reports. It had the potential to take action, but it failed to implement effective accounting practices to monitor its conservation directives.

Excessive emphasis on renewable energy objectives: The court prioritized India's commitment to green energy, particularly solar power, above the continuity of the GIB. This highlights a significant issue in India's environmental management: the need to save wildlife while also advancing green energy initiatives.

Insufficient consideration was given to alternative conservation measures. The Court, in its effort to preserve the GIB, neglected to explore other protective measures such as habitat restoration, enhancement of breeding programs, or the implementation of more stringent agricultural regulations. Typically, technological solutions such as subterranean cables and avian deterrents were used. Nonetheless, both of them possess constraints regarding finances and logistics.

Several significant conclusions may be derived from the verdict. The verdict underscores the need for judges to use sensitivity in balancing the protection of natural processes with the pursuit of developmental objectives. The courts are responsible for ensuring that environmental trade-offs do not adversely affect vulnerable species more than other species.

Integrating science and law: The Court solicited expert views about subterranean lines; nevertheless, a comprehensive scientific strategy for the protection of GIB has yet to be established. This is a lost opportunity to effectively integrate legal frameworks with environmental considerations.

Maintaining oversight of the courts: Maintaining legal oversight was essential to ensure compliance with the court's directives. Conversely, the initiative proved to be inadequate due to its insufficient tracking methodologies.

Species-focused conservation methods: The Supreme Court needs to prioritize species protection by enacting legislation that is both regionally tailored and feasible. These possibilities include breeding initiatives, habitat protection, and the construction of escape-proof enclosures.

Enhanced mechanisms for ensuring individual accountability: To ensure adherence to instructions, the decision-making process must have enforceable schedules and systematic reviews.

Cooperative environmental management: To achieve comprehensive conservation outcomes, the legal framework must compel various government officials, activists, and local communities to collaborate.

Despite acknowledging the conflict between biodiversity preservation and the creation of renewable energy, the Supreme Court's finding falls short because of its reductionist stance. By ignoring serious environmental concerns in favour of installing underground power lines across the GIB's natural habitat, the Court has shown a bias in support of India's renewable energy goals. Even while renewable energy is essential in the fight against climate change, the choice misses the complex balancing act required for true environmental justice, which entails appropriately addressing both ecological and developmental imperatives.

Simple legislation that pits renewable energy against biodiversity runs the danger of making environmental protection a zero-sum game. The delicate ecology of the GIB will suffer greatly from more habitat loss. The Court unintentionally ignores its constitutional duty to protect biodiversity under Articles 21 and 48A in favour of technical and business considerations while defending overhead electricity lines. By limiting the matter to a straightforward yes-or-no decision rather than recognizing the complexity of the situation and the necessity for creative solutions that strike a balance between growth and conservation, the ruling goes against the principles of comprehensive environmental justice. By putting immediate gains ahead of the longer-term constitutional responsibility to balance human needs with the purity of nature, the ruling creates a concerning precedent that endangers both animals and vulnerable humans.

A big step forward in the M.K. Ranjitsinh case was the court's decision that climate change is against the Constitution. Still, the fact that they weren't able to save the GIB is disheartening; it shows how hard it is to balance economic growth with protecting a critically endangered species. In future court cases, the courts will need to be more fair and focus on the species. We can make sure that bigger environmental goals don't get in the way of our efforts to protect the environment in this way.

Also, this case shows how animal care and the switch to green energy sources are linked in a very complicated way. It also gives important information for rules around the world that protect the earth.

15.11 Conclusion

As climate change jurisprudence develops in India, the courts are becoming more and more crucial in juggling environmental preservation against development. Historic decisions emphasizing the crucial need of including sustainable practices into national legislation include Article 21's recognition of a right to be free from the consequences of climate change. Courts and specialist tribunals such as the NGT have proactively advanced environmental justice by closing enforcement gaps and promoting public accountability.

Still, challenges abound. When the court fails to give environmental preservation first priority over developmental goals, compromising concessions can result. As historic instances like M.K. Ranjitsinh demonstrate – which recognize climate change consequences as constitutional violations – aligning conservation efforts with renewable energy ambitions and other developmental imperatives is difficult.

Going forward, the courts have to adopt a scientific approach if they are to ensure that conservation decisions are based on extensive ecological research. There should be strong systems of responsibility to track respect of court orders. Comprehensive conservation plans need cooperative governance including government, civic society, and people living in a given area.

Strategic climate litigation in India might find inspiration in global trends, thereby improving enforcement systems and driving further policy development. Working towards a sustainable economic model helps Indian courts to guarantee the welfare of the present and future generations of their people as well as biodiversity. On this journey, we have to be alert, imaginative, and tenacious in our commitment to justice and equality.

Notes

1 Thomas F. Stocker, Dahe Qin, et al. (eds.), *Climate Change 2013: The Physical Science Basis. Contribution of Working Group I to the Fifth Assessment Report of the Intergovernmental Panel on Climate Change* 1535 (Cambridge University Press, New York, 2013).
2 J.T. Houghton, Y. Ding, et al. (eds.), *Climate Change 2001: The Scientific Basis, Contribution of Working Group I to the Third Assessment Report of the Intergovernmental Panel on Climate Change*, 881 (Cambridge University Press, New York, 2013).
3 United Nations Climate Action, *What Is Climate Change*? Available at: https://www.un.org/en/climatechange/what-is-climate-change (Last visited on January 06, 2024).
4 *State of Bihar* v. *Murad Ali Khan* (1988) 4 SCC 655.
5 *M.C. Mehta* v. *Kamal Nath* (1997) 1 SCC 388.
6 *M.K. Ranjitsinh & Ors.* v. *Union of India & Ors.*, 2024 SCC Online SC 570.
7 *Vellore Citizens Welfare Forum* v. *Union of India* (1996) 5 SCC 647.
8 *Gene Campaign and Others* v. *Union of India*, 2024 SCC Online SC 1793.
9 J. Muir, *Original Sanskrit Texts on the Origin and History of the People of India Their Religion and Institutions*, 23 (Trubner & Co., London, 5th ed., 1872).
10 Muir, above note 10 *Original Sanskrit Texts.*

11 Muir, above note 10 *Original Sanskrit Texts*.
12 Christina Voigt, "Equity in the Legal Context: Neglected Principles in the Climate Change Regime" 22 *Review of European, Comparative & International Environmental Law* 111 (2013).
13 Constitution of India, Art. 21.
14 *Subhash Kumar* v. *State of Bihar*, AIR 1991 SC 420.
15 *Virender Gaur* v. *State of Haryana* (1995) 2 SCC 577.
16 Jai Garg, "Climate Change and International Human Rights Law: Intersections and Challenges", *Journal of Human Rights* 127 (2012).
17 *M.K. Ranjitsinh & Ors.* v. *Union of India & Ors 2024* INSC 280.
18 Constitution of India, Art. 14.
19 Constitution of India, Art. 32.
20 Lavanya Rajamani, "Public Interest Environmental Litigation in India: Exploring Issues of Access, Participation, Equity, Effectiveness and Sustainability", 19 *Oxford Journal of Environmental Law* 295 (2007).
21 Rajamani, "Public Interest Environmental Litigation in India", above note 20.
22 The Constitution of India art. 21; See also *Subash Kumar* v. *State of Bihar* (1991) 1 SCC 598.
23 *Intellectuals Forum* v. *State of A.P.*, AIR 2006 SC 1350.
24 See also *Virender Gaur* v. *State of Haryana* (1995) 2 SCC 577. See, *Indian Council for Enviro-legal Action* v. *Union of India (Bichhri Case)* (1996) 3 SCC 212 (describing polluter pays principle); See also *M.C. Mehta* v. *Kamal Nath* (2000) 6 SCC 213, 220. 27. See, *Vellore Citizens' Welfare Forum* v. *Union of India* (1996) 5 SCC 647 (establishing precautionary principle); *Narmada Bachao Andolan* v. *Union of India* (2000) 10 SCC 664, 727 (shifting the burden of proof to the industry).
25 *Narmada Bachao Andolan* v. *Union of India*. MANU/SC/0640/2000.
26 IUCN World Commission on Environmental Law, "IUCN World Declaration on the Environmental Rule of Law" Principle 12, 2016.
27 Shinsa P. Mathew, "National Green Tribunal: Drifting Singularity" 2(2) *GLS Law Journal* 35–41 (2021).
28 Mathew, "National Green Tribunal", above note 27.
29 Setzer Joana and Lisa C. Vanhala, "Climate Change Litigation: A Review of Research on Courts and Litigants in Climate Governance" 10 *Wiley Interdisciplinary Reviews: Climate Change* 580 (2019).
30 Sabin Centre for Climate Change Law, *Climate Change Litigation Databases*, Available at: http://climatecasechart.com (last visited on January 10, 2025).
31 Joana Setzer and Rebecca Byrnes, "Global Trends in Climate Change Litigation: 2020 Snapshot Report" *Grantham Research Institute on Climate Change and the Environment* 4 (2020).
32 Setzer and Byrnes, "Global Trends in Climate Change Litigation", above note 31.
33 Global Climate Litigation Report: 2023 Status Review, Available at: https://www.unep.org/resources/report/global-climate-litigation-report-2023-status-review (Last visited on January 11, 2025).
34 Global Climate Litigation Report: 2023 Status Review, above note 33.
35 *Urgenda Foundation* v. *The State of Netherlands*, Case No. C/09/456689 / HA ZA 13–1396 (District Court of The Hague, June 24, 2015).
36 *Juliana* v. *United States* [9th Cir. 2020] 947 F.3d 1159.
37 *Client Earth* v. *European Investment Bank* [2021] Case T-393/19.
38 *Massachusetts* v. *Environmental Protection Agency* [2007] 549 U.S. 497.
39 *Urgenda Foundation* v. *The Netherlands* [2015] HAZA C/09/00456689.
40 *Milieudefensie and Others* v. *Royal Dutch Shell PLC and Others* [2021] C/09/571932.

41 Convention on Biological Diversity, 1992, Available at: https://treaties.un.org/pages/ ViewDetails.aspx?src=TREATY&mtdsg_no=XXVII-8&chapter=27 (Last visited on January 12, 2025).
42 The Biological Diversity Act, 2002 ACT NO. 18 OF 2003.
43 The Biological Diversity Act, 2002 ACT NO. 18 OF 2003, above note 42 sec. 8, 22, 41.
44 Biological Diversity Act, 2002, sec. 23.
45 Biological Diversity Act, 2002, sec 37.
46 *M.C. Mehta* v. *Union of India 1986* SCC (2) 176.
47 *Vellore Citizens' Welfare Forum vs Union of India and Ors*, 1996 5 SCR 241.
48 *Indian Council for Enviro-legal Action* v. *Union of India and Ors* [1996] Supp. (1) S.C.R. 507 A.P. Pollution Control Board vs Prof. M.V. Nayudu (Retd.) & Others 1999 (2) SCC 718.
49 Law Commission of India, "108th Report on Proposal to Constitute Environment Courts" (September, 2003).
50 The National Green Tribunal Act, 2010, ACT NO. 19 of 2010 sec 3,4,5,6.
51 The National Green Tribunal Act, 2010, sec 14.
52 Gitanjali Nain Gill, "Mapping the Power Struggles of the National Green Tribunal of India: The Rise and Fall?" *Asian Journal of Law and Society*, 7 (2020).
53 *M.K. Ranjitsinh & Ors.* v. *Union of India & Ors.*, 2024 SCC Online SC 570.
54 *M.K. Ranjitsinh & Ors.* v. *Union of India & Ors*, above note 53.

Part IX

Implementation Aspects

16 Implementation of Sustainable Development Goals in Nepal

Amber Prasad Pant

16.1 From Sustainable Development to SDGs in Historical Context

The draft of Sustainable Development Goals (SDGs) was prepared by Open Working Group (OWG) formed by United Nations General Assembly resolution (UNGA Reso.) 67/555 dated January 22, 2013, as mandated by the United Nations Conference on Sustainable Development 2012 to develop a set of goals that would build upon Millennium Development Goals (MDGs). Its draft was followed by member state negotiations before adoption by UN on August 25, 2015.

The idea of sustainable development was discussed in the Founex Meeting of Experts held in Switzerland on June 4–12, 1971. Founex meeting tried to clarify the relationship between environment and development. The meeting was convened by the organizing committee in the run-up to the Stockholm Conference.[1] Thereafter, the UNGA Reso. of December 20, 1971, was adopted that focused on integrated and rational development.[2] These initiatives assured developing countries to attend the United National Conference on Human Environment 1972 held from June 5 to 16, 1972, which laid down the foundation of the concept of sustainable development. Its declaration stressed on environmental quality and the creation of good life, a life of dignity and well-being, maintenance of the earth to produce vital renewal resources, not to exhaust non-renewable resources, pollution control, economic and social development, better living conditions for present and future generations, and the means to orderly work; careful, rational, and wise planning or management; integrated and coordinated approach and international aid including transfer of financial and technological assistance taking into account of circumstances and requirement of developing countries; environmentally sound science and technology, education, mass-media; promotion of research and international cooperation; and so forth.[3] Many of these aspects are included even in the SDGs.

The outcome of the conference included 109 action plans and the establishment of UNEP for monitoring and implementation. Thereafter the concept entered mainstream, and a process of redefining and refining began globally and at the domestic level.

DOI: 10.4324/9781003733751-25

In 1974, the Cocoyoc Declaration redefined that "development should not be limited to the satisfaction of basic needs. There are other needs, other goals and other values. Development includes freedom of expression and impression, the right to give and receive ideas and stimulus."[4]

"In 1980, IUCN World Conservation Strategy" defined sustainable development as improving and maintaining the well-being of the people and ecosystem . . . integrating economic, social and environmental objectives by changing or strengthening their values, technologies and institutions.[5] Supplementing this strategy, the general principle of UN's World Charter for Nature 1982 was adopted, which stressed that sustainable development requires respecting nature and its essential processes and using natural resources for achieving optimum sustainable productivity through laws and practices of country and inter-governmental and national NGOs as well as by disseminating the knowledge of nature and having arrangement of funds, programs, and administrative structure.[6]

The definition of sustainable development was an important addition in the Brundtland Report 1987: "Our Common Future" was the report of the UN World Commission on Environment and Development (UNWCED). It defined sustainable development as "development that meets the needs of the present without compromising the ability of future generations to meet their own needs which contains the concept of needs and the idea of limitations."[7] The report contained 22 legal principles comprising substantive and procedural aspects covering environmental standards and monitoring and assistance to developing countries.[8]

The meaning was further refined in Caring for the Earth as a successor to the first World Conservation Strategy, which defined sustainable development to mean "improving the quality of human life while living within the carrying capacity of supporting ecosystems." We can achieve this without rejecting the many benefits that modern technology has brought provided that technologies work within those limits.[9]

A landmark United Nations Conference on Environment and Development, also called Rio Conference, was held on June 3–14, 1992, attended by 178 countries comprising 115 Heads of State and Governments. Its outcome was the Rio Declaration consisting of 27 principles, framework convention on climate change, convention on biological diversity, forest principles and Agenda-21, which contained qualitative and quantitative as well as substantive and procedural aspects. It developed a "global partnership norm" with greater responsibilities of developed countries (common but differentiated responsibilities), precautionary principles, principles of integration, no significant harm, and so forth. Its declaration emphasized a healthy and productive life, improving the living standards and better protection of the ecosystems for prosperity, to equitably meet the environmental and development needs of present and future generations, eradication of poverty, to reduce unsustainable patterns of production and consumption for higher quality of life, and so forth as elements of sustainable development, which are carried out in SDGs in some forms. The task of monitoring and implementation was entrusted to the UN Commission on Sustainable Development.

While Rio marked the formal acceptance of sustainable development as a goal, the concept was also discussed in the 1993 Copenhagen summit on Social Development, the 1994 Cairo Summit on Population, the 1995 Beijing Conference on Women, and in the 1997 Kyoto Protocol (Rio +5) aimed at controlling the greenhouse gases.

The number of past instruments had stressed for new external sources of financing for which the Global Environment Facility (GEF) was established by UN as a 3-year pilot program in 1991 but after Rio Conference 1992, it was made a permanent program for assistance to global projects and access to environmentally benign technologies to developing countries.

Unlike Agenda-21 and Rio declaration as a set of principles, the September 2000 UN Summit produced MDGs to be attained by 2015, which were altogether eight goals with specified indicators to monitor.[10]

Immediately, the World Summit on Sustainable Development 2002 issued additional goals on Water and Sanitation, Energy, Health, Agriculture and Biodiversity (WEHAB) agenda. Its outcomes included the Johannesburg Declaration as political commitment and a plan of implementation that intended to achieve the goals through public–private partnership submissions, which would be officially sanctioned as the central part of the summit's work.

In 2012, UN Conference on Sustainable Development (Rio+20) was held, which was for promoting a green economy as an engine for sustainable development and strengthening the institutional framework for sustainable development for which public–private initiatives as key vehicles were fully embraced by the UN, for sustainable development and launched a Universal Inter-governmental High Level Political Forum, replacing the CSD. The outcome document entitled "The Future We Want" was a government commitment for sustainable development. Finally, the Rio+20 Conference became a stepping stone by establishing a process for the SDGs 2015 and UNGA Reso. No. A /Res./70/1 (2015), known as Transforming Our World: The 2030 Agenda for Sustainable Development, was adopted on September 25, 2015, establishing the 2030 Agenda for sustainable development, a plan of action for people, the planet and prosperity, which contains 17 SDGs comprising environmental, economic, and social dimensions of sustainable development.[11] These 17 SDGs are structured into 5 pillars: people, planet, prosperity, peace and partnerships. On January 1, 2016, SDGs officially came into force, which are not legally binding, but countries are committed to implementing them through policies, plans, programs, laws, institutions, and judicial decisions.

Sustainable development has received nearly universal acceptance, and its general and wide recognition were found in several international instruments including in the SAARC documents. The Eighteenth Summit held on November 26–27, 2014, recognized that the SDGs would present opportunities to complement national and regional efforts on sustainable development. They directed to initiate an inter-governmental process to appropriately contextualize the SDGs at the regional level. It also required clean energy, poverty alleviation, agriculture and food security, environment, blue economy, health, education, and good governance for sustainable development.[12]

Prof. David Hunter and others emphasize that "sustainable development has been incorporated in International Declaration and treaties not only on the environment but increasingly in corporate Charter of Business Strategies, national policies and laws, and action plans for governments at every level-from international institution to city council. It is being taught in schools and exalted by civic organizations throughout the world."[13]

The 1997 decision of ICJ on Gabcikovo-Nagymaros Project (*Hungary* v. *Slovakia*) case was related to the 1977 treaty between Hungary and Czechoslovakia (to which Slovakia was the successor) to build and jointly operate a system of Locks on the Danube River for mutual benefits. Both the countries had breached the 1977 treaty and thus the judgement delivered by 15 judges with President Stephen M. Schwebel by a majority directed for reconciliation between environment and development and to consider the concept of sustainable development. Despite the use of the word "concept," however, the court recognized sustainable development by acknowledging the existence of environmental norms as part of the general corpus of international law. The dissenting opinion of Vice-President Judge Christopher Gregory Weeramantry, however, explicitly recognized sustainable development to be more than a mere concept and specifically said that it is a principle with normative value.[14]

Philippe Sands highlights principles of inter-generational equity, principles of sustainable use, principle of equitable use or intra-generational equity, and principles of integration as an element of sustainable development. Paricia Birnie and Alan Boyle on the other hand state substantive and procedural elements of sustainable development. According to them, substantive elements are set out in principles 3–8 and 16 of the Rio declaration, which include the sustainable utilization and conservation of natural resources, the integration of environmental protection and economic development, the right to development, the equitable allocation of resources both within the present generation and between the present and future generation (intra- and inter-generational equity), and the internalization of environmental costs through the application of the polluter pays principle. The principal procedural elements are found in environmental impact assessments. Alexander Kiss has taken a mid-view, suggesting that sustainable development should be considered as a concept while several principles have been proposed to establish its concrete content. Such proposed principles for Kiss include the following seven Principles, which were the Declaration on the Principles of International Law in the field of Sustainable Development adopted at the April 2002 Conference of the ILA held in New Delhi[15]:

1 The duty of states to ensure sustainable use of natural resources.
2 The principle of equity and the eradication of poverty.
3 The principle of common but differentiated responsibility.
4 The principle of the precautionary approach.
5 The principle of public participation and access to information and justice.

6 The principle of good governance.
7 The principle of integration and interrelationship, in relation to human rights and social, economic, and environmental objectives.

The view of David Hunter, James Salzman, and Durwood Zaelke is that they consider it as a concept. They also, however, agree that sustainable development is a principle accepted by the developing countries with worldwide acceptance.[16]

Assistance from developed countries to developing countries is essential for sustainable development. In this regard, it is stated that international environmental treaties now commonly include provisions on economic and technical assistance. The provisions obligate developed countries to transfer environmentally friendly technology to developing countries on fair and most favourable terms. Although these provisions may not be enforceable, aid and assistance including technology transfer is viewed as a necessary concession from the industrialized North to gain developing country participation in the implementation of global environmental treaties.[17]

The nature of such technology is, however, mentioned in general terms only. The World Resources Institute (WRI) identified a wide range of emerging technologies and practices whose broader adoption would promote sustainable development.[18] Their lists include[19] the following:

- Alternative agricultural practices
- Energy efficiency programs
- Industrial waste and recycling
- Industrial bio-products
- Light-weighting and material design
- Information technology
- Dematerialization
- Renewable energy
- Nanotechnology

Which technology is useful in a country depends upon economic and developmental needs and situations. The production, distribution, import, and export of technology has to be regulated by law of the country; such technology should be innovative and environmentally sound.

16.2 Sustainable Development in Nepal

16.2.1 Sustainable Development Before SDGs 2015 in Nepal

16.2.1.1 Plans and Policies

16.2.1.1.1 PERIODIC PLANS

Nepal began to incorporate provisions related to sustainable development in the periodic plans more specifically after the Stockholm Declaration 1972, namely from the fifth five-year plan for halting the pace of environmental degradation. In the sixth five-year plan of 1980–1985, some sectoral aspects, such as soil and water conservation, forestry, conservation of natural resources, and environmental concerns, were incorporated. The seventh plan (1985–1990) emphasized fulfilling the basic needs of the people and to promote alternative energy.

The brown sector and the integrated approach were started from the eighth five-year plan (1992–1997). It also laid down provisions of sustainable economic growth, poverty alleviation, and reduction of regional disparity. In the ninth five-year plan (1997–2002), the objective of poverty alleviation and the classification and zoning of agricultural land was included.[20] The tenth plan (2002–2007) also included poverty alleviation and sustainability. In the 11th plan (2007–2010), poverty alleviation, agriculture, and forest conservation were covered. It was in the 12th plan (2010–2013) that the goals were set to make people feel satisfied through economic growth, resulting in reduction of poverty and sustaining peace and for renewable energy. Before the implementation of SDGs, the last plan was the 13th three-year plan (2013–2015), which also had the objective to alleviate poverty. The priorities were set in energy, tourism, industries, trade, basic education, health, drinking water, infrastructure development, and research and development along with the technology transfer on renewable energy.

16.2.1.1.2 NATIONAL CONSERVATION STRATEGY 1987

As a result of IUCN's World Conservation Strategy 1980, Nepal prepared the National Conservation Strategy (NCS) in 1987, which was endorsed by the government as policy in 1988. It emphasized that sustainable development and conservation must be firmly linked with an objective to meet the needs and improve the quality of life of present and future generations.[21] The objective of the strategy was to satisfy people's basic needs and maintain life support systems, which are closely related to the safe drinking water and effective sanitation facilities.[22] In fact, ensuring sustainable use and maintaining essential life support systems are important objectives of the NCS for Nepal.[23]

This strategy has identified areas of food, shelter, clothing, education, health and security as an area of basic needs programme.[24]

Further, it was also emphasized that "the integration of the conservation with development is a salient feature of the NCS for Nepal."[25]

Regarding the industrial arrangement, the NCS makes a provision to establish a separate policy-making body named the National Council for the Conservation of National and Cultural Resources (NCCNCR), which replaced the NCCNR.[26]

The focus of the NCS was on wise use, protection, preservation and restoration and provision dealing with biological diversity, soil conservation, watershed management, national park, protected areas, wildlife conservation, and natural heritage.

16.2.1.1.3 NATIONAL ENVIRONMENTAL POLICY AND ACTION PLAN INTEGRATING ENVIRONMENT AND DEVELOPMENT 1993

Nepal prepared this policy and action plan to respond and acknowledge the Rio outcomes in the year 1993, which incorporated the provision of integrating environment and development with following main aims and objectives[27]:

- To manage natural and physical resources efficiently and sustainably
- To balance development efforts and environmental conservation for the sustainable fulfilment of the basic needs of the people
- To safeguard national heritage
- To mitigate the adverse environmental impacts of development projects
- To integrate environment and development through appropriate institutions, adequate legislation, economic incentives, and sufficient public resources

These five policies and action plans were intended to cover the quantity and quality aspects to maintain and achieve sustainable development.[28]

16.2.1.1.4 ENVIRONMENTAL IMPACT ASSESSMENT GUIDELINES NATIONAL EIA GUIDELINE 1992

The Guideline makes clear that project proponents have to carry out environmental assessments. It is stated that if there is no certainty on whether any projects require EIA or not, IEE must ascertain it.[29] Later, sector-specific EIA guidelines were also issued, namely EIA Guidelines for Industry Sector 1995 and EIA Guidelines for Forestry Sector 1996.[30]

16.2.1.1.5 NEPAL ENVIRONMENTAL PLANNING AND GUIDELINES 1998

This guideline has set four levels of planning at local, district, regional, and national levels, with more attention to rural planning and urban planning. It states that environmental planning aims to enhance conservation and promote wise use of resources to improve the quality of life and to promote sustainable development. Effective environmental planning requires inter-sectional collaboration, participatory approach, decentralized decision making, and the genuine devolution of power over resource management in local communities.[31]

16.2.1.1.6 SUSTAINABLE DEVELOPMENT AGENDA FOR NEPAL 2003

This Agenda is to respond and comply with Nepal's commitment to Rio Conference, Millenium Development Goals 2000, World Summit on Sustainable Development held in 2002, and many other environmental instruments. According to this Agenda for Nepal, "the over-arching goal of sustainable development in Nepal is to expedite a process that reduces poverty and provides to its citizens and successive generations not just the basic means of livelihood, but also the broadest of opportunities in the social, economic, political, cultural, and ecological aspects of their lives."[32]

It is further stated that a corollary inherent in viewing sustainable development in Nepal in these broad terms is a national resolve to pursue happy, healthy, and secure lives as citizens who lead a life of honour and dignity in a tolerant, just, and democratic nation. This Agenda illustrates broad goals.[33]

16.2.1.1.7 NATIONAL BIODIVERSITY STRATEGY 2002

This strategy was endorsed in 2002 for the protection and sustainable use of biological resources, ecological processes, and ecological systems. It required the creation of the National Biodiversity Unit, a National Biodiversity Coordination Committee, and a thematic committee for the effective implementation of this strategy.[34]

16.2.1.2 Institutional Set-Up

In fact, all the ministries, departments, and offices are concerned with the environment. The names of ministries are time and again changing according to the restructuring of the ministries by the decisions of the government. The other prominent institutions were the Nepal Planning Commission, Environmental Council, and Parliamentary Committee on Environment Conservation. In the Department of National Parks and Wildlife Conservation, there was CITES Unit.

16.2.1.3 Constitutional and Legal Provisions

The Constitution of the Kingdom of Nepal 1990 and the Interim Constitution of Nepal 2006 had a provision of right to life, right to live with dignity, and personal liberty under the fundamental rights and provisions.[35]

The Environment Protection Act 1997 intended to provide for the protection of the environment and to maintain a clean and healthy environment by minimizing as far as possible adverse impacts likely to be caused by environmental degradation and to achieve sustainable development. This Act and the Environment Protection Rules 1997 contained detailed provisions on pollution control, environmental standards, environmental impact assessment, and environment protection council headed by the Prime Minister.[36]

The Local Self-Governance Act 1999 was another Act that empowered the local bodies (urban and rural) to protect forests and other public resources and places. The protection of the environment, cleanliness, and removal of waste, conservation of forests, vegetation, and planting of trees, and so forth was within their jurisdiction.

16.2.2 SDGs in Nepal: 2016–2030

16.2.2.1 Nature of SDGs

History shows that some forms of sustainable development are included in different international environmental instruments. The SDGs contain 17 goals, 169 targets, and 244 correlative indicators (which became 232 after deduction of duplication). In the context of Nepal, 247 indicators were added (making it 479 indicators to track the actual status of SDGs in Nepal). This figure of Nepal was also reduced to 301 due to unavailability of data in the past several years from the date of enforcement in 2016.[37] These SDGs, targets, and indicators are integrated and indivisible, global in nature, and universally applicable paying attention of national realities, capacities, levels of development, and national policies and priorities.

Nepal has classified 17 SDGs into 5 dimensions, namely economic development in SDG numbers 1, 2, 8, 10; social development in SDG numbers 3, 4, 5, 6; infrastructure development in SDG numbers 7, 9, 11; environment management in SDG numbers 12, 13, 15; and institutional development and good governance in SDG numbers 16 and 17.[38] In these dimensions, environment-specific goals were stated only in SDG numbers 12, 13, and 15. However, David Hunter, James Salzman, and Durwood Zaelke have written in their recent book that environment specific goals include SDG numbers 6, 7, 13, 14, 15.[39] These five SDGs, according to these authors, are environment specific, which means the remaining SDGs are general goals which have direct and indirect bearing on environment in developing countries.

Regarding the provisions of the SDGs, we find the following goals:

Goal Number	*Goal (Ends)*	*Goal (Means/Technology)*
1	End poverty in all its form everywhere	Nationally appropriate social protection system, appropriate new technology and financial services, and enhanced development cooperation for developing countries
2	End hunger, achieve food security and improved nutrition, and promote sustainable agriculture	Resilient agriculture practices, enhanced cooperation including technology development and food reserves
3	Ensure healthy lives and promote wellbeing for all age groups	Through prevention and treatment, access to sexual and reproductive health care services, research for vaccine and medicine, increase self-financing, early warning
4	Ensure inclusive and equitable quality education and promote lifelong learning opportunities for all	Free, equitable, and quality primary and secondary education, technical and vocational skills, acquire the knowledge and skills to promote sustainable development, expand global scholarship for vocational training and information and communication technology

(*Continued*)

(Continued)

Goal Number	*Goal (Ends)*	*Goal (Means/Technology)*
5	Achieve gender equality and empower all women and girls	Through public services, infrastructure and social protection policies, participation, and opportunities, enhance the use of enabling technology, in particular information and communication technology
6	Ensure availability and sustainable management of water and sanitation for all	Adequate and equitable sanitation and hygiene and safe drinking water, integrated water resource management, including transboundary cooperation, expand international cooperation and capacity building support to developing countries in water and sanitation including water harvesting, desalination, water efficiency, wastewater treatment, recycling and reuse technology
7	Ensure access to affordable, reliable, sustainable, and modern energy for all	Enhance international cooperation to facilitate access to clean energy, research and technology including renewable energy, energy efficiency and advanced and cleaner fossil-fuel technology, promote investment in energy infrastructure and clean energy technology
8	Promote sustained, inclusive, and sustainable economic growth, full and productive employment and decent work for all	Higher economic productivity through technological upgrading and innovation
9	Build resilience infrastructure and promote inclusive and sustainable industrialization and foster innovation	Upgrade infrastructure, retrofit industries to make them sustainable with increased resource use, efficiency and greater adoption of clean and environmentally sound technologies and industrial processes with all countries taking action in accordance with their respective capabilities, enhance scientific research, upgrade the technological capabilities of industrial sectors, include domestic technology development to increase access to information and communication technology

(*Continued*)

(Continued)

Goal Number	*Goal (Ends)*	*Goal (Means/Technology)*
10	Reduce inequality within and among countries	Implement the principle of special and differential treatment, encourage official development assistance and financial flow where the need is greatest
11	Make cities and human settlements inclusive, resilient, and sustainable	Reduce the adverse per capita environmental impact of cities by paying special attention to air quality and waste management
12	Ensure sustainable consumption and production patterns	Achieve the environmentally sound management of chemicals and all wastes in accordance with agreed international frameworks, reduce their release to air, water, and soil by strengthening technological capacity of developing countries.
13	Take urgent action to combat climate change and its impact	Improve institutional and human capacity and fully operationalize the green climate fund
14	Conserve and sustainably use the oceans, seas, and marine resources for sustainable development	Through enhanced scientific cooperation at all levels, implement science-based management plans and best available scientific knowledge and research capacity, transfer marine technology
15	Protect, restore, and promote sustainable use of terrestrial ecosystems, sustainably manage forests, combat desertification, and halt biodiversity loss	Conservation, restoration, sustainable use, increase afforestation and reforestation, enhance capacity, fair and equitable sharing of benefits and appropriate access
16	Promote peaceful and inclusive societies for sustainable development, provide access to justice for all and build effective, accountable, and inclusive institutions at all levels	Promote rule of law, ensure access to justice, reduce corruption and bribery, substantially strengthen national institutions through international cooperation
17	Strengthen the means of implementation and institutionalize the global partnership for sustainable development	International support to developing countries on finance, technology, and capacity building, trade, etc., cooperation on and access to science, technology, and innovation, enhance knowledge sharing on mutually agreed terms through a global technology facilitation mechanism, promote transfer of environmentally sound knowledge to developing countries

Several means and techniques are found in the SDGs, out of which a few are related to international aid, assistance, and cooperation including the technology and its transfer. David Hunter and others in their book also mention that many of the SDGs have an explicit focus on such areas and capacity strengthening. These authors specifically refer to goals 6.a, 7.a, and 9.4 for having such explicit focus on technology.[40]

Goal 6: Clean Water and Sanitation, 6.a:

> By 2030, expand international cooperation and capacity building support to developing countries in water and sanitation-related activities, water efficiency, wastewater treatment, recycling and reuse technologies.

The goal on Affordable and Clean Energy is provided in goal 7.a:

> By 2030, enhance international cooperation to facilitate access to clean energy research and technology, including renewable energy, energy efficiency and advanced and cleaner fossil-fuel technology, and promote investment in energy infrastructure and clean energy technology.

Similarly, the goal on Industry, Innovation and Infrastructure is in goal 9.4:

> By 2030, upgrade infrastructure and retrofit industries to make them sustainable, with increased resource-use efficiency and greater adoption of clean and environmentally sound technologies and industrial processes, with all countries taking action in accordance with their respective capabilities.

Innovative (new or improved) technology is an important technique in the present global regime, which will yield effective results in the implementation of SDGs in Nepal.

16.2.2.2 Implementation of SDGs in Nepal

16.2.2.2.1 IMPLEMENTATION THROUGH LAWS

The SDGs that came into force on January 1, 2016, have the objectives to "leave no one behind" (LNOB). While these global goals are not legally binding, governments are expected to take ownership and fulfil commitments through law, plans, policies and programs, and institutional framework in addition to the contributions of judicial pronouncements for the implementation.

The Constitution of Nepal 2015 was promulgated and enforced on September 20, 2015. The commitment of Nepal to the integration of social, economic, and environmental dimensions into development frameworks at national and sub-national levels were accordingly embraced into the Constitution of Nepal.

The Constitution serves as the cornerstone of the nation's legal framework, which emphasizes for LNOB: echoing the core principles of SDGs.

The Constitution of Nepal has provided the exclusive power of local government and concurrent power of federal, provincial and local levels with respect to SDGs under five dimensions, as stated in following table[41]:

Serial Number	*Dimensions of SDGs*	*Exclusive Power of Local Governments (Schedule 8)*	*Concurrent Power of Three Tiers of Government (Schedule-9)*	*Sustainable Development Goals*
1	Economic development	• Cooperative institutions • Local market management • Rural roads, argo roads, irrigation • Agriculture and animal husbandry, agro products management • Management, operation and control of agricultural extension	• Cooperatives • Agriculture • Poverty alleviation • Landless squatters' management	• End poverty • Zero hunger • Decent job and economic growth • Reduced inequality
2	Social development	• Basic and secondary education • Basic health and sanitation (drainage) • Management of senior citizens, persons with disabilities and the incapacitated • Collection of statistics of the unemployed • Water supply • Protection and development of languages, culture and fine arts	• Education • Health • Services such as water supply • Social security	• Good health and wellbeing • Quality education • Gender equality • Clean water and sanitation

(*Continued*)

(Continued)

Serial Number	*Dimensions of SDGs*	*Exclusive Power of Local Governments (Schedule 8)*	*Concurrent Power of Three Tiers of Government (Schedule-9)*	*Sustainable Development Goals*
3	Infrastructure development	• Local level development plans and projects • Local roads • Distribution of house and land ownership certificates • Small hydropower projects, alternative energy	• Services such as electricity, irrigation, • Motor vehicle permits	• Affordable and clean energy • Industry, innovation, and infrastructure • Sustainable cities and communities
4	Environment management	• Basic sanitation (waste management) • Environment protection and biodiversity • Disaster management • Protection of watersheds, wildlife, mines and minerals.	• Forests, wildlife, birds, water uses, environment, ecology and biodiversity • Mines and minerals • Disaster management	• Responsible consumption and production • Climate action • Life on land
5	Institutional development and good governance	• Town police • Operation of FM • Local taxes, service charge, fee, tourism fee, penalty • Management of the local services • Collection of local statistics and records • Management of village assembly, municipal assembly, district assembly, local courts, mediation and arbitration • Local records management	• Newspapers • Service fee charge, penalty and royalty from natural resources • Personal events, births, deaths, marriages, and statistics • Archaeology, ancient monuments and museums	• Peace, just and strong institutions • Global partnership

All the 17 SDGs classified into 5 dimensions are well embodied in the Constitution.

In the preamble, the Constitution emphasizes protecting and promoting social and cultural solidarity and ensuring economic equality, prosperity, and social justice by eliminating discrimination and committing socialism. Specifically, the Constitution also incorporates provisions in fundamental rights related to social-economic rights such as right to social justice, right to social security, right to education, and right of women, children, Dalits, senior citizens and economic rights through right to employment, labour, health, food, housing, and so forth.[42] Such social and economic rights are written in detail in the directive principles and state policies as well.[43]

Regarding environmental rights, the Constitution includes the right to live with dignity and right to freedom including personal liberty.[44]

Art. 30 entitled "Right to Clean Environment":

1. Every citizen shall have the right to live in a clean and healthy environment.
2. The victim shall have the right to obtain compensation, in accordance with law, for any injury caused by environmental pollution or degradation.
3. This Article shall not be deemed to prevent the making of necessary legal provision for a proper balance between the environment and development in development works of the nation.

Integration, balance, and environmentally sustainable development is the guiding norm in the Constitution.

Article 51(g) entitled "Policies Relating to Protection, Promotion and Use of Natural Resources" reads:

1. To protect, promote, and make environmentally friendly and sustainable use of natural resources available in the country in consonance with national interest and adopt the concept of inter-generational equity and make equitable distribution of fruits, according priority and preferential right to the local communities;
2. To make multi-purpose development of water resources, while according to priority to domestic investment based on public participation.
3. To ensure reliable supply of energy in an affordable and easy manner and make proper use of energy, for the fulfillment of the basic needs of citizens by generating and developing renewable energy.
4. To develop sustainable and reliable irrigation by making control of water induced disasters and river management.
5. To conserve promote and make sustainable use of forests, wildlife, birds, vegetation and biodiversity, by mitigating possible risks to environment from industrial and physical development, while raising awareness of public about environmental cleanliness.

6 To maintain the forest area in necessary lands for ecological balance.
7 To adopt appropriate measures to abolish or mitigate existing or possible adverse environmental impacts on nature, environment or biological diversity.
8 To pursue the principles of environmentally sustainable development such as the principle of polluter pays, of precaution in environmental protection and of prior informed consent.
9 To make advance warning, preparedness, rescue, relief, and rehabilitation to mitigate risks from natural disasters.

Provisions of environmental rights and sustainable development, Art 51(b), stress on "best interest and prosperity of the people through economic, social and cultural transformation"; Art 51(d) enumerates policies related to economy, industry, and commerce including "obtaining of foreign assistance transparent "and" to utilize knowledge, skill, technology and capital of non-resident Nepalese."

Policies relating to development are found in Article 51(f):

- To formulate sustainable socio-economic development strategies and programs.
- To develop balanced, environmentally friendly, quality, and sustainable physical infrastructures.
- To enhance investment in scientific study, research works, and science and technology.
- To ensure easy and simple access of the public to information technology by developing and expanding information technology to the tune of national needs and make optimum utilization of information technology in national development.

In line with the mandates of the Constitution, the Environmental Protection Act 2019 was enacted, repealing the earlier Act of 1997.

The preamble of the 2019 Act states the objective to maintain a proper balance between environment and development and to protect the fundamental right of the citizen to live in a clean and healthy environment, provide the victim with compensation by the polluter for any damage resulting from environmental pollution or degradation, mitigate environmental impacts on environment and biodiversity, and face the challenges posed by climate change. Unless the environmental study report is approved after completing the legal processes and the environmental standards are complied with, no project can be permitted to be implemented. The Act also requires specified policies, programs, or projects to be carried out post strategic environmental analysis. In addition, prior to the implementation of a project, the proponent shall have to prepare an environmental management plan as required. The Act also empowers the Government of Nepal or provincial Government to make an environmental study. The Act further makes a provision for pollution control, climate change, and national heritage and environment protection areas. A high-level Environmental Protection and Climate Change Management National Council chaired by the Prime Minister and comprising Chief Ministers of all the provinces is created by the Act for environmental protection and climate change. The Act has made a provision of fines, compensation, and order to stop acts if found that their works are executed in a manner inconsistent with the laws or against the terms and conditions of the approval.

As per the power under Sec. 15 of the Act, the government is empowered to make environmental standards. It is informed by the officials in the Ministry of Forest and Environment that there are about 28 environmental standards, including 19 in water.

For the air quality, Nepal has endorsed National Ambient Air Quality Standards 2012, incorporating nine air quality parameters. The status of air quality and compliance of other standards are monitored by the Department of Environment under the Ministry of Forests and Environment. For air quality, there are 27 air quality monitoring stations, with the highest number in the Bagmati Province.[45]

Another Act is the Local Government Operation Act 2017, which came into effect on October 15, 2017. It provides the criteria to divide a state into municipalities or rural municipalities and respective rights, duties, and responsibilities in development and conservation sectors. Further, it empowers the local bodies to work for conservation of the environment, protected areas and species, and controlling environmental pollution, hazard, solid waste, and so forth.

The Mines and Minerals Act 1985 and Petroleum Act 1983 are also equally supporting, as both Acts prohibit the mining activities affecting conservation of soil, forests, and sustainable land use. The National Parks and Wildlife Conservation Act 1973 prevents activities such as hunting, mining, cutting and removal of trees, and cultivating inside the protected areas without permission of the authorities. Its 1994 amendment added provisions of buffer zones and imposed obligation to local people for the conservation and management of protected areas. This amendment also required that the 30–50% income from protected areas must be spent through local buffer zone management committees for the community development of local people and for conservation of biodiversity and livelihood of local people.

The Forests Act 2019 notes two types of forests, namely private forests and government-managed forests (forest conservation area, community forest, partnership forest, religious forests, lease hold forest, national forest within province, or inter-provincial forest). The Act restricts any act that harms the environment.

The Criminal Code of 2017 prohibited food adulteration, water pollution, and environmental pollution and provided punishment to those who commit such crime.[46]

The Foreign Investment and Technology Transfer Act 2019 was enacted to attract foreign investment and technology for the economic prosperity of the country. The Act defines the technology transfer in section 2(f) "to mean any transfer of technology to be made under an agreement between an industry and a foreign investor on the following matter:

(1) Patent, design, trademark, goodwill, technology specificity, formula, process
(2) User's license, technological know-how sharing or use of technological knowledge (franchise)
(3) Provision of foreign technical advisor, management and marketing service of other technological skill or knowledge."

This Act makes the provisions of technology transfer for foreign investment (Sec. 7 entitled "investment may be made through technology transfer") with the provision of the terms of technology transfer to be made for foreign investment as specified in the technology transfer agreement entered between the industry concerned and the foreign investor. For implementing the SDGs, and entering into agreement, the Government of Nepal must follow the provisions of the Art. 278 of the Constitution of Nepal 2015 and provisions of the Treaty Act 1990.

In the area of energy related to SDG number 7, Renewable Energy and Energy Efficiency Bill was approved by the Federal Government and submitted to National Assembly in 2024. As of the time of writing, the legislative management committee is preparing the report based on such discussions. Once the bill is enacted, it will pave the way for transformation of the Alternative Energy Promotion Center (AEPC) of the Ministry of Energy, Water Resources and Irrigation of Government of Nepal into a new organization with an extended mandate.

16.2.2.2.2 IMPLEMENTATION THROUGH PLANS, POLICIES, AND PROGRAMS

Nepal has embraced the transformative agenda of the SDGs aligning its national plans, policies, and programs with goals to ensure sustainable development. Immediately after the enforcement of SDGs, Nepal implemented the three-year development plan from 2016 to 2018/19 and thereafter implemented a five-year plan from 2019 to 2023–24, both of which mainstreamed the SDGs into the country's development strategies. These plans outlined Nepal's approach to SDGs, focusing on poverty reductions, health, education, gender equality, infrastructure development, environmental sustainability, and good governance. Currently, the 16th development plan is in execution from 2024 to 2028/29, which demonstrates Nepal's commitment to the SDGs.[47]

The overarching theme of the 15th plan was a comprehensive 25-year long-term vision till 2043, which aims to achieve "Prosperous Nepal, Happy

Nepali," focusing on building strong economic, social, and physical infrastructure to accelerate economic growth and promote prosperity and happiness. The 25-year vision aims to boost production and productivity by promoting the diversification, modernization, and commercialization of the agricultural sector, utilizing clean, modern, and high technology in the industrial and service sectors and advancing research and development. The current 16th plan has carried this 25-year plan vision with an intention to achieve good governance, social justice, and prosperity.[48]

Additional long-term sectoral perspective plans that are in implementation include water supply, sanitation, and hygiene sector development plan (2016–2030), school education sector plan (2021–2050), Nepal's long-term strategy for net-zero emission 2021,[49] Agricultural Development Strategy (2015–2035), the National Agro-Forestry Policy 2019, Environment Policy 2019, Climate Change Policy 2019, and Forest Policy 2018.

The most recent policy is the Green Hydrogen Policy 2024, which intends to meet the commitment to maintain zero-carbon emissions goal by 2045 with the help of renewable energy, including hydropower. The policy aims to reduce reliance on petroleum products by promoting the generation, storage, and transportation of green hydrogen. It further focuses on utilizing green hydrogen in the production of chemical fertilizers and other industrial processes while also fostering research and innovation in green hydrogen technologies.[50]

The Government of Nepal developed a National Framework for Leave No One Behind in 2022, which implements the LNOB 2030 Agenda. Similarly, the UN Sustainable Development Cooperation Framework 2023–2027 is under implementation, aiming to achieve sustainable development as well as improved governance and participation.[51]

16.2.2.2.3 IMPLEMENTATION THROUGH INSTITUTIONAL ARRANGEMENTS

The Committee on Development, Economic Affairs, and Good Governance of the National Assembly of the Parliament reviews and scrutinizes the implementation and localization of sustainable development initiatives through reviews of National Planning Commission (NPC) and line ministries. The National Steering Committee Chaired by the Prime Minister oversees the implementation and coordination of plans and policies. The main operating body is the Implementation and Monitoring Committee chaired by the Vice-Chairman of NPC, which serves as Central body. Thereafter, there are seven thematic committees each chaired by a member of the NPC. In the provinces, the seven Provincial Planning Commissions are tasked with the localization of SDGs. Mayors or chairpersons work at the grassroots level.[52]

In this way, each level of government has been entrusted with incorporating SDGs into their plans, policies, and programs. Nepal has been at the forefront of localizing the SDGs within its planning and budgetary processes and capacity-enhancing programs across each level of government.[53]

The AEPC, established in 1996, is playing a crucial role in the implementation of renewable energy and energy efficiency, accountable to Ministry of Energy.

16.2.2.2.4 IMPLEMENTATION THROUGH JUDICIAL PRONOUNCEMENT IN NEPAL

The Supreme Court Observed in Puodwar hot water pond case that the construction of Mistry Stream-II hydropower through underground channel without IEE report of geologist as per EPA 2019 and EPR 2020 is against balancing the environment and development and neglecting the touristic, cultural, religious and economic importance and archaeological value.[54]

In *Nijgad International Airport* case, the supreme court full bench comprising five justices, by a majority, said that the sustainable forest management will fall under the concept of sustainable development and thus the projects should be implemented only if the EIA norms, procedure and processes are objectively followed in accordance with the law.[55]

In another case, involving the issue of the constitutionality of clause 199 of the annual budget of 2020–21 (2078–79 B.S.) fiscal year that permitted excavation, exploitation, and export of natural resources, namely stone, gravel, and sand, the constitutional bench of five justices acknowledged the constitutional provision of clean and healthy environment and emphasized that the conservation, sustainable use, and equitable sharing of benefits of natural resources, energy, and biodiversity for present and future generations is a sustainable development mandated by the Constitution. The court stated that sustainable development is a principle in several cases with its intrinsic value: The court said that this principle is being developed as a basic principle for environmental justice and climate justice and this principle involves many other principles, namely intergenerational equity principle, precautionary principle, polluter-pays principle, prior informed consent, and ecosystem service. Other responsibilities include environmental integrity for present and future generations, non-regression, public trust principle, sovereignty over natural resources, common but different responsibility principle, and regional cooperation. On the issue of irreparable damage to the environment, the court further stated that it should ascertain with scientific facts and public hearings.[56]

The court concluded that sustainable development principles should be recognized as fundamental principles. In view of legislative powers of federal province and local governments and when there is uncertainty of excavation place and impacts of excavation, budget provision of clause 199 in such situation was against the constitutional, legal and policy provisions but now, as the clause is deleted in the supplementary budget, no specific order is required. Further the court stressed that without federal law, no excavation, exploitation and export work can be allowed and the court therefore issued mandamus to make arrangement for effective environmental protection laws.[57]

The Supreme Court held[58] that no citizen's social, economic, and cultural rights can be protected without clean and healthy environment. It also held that a clean and healthy environment is a right to life, for which necessary equipment needs to be installed.[59]

In this case, the court was satisfied with the installation of venturi wet-scrubber in a VSR plant for controlling dust and smoke.[60]

In number of other cases, the court stated that sustainable utilization of natural resources is the sustainable development in real sense,[61] natural resources belong to present and future generations,[62] historical heritage is a sign of civilization and priceless identity,[63] haphazard exploitation of natural resources that endanger its existence is not sustainable development,[64] and environment and biodiversity conservation must be respected (obligation to respect), protected (obligation to protect), and fulfilled (obligation to fulfil). So the government should fulfil its obligation and hand over a clean environment to future generation.[65] The court also spoke that sustainable development cannot be imagined when environment is adversely affected.[66]

16.2.2.2.4.1 Goal-Wise Implementation in Nepal Nepal has so far submitted three Voluntary National Reviews (VNRs), in 2017, 2020, and 2024 to the United Nations on the implementation of the SDGs, which highlight its legal, policy, and institutional framework. Therein, I have also added some landmark judicial decisions on sustainable development that also help to implement SDGs. Goal-wise implementation of SDGs is briefly outlined now.

Goal 1: the Social Security Allowance program, foreign employment opportunities, increase of tangible property in the name of woman are in progress.[67]

Zero hunger of goal 2: It is addressed by the Zero Hunger Challenge Nepal Initiative, providing agricultural training, supporting smallholder farmers, promoting nutritious crops, and distributing food to vulnerable communities. But under-nourishment, anaemia, and lack of food storage facilities in remote areas are still prevalent.[68]

Good health and well-being of goal 3: It is achieved through reducing new HIV infections and maternal mortality ratio (succeeded in eye treatment but reducing tuberculosis, communicable diseases is still a challenge).[69]

Quality education of goal 4: The enrolment rate in primary education and gender parity has increased but sufficient physical facilities and disparity in access to quality education for all exist.[70]

Goal 5: Reduction in physical or sexual violence against women and child marriage has occurred (but achieving gender equality in job opportunity, labour, and representation in parliament is still in progress).[71]

The wash initiative: it has helped to achieve goal 6 for clean water, sanitation, and hygiene education but still faces the challenge of disparities in access to safely managed drinking water services between urban and rural areas.[72]

Goal 7: Affordable and clean energy is in progress in renewable technologies such as solar, wind, hydro, and biogas as spearheaded by

AEPC. However, Nepal is behind in strengthening institutional capacities, fostering private sector participation, and promoting regional energy cooperation.[73]

Goal 8: It is related to decent work, employment, and sustainable economic growth, which has made limited progress through training, access to microfinance, and business support services. But COVID-19 destroyed the economy, leading to losses of jobs and economic activity.[74]

Improvement in road density: The establishment of the National Innovation Centre (NIC) is a remarkable achievement for goal 9. NIC is considered a catalyst for a prosperous and technologically advanced Nepal. But the lack of required investment and enhanced connectivity is troubling.[75]

Goal 10 reduced inequality: The goal is to provide access to education, healthcare, and economic opportunity to marginalized and vulnerable groups, including women, Dalits, and indigenous communities, and balance development across the provinces and local levels. Now the Dalit welfare programs empower Dalits to participate fully in social economic sphere.[76]

Goal 11: It requires developing resilient infrastructure, including transport, housing, public places, and clean environment for sustainable cities and communities, which is in progress after the Earthquake Resilient Reconstruction and Rehabilitation Program of 2015.[77]

Responsible consumption and production: Goal 12 has moderately progressed through the Plastic Bag Free Kathmandu Valley Campaign. Improvement in resource efficiency and waste management across each sector requires policy and technological shifts and collaborations among stakeholders.[78]

Climate action: Goal 13 is in progress, with reduced greenhouse gas emissions and preparations of 241 local adaptation plans of action and community adaptation plans. Renewable energy for rural livelihood has helped to reduce reliance on fossil fuels.[79]

Goal 15: Terrestrial ecosystems, forest management, combating desertification, halting land degradation, and biodiversity loss are in progress. The National Parks and Wildlife Conservation Program has shown progress through habitat restoration, anti-poaching measures, and community engagement but requires scientific management.[80]

Goal 16: Establishment of judicial committees at all local levels and free legal aid in court cases for the needy people have been implemented so far in 63 districts (provided access to justice for all).[81]

Strengthening the means of implementation and revitalize the global partnership for sustainable development: Implementation of Goal 17 is collaborated and coordinated through the Development Partners Coordination Committee in Nepal, which bring together government agencies and international organizations and ensure alignment of development efforts. The government budget, revenue, official development assistance, foreign direct investment, and remittance flow have a mixed impact, which requires refinement in policies and enabling environment and institutional arrangements.[82]

16.3 Successes, Challenges, and Conclusions

There have been remarkable strides in electricity access, energy consumption, cleaner cooking methods, and road transportation. Under climate action, notable achievements were observed in forest and species conservation. Access to basic water supply and sanitation has also increased. Some advances are also found in environmental impact minimization.[83] On the whole, Nepal's average progress reached 58.6% against mid-term 2022 targets and 41.7% against long-term 2030 targets.[84]

Despite Nepal's dedication and sincerity in implementing the SDGs, several challenges and constraints remain. One of them is financial. Budgets and annual program coding by SDG categories have helped track SDGs' spending progress, and intra-government management information system has helped to facilitate data collection and its sharing; however, allocation of annual budget is not sufficient. It is found that the percentage of annual budget allocation for SDG numbers 6, 7, 9, 13, and 15 has seen a decreasing trend; there was no budget allocation for SDG 14 and data is unavailable for SDG 5. Nepal seeks international support to financing its SDG programs and to address other challenges and constraints.

SDGs are changing according to time and place, which indicates that by the completion of the 2030 target, the United Nations will bring another global instrument in the light of international need in a different shape. The aid and assistance in financing, expertise, and innovative technology will hopefully get additional focus. The issue of sustainable development is still a *concept* for many, *principle* for some, and *mid-way* for others, who treat it as a concept supported by principles. The SDGs are still wide and aspirational. The High-Level Political Forum is under the control of UNGA and UN Economic Social Council (ECOSOC). The UN Division for Sustainable Development Goals, a UN Department of Economics Social Affairs, serves as the Secretariat for the SDGs, tasked with monitoring and providing assistance for SDG implementation. The issue of monitoring and assistance as stated above could be resolved if the UN takes a clear stand in the next instrument.

Nepal is recognizing sustainable development as a *principle*. The regional body of SAARC has also recognized the SDGs. It can be assumed that a Global Agreement of updated SDGs will be adopted by the United Nations while SAARC will also move ahead with its course in addressing socio-economic and environmental problems for a *Prosperous and Happy South Asia*. It can be suggested that the next global summit on sustainable development in 2030 should be used as an opportunity for the adoption of a global pact on SDGs.

Notes

1 UNEP Secretariat, "Environment and Development," 1 *Environment International* 141–146 (Pergamon Press, 1978).
2 UN General Assembly, *Development and Environment*, GA Res 2849, GAOR, UN Doc A/RES/2849 (December 20, 1971).

3 Stockholm Declaration, 1972, UN Doc A/Conf 48/14/Rev.1 (December 15, 1972), Preamble and Principles 1–26.
4 UNEP, the Cocoyoc Declaration adopted by the participants in the UNEP/UNCTAD Symposium on "Patterns of Resource Use, Environment and Development Strategies" held at Cocoyoc, Mexico, from 8 to 12 October 1974, UN Doc A/C.2/292 (Nov 1, 1974).
5 Ben Boer, "Institutionalizing Ecologically Sustainable Development: The Roles of National, State and Local Governments in Translating Grand Strategy into Action," 31/2 *Willamette Law Review*, 308 (1995).
6 UN World Charter for Nature 1982, Art. 1, 4, 14, 15, 17, 21.
7 WCED, *Our Common Future* 43–46 (Oxford University Press, Delhi, 1989).
8 WCED, *Our Common Future*, at 348–351, above note 7.
9 *Caring for the Earth: A Strategy for Sustainable Living* 3, 4 (IUCN/UNEP/WWF, Gland, Switzerland, 1991).
10 The eight goals were as follows:

1 Eradicate extreme poverty and hunger
2 Achieving universal primary education
3 Promote gender equality and empower women
4 Reduce child mortality
5 Improve maternal health
6 Combat HIV/AIDS, malaria and other diseases
7 Ensure environmental sustainability
8 Developing a global partnership

11 These 17 SDGs are discussed in section 2 in this chapter.
12 David Hunter James Salzman, et al., *Part A International Environmental Law and Policy* 141, 159–161 (6th ed., Foundation Press, 2022). See also *Kathmandu Declaration Eighteenth SAARC Summit Kathmandu* (26–27 November 2014).
13 *Kathmandu Declaration Eighteenth SAARC Summit Kathmandu*, at 161.
14 *Kathmandu Declaration Eighteenth SAARC Summit Kathmandu*, at 333–334, above note 13.
15 *Kathmandu Declaration Eighteenth SAARC Summit Kathmandu*, at 196–198, 215, above note 13. See also Philippe Sands, *Principles of International Environmental Law*, vol. I, 199 (Manchester University Press, 1995); Patricia Birnie and Alan Boyle, *International Law, and the Environment* 86–95 (Oxford University Press, 2002).
16 Birnie and Boyle, *International Law, and the Environment*, above note 15 at 336.
17 Birnie and Boyle, *International Law, and the Environment*, above note 15 at 77.
18 Birnie and Boyle, *International Law, and the Environment*, above note 15 at 74–75.
19 Birnie and Boyle, *International Law, and the Environment*.
20 *Tenth Plan* National Planning Commission (2002–2006). See also Amber Prasad Pant, "Nepal's Legal Initiatives on Land Use for Sustainable Development," in Nathalie J. Chalifour Patricia Kameri-Mbote et al. (eds.), *Land Use Law for Sustainable Development* 423 (Academy of Environmental Law Research Studies, Cambridge University Press, 2007).
21 *Building on Success: The National Conservation Strategy for Nepal* xv (HMG/IUCN, 1988).
22 *Building on Success: The National Conservation Strategy for Nepal*, above note 21 at 34.
23 *Building on Success: The National Conservation Strategy for Nepal*, above note 21 at 85.
24 *Building on Success: The National Conservation Strategy for Nepal*, above note 21.
25 *Building on Success: The National Conservation Strategy for Nepal*, above note 21 at 87.
26 *Building on Success: The National Conservation Strategy for Nepal*, above note 21 at 90, 121.
27 *Nepal Environmental Policy and Action Plan, Integrating Environment and Development* XI (HMG Environment Protection Council, August 1993).

28 A. P. Pant, "Sustainable Infrastructure Development: Principles and the Nepalese Experience," 12/1 *Nepal Law Review* 9 (1998).
29 National EIA Guideline 1992, sec. 9; Dr. Amber Prasad Pant, "Environmental Law in Nepal: An Overview," 1 *Survey of Nepalese Law* 12 (2000).
30 Pant, "Environmental Law in Nepal," above note 29 at 12–14.
31 Dr. Amber Prasad Pant, "International Environmental Governance and the Implementation of Environmental Treaties in Nepal: An Overview," 15/1 & 2 *Nepal Law Review* 26 (2002).
32 Sustainable Development Agenda for Nepal, HMG of Nepal 2 (2003).
33 Sustainable Development Agenda for Nepal, above note 32.
34 Dr. Amber Prasad Pant, "Climate Change and Forest Management in Nepal," in Ed. Couzins, Alexander Paterson et al. (eds.), *Protecting Forest and Marine Biodiversity: The Role of Law* 204 (Edward Elgar, 2017).
35 Constitution of the Kingdom of Nepal 1990, Art. 12(1) & (2), 26(3) & (4) and Interim Constitution of Nepal 2006 Art. 12(1) & (2), 35(4) & (5).
36 Pant, above note 34 at 209.
37 *Nepal Sustainable Development Goals: Revised Indicators*, 1–2 (Government of Nepal National Planning Commission, Kathmandu, 2023).
38 *Voluntary National Review of Sustainable Development Goals*, 22 (Government of Nepal National Planning Commission, Kathmandu, 2024).
39 Salzman, et al., above note 12 at 173.
40 Salzman, et al., above note 12 at 79–80.
41 *Voluntary National Review of Sustainable Development Goals*, above note 38 at 22.
42 Constitution of Nepal, Art. 18, 24, 31, 33–44.
43 Constitution of Nepal, Art. 50(2) and (3), Art. 51(b), (c), (d), (e), (f), (h), (i), (j). See also the *Social Security Act 2018* which gives effect to the provisions of the Constitution by providing social security allowance to Nepali citizens, namely senior citizens, indigent, incapacitated and helpless persons, helpless single women, citizens with disabilities, children, and citizens unable to take care of themselves.
44 Constitution of Nepal, above note 42, Art. 16 and 17.
45 Govind Prasad Lamichhane, Nabina Maharjan, et al., *Status of Air Quality in Nepal Annual Report, 2021* 17, 18, 22 (2023).
46 National Criminal Code 2017, sec. 107, 111, and 112.
47 *Voluntary National Review of Sustainable Development Goals*, above note 38 at 1.
48 *Voluntary National Review of Sustainable Development Goals*, above note 38 at 11.
49 *Voluntary National Review of Sustainable Development Goals*, above note 38 at 6–7.
50 *Progress at a Glance: Year in Review (2023/24)*, 9–10 (Alternative Energy Promotion Centre, 2024).
51 *Voluntary National Review of Sustainable Development Goals*, above note 38 at 10.
52 *Voluntary National Review of Sustainable Development Goals*, above note 38 at 20.
53 *Voluntary National Review of Sustainable Development Goals*, above note 38 at 5.
54 *Rumlal Pun and others* v. *Ministry of Energy, Water Resources and Irrigation*, Case no. 079-wo-1520, decided on Jan. 28, 2024 (unpublished).
55 *Senior Advocate Prakash Mani Sharma Representing Pro Public and Others and Senior Advocate Krishna Prasad Bhandari* v. *Govt. of Nepal*, Writ No. 076-wf-0006 and 076-wf-0007, decided on May 26, 2022 (unpublished).
56 *Advocate Shailendra Prasad Ambedkar and Others* v. *Govt of Nepal, Prime Minister and Office of the Council of Ministers and Others*, Writ No. 077-wc-0099 and other 7 writs, decided on April 20, 2022 (unpublished).
57 *Advocate Shailendra Prasad Ambedkar and Others*, above note 56.
58 NKP 2021, no. 7, p. 1278.
59 NKP 2022, no. 7, p. 1315.
60 NKP 2022, no. 7, p. 1315.
61 *Advocate Sunil Ranjan Singh* v. *Government of Nepal, Prime Minister and Council of Minister*, NKP 2020–21, no. 9, p. 1727.

62 *Advocate Ram Chandra Simkhada* v. *Government of Nepal, Prime Minister and Council of Ministers*, NKP 2019, no. 2, p. 367.
63 *Rato Kumari Bikash Sewa Samiti* v. *Prime Minister and Council of Ministers*, NKP 2018, no. 4, p. 699.
64 *Advocate Khagenedra Subedi and Others* v. *Government of Nepal, Prime Minister and Council of Ministers*, NKP 2018–19, no. 9, p. 1626.
65 *Advocate Tulasi Simkhada* v. *Government of Nepal, Prime Minister*, NKP 2019, no. 10, p. 1818.
66 *Dharam Bahadur Lamichhane and others* v. *Ministry of Physical Infrastructure and Transport and others, NKP 2018–19*, no. 9, p. 1615.
67 *Voluntary National Review of Sustainable Development Goals*, above note 38 at 39.
68 *Voluntary National Review of Sustainable Development Goals*, above note 38 at 30.
69 *Voluntary National Review of Sustainable Development Goals*, above note 38 at 30, 31.
70 *Voluntary National Review of Sustainable Development Goals*, above note 38 at 31, 32.
71 *Voluntary National Review of Sustainable Development Goals*, above note 38 at 32, 33.
72 *Voluntary National Review of Sustainable Development Goals*, above note 38 at 33, 34.
73 *Voluntary National Review of Sustainable Development Goals*, above note 38 at 34, 35.
74 *Voluntary National Review of Sustainable Development Goals*, above note 38 at 35, 36.
75 *Voluntary National Review of Sustainable Development Goals*, above note 38 at 36, 37.
76 *Voluntary National Review of Sustainable Development Goals*, above note 38 at 37.
77 *Voluntary National Review of Sustainable Development Goals*, above note 38 at 38.
78 *Voluntary National Review of Sustainable Development Goals*, above note 38 at 39.
79 *Voluntary National Review of Sustainable Development Goals*, above note 38 at 39, 40.
80 *Voluntary National Review of Sustainable Development Goals*, above note 38 at 40, 41.
81 *Voluntary National Review of Sustainable Development Goals*, above note 38 at 41, 42.
82 *Voluntary National Review of Sustainable Development Goals*, above note 38 at 42, 43.
83 *Voluntary National Review of Sustainable Development Goals*, above note 38 at 8.
84 *Voluntary National Review of Sustainable Development Goals*, above note 38 at 2.

17 Carbon Dioxide Capture, Utilization, and Storage as an Implementation of ESG

Moon-hyun Koh

17.1 Introduction

Environmental issues are not bound by borders. As the gravity of environmental issues is becoming even serious, worldwide interest is growing upon ESG. CCUS (carbon dioxide capture, utilization, and storage) technology, a representative technology for large-scale greenhouse gas reduction, can be said to be one of the important technologies for implementing ESG (Environmental, Social, Governance). The importance of CCUS is growing as the dependence on fossil fuels does not turn down and as development of efficient eco-friendly energy is slower. CCUS is considered the most promising technology, which provides a sustainable solution to supply abundant energy sources while meeting the planned reduction target. Thus, each state is active in research and development, not to fall behind with the CCUS technology.

The Korean government is not an exemption in supporting CCUS and sustainable development. Yet the development of CCUS technology and the implementation of CCUS projects cannot be fully achieved without institutional and legal support.

A legal framework is indispensable as it functions as a vessel to systematically support the technology. Nonetheless, worldwide CCUS legislation does not have a coherent framework, resulting in a wide gap between technological development and its policies. Therefore, necessity of legislative efforts to help the development of CCUS is continuously being highlighted. In South Korea, the legal research team of K-COSEM (Korea-CO_2 Storage Environmental Management) created a draft of the integrated CCS Act to oversee CCS through industry–academia cooperation. As a result, it was passed through the Congress in 2024; review on the CCUS Framework of Korea and its proposed Integrated Draft Act shall shed light on future CCUS legislation and its positive implication on the global ecosystem.

DOI: 10.4324/9781003733751-26

17.2 Concept of CCUS and Role of CCUS in Clean Energy Transitions

17.2.1 Concept of CCUS

The issue of reducing carbon dioxide, the main cause of global warming, is a hot issue around the world, and the 2015 Paris Climate Agreement also focuses on this issue. The Paris Agreement has set a universal target of holding global warming to well below 2°C and urged nations to pursue efforts to limit the rise to 1.5°C. Carbon dioxide capture, utilization, and storage (CCUS), a representative technology for massive reduction of carbon dioxide, has recently been receiving attention as part of the Moon Jae-in administration's Green New Deal policy.

CCUS involves the capture of CO_2, generally from large point sources like power generation or industrial facilities that use either fossil fuels or biomass as fuel. If not being used on-site, the captured CO_2 is compressed and transported by pipeline, ship, rail, or truck to be used in a range of applications (CCU), or injected into deep geological formations such as depleted oil and gas reservoirs or saline aquifers (CCS).[1] CCS (carbon dioxide capture and storage) technology has been defined as a technology that captures carbon dioxide from sources that emit large amounts of carbon dioxide using fossil fuels, such as thermal power plants, steel plants, and cement plants, before it is emitted into the atmosphere, transports it, and stores it in safe underground geological layers to isolate it from the atmosphere. Recently, the concept has been expanded to include processing technology, which is a technology that utilizes the captured carbon dioxide to increase the recovery of oil or natural gas, or converts carbon dioxide into useful substances such as polymers or biodiesel using chemical or biological methods (CCU).[2]

Carbon capture and utilization (CCU) refers to a range of applications through which CO_2 is captured and used either directly (i.e. not chemically altered) or indirectly (i.e. transformed) into various products. Around 230 Mt of CO_2 is currently used each year, mainly in direct use pathways in the fertilizer industry for urea manufacturing (~130 Mt) and for enhanced oil recovery (~80 Mt).[3]

New utilization pathways in the production of CO_2-based synthetic fuels, chemicals, and building aggregates are gaining momentum. The current project pipeline shows that just under 15 Mt of CO_2 per year could be captured for these new uses by 2030, including around 8 Mt CO_2 in synthetic fuel production. If all announced projects are commissioned, they could reach around two-thirds of the level of CO_2 utilization for synthetic fuel production by 2030 envisaged in the Net Zero Emissions by 2050 (NZE) scenario. In addition, to be compatible with the NZE scenario, all the CO_2 would need to come from air or biogenic sources, which is currently only the case for just over 4 Mt CO per year of planned CCU to fuels capacity for 2030.[4]

CCS technology is largely divided into three stages: the process of capturing carbon dioxide generated during the combustion of fossil fuels, the process of transporting it to storage, and the process of permanently injecting it underground. Capture refers to the process of technically filtering out only carbon dioxide from greenhouse gases generated from large-scale energy complexes such as power plants, industrial processes, and natural gas wells and compressing it into a liquefied state. "Transport" refers to transporting the captured carbon dioxide to a area where it will be stored deep underground using pipelines or ships. "Storage" refers to the process of capturing and transporting carbon dioxide by drilling deep underground (about 800 metres or more) and injecting it under high pressure into brine layers, fully developed oil and gas fields, coal seams, and so forth.[5] The main sources of carbon dioxide in many countries, which are the target of CCS technology, are the power generation sector that utilizes gas, coal, geothermal heat, and biomass. The capture process can be divided into pre-combustion capture and post-combustion capture, and it is the most expensive stage in CCS technology, requiring R&D to reduce costs. The capture stage generally does not raise significant legal issues beyond the scope related to the construction and operation of industrial plants. The transportation stage includes purifying and compressing the captured carbon dioxide to the supercritical phase in order to transport it. The transportation process is usually done through pipelines, which is more economical than using transportation means such as rail or ship. The storage stage is to inject it into a geologically feasible storage layer, which must be deep enough as a caprock structure, physically impermeable, and have enough space to permanently trap CO_2.

The success or failure of CO_2 underground storage depends on finding a geological structure that can store large amounts of CO_2 stably for a long period of time, and the ideal geological structure for underground storage is an oil or gas field that has held fossil fuels for a geological period.[6] However, these geological structures are distributed regionally, so their utilization is inevitably limited in regions like Korea that do not have them. As an alternative, marine underground storage, that is, deep saline aquifers, is being studied. In addition, geological structures such as coal seams, salt beds, and underground cavities are being studied as candidates for underground storage.

Meanwhile, CO_2 can be stored in porous rock types that can contain crude oil or natural gas. Likewise, impermeable cap rocks, which store oil and gas underground and prevent it from moving to the surface, can also trap CO for long periods of time on a geological scale. The key issue raised here is unpredictability.[7]

In other words, it is impossible to completely resolve concerns about whether the carbon dioxide injected underground will continue to remain sealed underground, rise vertically and leak to the ground, or move in other unpredictable ways, causing harm to the natural world.[8]

Therefore, the key legal issues related to CCUS include regulations to ensure the safety of CCUS technology, as well as raising the public's environmental awareness of climate change, cooperation and communication with ordinary citizens, legal compensation for landowners or stakeholders, and a balance between public and private interests.

CCUS has a convergent nature that requires a high level of technicality due to the nature of its concept, and thus requires sensitive consideration of application even to a single legal term. For example, in the process of promoting legislation related to CCUS, factors such as which process to set as the capture or storage stage must be determined, but this cannot be determined solely by legal theory, and legislation must be made based on the technical results after technical research. This is because even if a technical problem such as CO_2 leakage occurs, the business entity operating each process must be clearly identified in order to clearly distinguish responsibility. In addition, due to the nature of CCUS technology, which involves excavation deep underground, issues such as how far underground the landowner's ownership extends and how far a system should be established to protect the rights of stakeholders may arise. In this respect, the establishment of a flexible and timely system for CCUS experiments and verification must be supported, and revision of existing laws and systems related to the environment and resource development is also necessary.[9]

17.3 The Importance and Role of CCUS in Clean Energy Transitions

CCUS can be retrofitted to existing power and industrial plants, allowing for their continued operation. It can tackle emissions in hard-to-abate sectors, particularly heavy industries like cement, steel, or chemicals. CCUS is an enabler of least-cost low-carbon hydrogen production, which can support the decarbonization of other parts of the energy system, such as industry, trucks, and ships. Finally, CCUS can remove CO_2 from the air to balance emissions that are unavoidable or technically difficult to abate.[10]

17.4 CCS Policy and CCS Projects in South Korea

17.4.1 CCS Policy in South Korea

The South Korean government has been implementing various policies to achieve the targets together with industrial and ordinary citizens, after setting the national greenhouse gas reduction target in 2009. This target has been updated prior to the Paris Agreement, and finalized in December of 2015. It is well known that the Paris Agreement has set a universal target of holding global warming to well below 2°C and urged nations to pursue efforts to limit the rise to 1.5°C.[11] In accordance with the principle of discriminatory accountability, the Agreement recognized the voluntary reduction targets, also known as Intended Nationally Determined Contributions (INCD), which was submitted by each nation under their own circumstance prior to the convention (further agreed on the principle that each national submit an upward target every five years from 2020). Korea also submitted INDC in

June 2015 to reduce greenhouse gas emissions by 37% compared to BAU in 2030, ahead of the Paris Agreement, which became the new long-term goal in setting the nationwide CCS framework.[12]

In July of 2011, the government established and announced the National CCS Comprehensive Plan to reduce greenhouse gas emissions and create new growth engines through efficient CCS technology development, and it is updated with the New Climate System worldwide under Paris Agreement under the new terminology Korea CCS 2020. It remains a national top-level CCS project plan that aims to promote R&D on CCS, establish infrastructure for CCS, and commercialize the international technology by 2020 with the vision of Leaping a Global Power of CCS Technology as an action plan for promoting the development and use of CCS technology.[13] The Ministry of Environment launched the Korean CO_2 Storage Environmental Management Research Center in April of 2014. It promotes the development of technologies to detect and manage the environmental impacts and risks associated with the storage of carbon dioxide in the ground. In accordance with the National CCS Comprehensive Implementation Plan and the Roadmap for Achieving the National Greenhouse Gas Reduction Target, the Ministry of Environment develops technologies to assess and manage the environmental impacts and risks of CCS technology, one of the key ways to reduce greenhouse gas emissions.

17.4.2 CCS Projects in South Korea

Major government offices that are responsible for CCS-related research are the Ministry of Science, ICT and Future Planning, the Industry, the Ministry of Trade, Industry and Energy, and the Ministry of Maritime Affairs and Fisheries. Affiliated organizations and research teams perform further research on specific topics. Under affiliation with the Ministry of Science, ICT and Future Planning, Korea Carbon Capture & Sequestration R&D Center (KCRC) conducts the studies. It serves as hub for CCS, to carry out the macro-plan for achieving the goal by 2020.[14] It basically plays a role of securing and verifying original technology for CCS, and of increasing the capability for trades through technology transfer and commercialization. It concentrates on all three: capturing, storing, and converting stage, major elements of CCS. From 2011 to 2013, it strove to verify the concept of the first stage CCS technology, and acquired more than four types of CO_2 capture technology through new material addition and process modelling. From 2014 onwards, it stepped into the second stage, under mission to develop the bench-scale CCS technology. From 2017 to 2019, it was in the midst of pilot-scale verification to tackle the large-scale emission sources on the basis of stages 1 and 2.[15] In fact, KCRC succeeded in demonstrating 150 Nm^3/h, which is the world's top CO_2 capture technology, in developing the world's first graphene-based next-generation CO_2 separator, in developing high-value biodegradable plastic manufacture technology with simultaneous disposal of greenhouse gases, in developing core source technology for algae selection and cultivation for CO_2 capture, and much more.

The Ministry of Environment is in charge of developing technologies necessary for establishing an environmental management system throughout the entire carbon dioxide capture and storage process for the National CCS Comprehensive Promotion Plan established by the Green Growth Committee in July 2010. Accordingly, the Korea Environmental Industry & Technology Institute launched the Carbon Dioxide Underground Storage Environmental Management Research Center (Korea-CO_2 Storage Environmental Management, K-COSEM) in 2014 and has been conducting research on leak monitoring and environmental risk assessment/management technology development and environmental management laws and regulations for carbon dioxide underground storage until the end of 2020. As the research director of the fourth sub-project of the K-COSEM research centre, I also studied the so-called "Draft Act on Carbon Dioxide Storage, etc." and the CCS Public Acceptance Guidelines.

17.5 "Draft Act on Carbon Dioxide Storage, etc." and "CCUS Act" in South Korea

17.5.1 Introduction

The national vision of achieving carbon neutrality by 2050 is closely related to the implementation of the nationally determined contribution (NDC) under Articles 3 and 4 of the Paris Agreement in the broad sense.

According to IEA data, CCUS technology is a representative technology for large-scale greenhouse gas reduction and is reported to contribute greatly to carbon neutrality by 2050[16] and reduce carbon dioxide in heavy industries such as cement, steel, and chemicals.[17]

In a broad sense, the content related to CCUS is in line with the core international response system for achieving carbon neutrality in response to the climate crisis as the "best available science" stipulated in Article 4, Paragraph 1, of the Paris Agreement, and ultimately corresponds to our country's NDC implementation obligation. The provisions related to CCUS under the Framework Act on Carbon Neutrality and Green Growth for Climate Crisis Response (Framework Act on Carbon Neutrality) include Article 33 (Expansion of Carbon Sink Sources, etc.) and Article 34 (Promotion of Carbon Capture, Utilization, and Storage Technology) of the Framework Act on Carbon Neutrality. In order to achieve our country's goal of carbon neutrality by 2050, it is very important to establish more systematic and specific laws and regulations to strengthen investment in and institutional support for CCUS technology development, which is in line with the "provision of incentives and promotion of greenhouse gas emission mitigation" of Article 6, Paragraph 4 (b) of the Paris Agreement. Therefore, the enactment of the so-called "Act on Carbon Dioxide Storage, etc." is urgent. As a result of my 15-year argument while proposing the so-called "Draft of Korean Integrated CCS Act," CCUS Act was enacted on February 6, 2024, and came into effect on February 7, 2025, by Law No. 20203.

In this chapter, this author first introduces the "Draft of Korean Integrated CCS Act" as an important means of South Korea's carbon dioxide capture, utilization, and storage policy, and then examines the CCUS Act.

17.5.2 *Outline of South Korean Integrated CCS Draft Act*

Following technological advance, the Ministry of Environment supported the efforts to conceive the CCS-related laws and policies. Despite efforts, the CCS-related Act did not exist in South Korea until 2024. In order to establish a stable settlement of CCS, necessity on establishment of realistic and applicable laws and institutions emerged. Thus, the K-COSEM operated the Task Force for Integrated CCS Act from April 2014 to the end of 2018, under cooperation with scientists and engineers in the field. On March of 2016, a provisional South Korean Integrated CCS Draft Act (CCS Single Act) was reported to the Ministry of Environment. After thorough review on the framework throughout April and May, a final draft was presented in 2018. The South Korean Integrated CCS Draft Act reflects a legal system that encompasses a variety of ministries and agencies, central and local, private and public sectors, the working and research communities, in consideration of wide spectrum of perspectives.

South Korean Integrated CCS Draft Act consists of overall ten chapters and addendum. An outline of the Integrated CCS Draft Act and its description is as follows.[18]

17.5.3 *Chapter 1: (General Rule) and Chapter 2: (Establishment of Basic Plans for CCS)*

Chapters 1 and 2 cover the basics, as a cornerstone to create systemic legal basis for CCS. It covers the basic plan for carbon dioxide capture and storage, budget plan for carbon dioxide capture and storage, and special committees of CCS. Article 7 envelops the items to be included in the basic plan. Not only carbon dioxide capture and storage but also the interest in utilization are stipulated in the law. It enables the development of science and technology and the change of society to be continuously reflected, in that the plan should be established and implemented every five years. Implementation plan to establish and implement the specific assignments shall be presented each year based on the basic plan, as stipulated in Article 8, with coordination between related ministries if unclear. The draft attempts to solve the issue concerning legal definition of carbon dioxide stream – whether it shall be deemed waste or other – through Article 2 (8) of the draft, which devolves the power to set the criteria to the Ministry of Environment, while defining the concept of carbon dioxide stream. Article 9 stipulates the committee of capture process to be under the supervision of prime minister, considering the characteristics of the CCS project (being large-scale and time-taking national government project). It includes experts in each sector for professional advisory support and gathers representatives from various institutions. To delegate partial discretion for practical assignments, subordinate provisions such as presidential decrees shall be stipulated to establish a secretariat. It shows that the legislation recognizes the importance and gravity of CCS in that it prescribes discretion to allow dispatch and advisory from related institutions. In respect to securing financial resources, Article 10 stipulates the necessity for the financial resources as well as demand for its necessary resources being sustainable.

17.5.4 Chapter 3: Facilities for Carbon Dioxide Capture

Chapter 3 regulates general carbon dioxide capture processes, including installation obligations and approvals of capture facilities in power generation companies, design, construction, and inspection of facilities, collection facilities, installation of joint collection facilities, management of facilities, safety management regulations, and so forth. The installation of capture facilities is ultimately aimed at reducing carbon dioxide emissions. It is important to impose installation obligations and regulations on CCS in a way to release the liability of individual companies. The capture facility takes considerable time and requires large sites, as preliminary measures such as storage exploration and vehicle securing is required beforehand.

Article 11 states the site acquisition obligation is enforced legally and installation is not the subject of sanction. To compel the installation obligation, penalties are imposed for under-consumption, and sanctions are taken in consideration of the principle of proportionality so that penalties are not excessive. In addition, approval for the installation of the capture facility and collection of the carbon dioxide stream while meeting safety standards is required.

Article 12 is the licensing clause. It supports the business which handles installation of the capture facilities. Article 13 puts emphasis on the technical requirements such as safety and efficiency. Article 14 introduces a start-up notification system to enhance safety. This chapter therefore considers both convenience and safety in the process. If an unqualified person constructs the facilities, the business will be cancelled and become subject to sanctions according to Article 13. At the same time, if one is qualified as in Article 12, the facilities constructed according to Power Development Promotion Act will also be subject to permission, based on the principle of one-stop administration, which the present government is introducing for the convenience of the people. This chapter recognizes both the safety of CCS and the convenient interests of operators. In general, it recognizes the risk of environmental problems as in Articles 17, 19, and 20 on environmental safety inspection; while Article 15 requires the provision of rules governing the joint installation of adjacent areas such as industrial complexes, in that common installation is efficient due to the nature of power plants and collection facilities, which are large-scale.

17.5.5 Chapter 4: Transportation of Carbon Dioxide

Chapter 4 covers installation obligations and approvals of carbon dioxide transportation operators, inspection of facilities, management of facilities, and safety management regulations. In accordance with Articles 24 and 25, the transporting authority is responsible for the transportation, and the authorization and agendas are laid down in accordance with Article 26. However, the Corporation may establish a transportation plan and issue change order under approval of the Minister of Environment, so that the autonomous discretion and discipline of the administrative office are harmonized. In addition, Article 27 stipulates the completion of inspection for leakage detection and other environmental and safety management regulations, and enforces discipline on transportation where environmental disputes and safety issues may arise. Transportation owes risk of leakage in a vast region. It is necessary to introduce a stopping order system as well as an improvement order system through Article 28, to submit the operation plan in accordance with Article 29, and to impose the obligation to comply with the guidelines.

17.5.6 Chapter 5: Management of Carbon Dioxide Storage

Chapter 5 covers carbon dioxide storage companies, securement and decision of storage location, approval of storage installation, approval of storage plan, announcement of storage information, cost of storage, measures to cope with industrial complexes, safety regulations for storage operation, notification of carbon dioxide leaks and committee reports, periodic inspections, leakage liability, closures of storage site and aftercare, and monitoring processes. In consideration of the public nature of the storage, the opacity of the business, and the large scale of initial investment, Article 31 requires the corporation to preserve the storage business. Similarly, as in the case of transportation, a storage plan should be established, presented, and approved by the Minister of Environment, while the Minister of Environment has the authority to change the plan as well. Article 32 stipulates the securement and exploration of storage facilities that are particularly problematic. In the case of South Korea, a small country, it is difficult to spot the storage facilities in Korea. For the fairness of the procedure, the location of the repository is determined according to the results of the exploration and the obligation to publish is also provided for its location and area. In addition, due to the geographical nature of Korea, there is a lot of marine storage, but it can be installed on land near the ocean for temporary storage. As the process is particularly costly, rules for financial resources management are important. In Article 36, transportation and storage expenses are partially charged to the entity that installs the collection facilities, which is a type of deposit to guarantee the fulfilment of the obligations. It also provides legal grounds for payment in advance, which can be costly and time-consuming. To ensure stability, various regulations such as periodic inspections, occasional inspections, and status reports are required in preparation to cover the risk of leakage during storage.

Article 41 stipulates the sale of emission permits. It specifies purchase of carbon credits for carbon dioxide leaks, which could be exerted following the process of purchasing the carbon credits and submitting them to the Minister of Environment at the time of leakage. This shows the coupling between the ETS and CCS. The ETS (Emission Trading Scheme), which corresponds to the Article 17 of the Kyoto Protocol, is a system that allows trading within the host nation to reduce greenhouse gas emissions. This system also allows the amount of greenhouse gas reductions to be bought and sold like commodities in the market, which helps to minimize the overall reduction cost. The introduction of the emission trading system, the recognition of the carbon dioxide captured by factories as credits, and linkage between them will contribute to the development of CCS technology and the activation of emission trading. In other words, it would allow CCS reductions to be converted into greenhouse gas emission credits. In addition to the sanctions, the approval system is operated in accordance Article 42 as part of the post-management system, and the obligation to monitor the leakage situation is imposed in Article 43.

17.5.7 Chapter 6: Korean Carbon Dioxide Storage Management Service, Chapter 7: Funds on Carbon Dioxide Storage Management, and Chapter 8: Promotion of CCS Industry

The Korean Carbon Dioxide Storage Management Service is established to effectively manage the overall CCS process including the storage of carbon dioxide, with its branch institutions and research centre as part of the Service. Chapter 6 stipulates the establishment, business management, human resources, finances, support, and guidance, monitoring of Korean Carbon Dioxide Storage Management Service. Chapter 7 considers the establishment and management of the funds. Provisions are explicitly written to have the penalty fine and costs for transportation or storage deposit in the funds. This is means procuring the efficiency of financial management and procedural transparency, as parties could use the expenses only after they deposit the amount in the funds and then withdraw it from the funds. Chapter 8 covers promotion and implementation of research and demonstration projects (Korean CO_2 Storage and Environment Center). Although specific issues are delegated to subordinate provisions, it does state the importance of government support for key policy issues in CCS projects such as the CCS R&D Center. Article 53 also states that Center shall also support the pilot project if needed, which implies the perspective of legislation to deem CCS project to be widely recognized and not be restricted to major tasks. Article

55 defines the Korean CO_2 Storage and Environment Center as a basic organization and special corporation for the implementation of the national CCS comprehensive plan.

17.5.8 Chapter 9: Public Communication of CCS

This chapter is organized under the idea that public communication is necessary, adopting various measures such as holding public relations training and seminars for experts, industry, NGOs, and students, in order to enhance public awareness related to CCS. Public acceptance had been one of the key issues, in establishment of CCS infrastructure. To improve public acceptance, information disclosure and publicity are highly demanded. Korean society generally lacks public understanding on overall CCS. It is very important to solve the problem of public acceptance, since CCS projects can only be successful if solid trust on environmental safety is formed. Through access to information, relevant education, and promotion, it would be easier to form public consensus on the necessity of CCS. Article 56 stipulates the announcement of information on capture, transport, and storage of carbon dioxide, and Article 57 stipulates the installation and operation of the public council. The council functions to form consensus through discussion and coordination of opinions and to ensure the sharing of information and the promotion of CCS. Regional councils are also established under the provision, meeting the principle of local autonomy. Article 58, on the other hand, stipulates the support plan on installation area and establishes and implements a support plan for the local resident. "This holds significance as it establishes a framework for public consultation, *although concerns may arise regarding its abstract nature and practical application.*"

Many Korean institutions and national research centres have implemented offshore and onshore CCS projects. The Korea Institute of Energy Technology Evaluation and Planning had implemented an underground offshore storage project in Pohang basin and succeeded the test injection in March 2017. But at that time, coincidentally, an earthquake occurred off the coast of Pohang. Research was halted due to claims by civic groups that the cause of the earthquake was the CCS project. Later, according to the results of the investigation by the Earthquake Incident Investigation Team, it was not the CCS research that caused the earthquake, but geothermal power research. Research can be unfairly halted like this, so it is necessary to further expand public communication on a regular basis. The Korea Institute of Geoscience and Mineral Resources has implemented the first onshore pilot project as part of CCS 2020 in Janggi basin.

Through the well-known case of Barendrecht in the Netherlands, countries are recognizing that public understanding is as much as important compared to technology research and development.[19]

A survey result by K-COSEM indicates that 55.5% of the public are unaware of CCS or climate change issue. Just as many countries have struggled before implementing any legal framework, Korea is also undergoing the phase of stabilizing the public understanding issue.[20]

17.5.9 CCS-Related Act on Public Understanding

Chapter 10: Supplementary, Chapter 11: Penalties, and Addendum. Chapter 10 covers miscellaneous yet important issues, such as report and inspections, environmental impact assessment, hearing, delegation, consignment, commission fee, penalty application, and public services. Article 60 regulates the cases of entering lands during the whole process of exploration, capture, transportation, and installation of storage reservoirs. With respect to safety enhancement, Article 59 regulates examination and inspection, and Article 61 regulates environmental effectiveness assessment; with respect to the enhancement of procedural accountability, Article 62 regulates hearing, and Article 63 regulates mandate and delegation. Chapter 11 includes penalties and penal regulations, while Addendum covers enforcement date, interim measures, establishment of Industrial Complex Committee, establishment cost of industrial complexes, and amendment of other laws.

17.6 CCUS Act in Korea

17.6.1 Reason for Enactment

The CCUS Act was enacted to prevent the serious impact of the climate crisis and contribute to the sustainable development of the national economy by efficiently capturing carbon dioxide emitted from industrial activities and storing it underground or by establishing the necessary controls for the development of technology and industrialization for industrial and living use.

17.6.2 Main Contents of the CCUS Act

The main contents of the CCUS Act are as follows.[21]

17.6.2.1 Define Capture, Transport, Storage, Utilization, and Carbon Dioxide Transport Pipelines (Article 2)

Article 2 (Definition). The terms used in this Act shall have the following meanings:

1 "Capture" means collecting carbon dioxide emitted from greenhouse gas emission sources such as industrial activities or in the atmosphere in containers or facilities for domestic or international storage or utilization.
2 "Transportation" means an act of using a means prescribed by Presidential Decree to deliver captured carbon dioxide (carbon dioxide streams under Article 10, Paragraph 1 of the Marine Waste and Marine Pollution Sediment Management Act and carbon dioxide prescribed by Presidential

Decree; hereinafter the same shall apply) to a storage facility or utilization facility for domestic or international storage or utilization.

3 "Storage" means permanently isolating captured carbon dioxide from the atmosphere by injecting it into domestic or international land or marine soil so that it does not leak into the atmosphere.

4 "Utilization" means converting captured carbon dioxide into a substance useful for human life or economic activities and utilizing it as a resource in industry or directly utilizing exhaust gas containing carbon dioxide.

5 "Carbon dioxide transport pipeline" refers to a pipeline and related facilities for transporting captured carbon dioxide, as determined by Presidential Decree.

6 "Storage facility" refers to a place for storing captured carbon dioxide in land or sea soil to prevent it from leaking into the atmosphere or the ocean, and is a place that meets the standards determined by Presidential Decree.

B. The government shall establish and implement a basic plan on carbon dioxide capture, etc. every five years, and the heads of relevant central administrative agencies shall establish and implement an implementation plan every year in accordance with the basic plan (Articles 5 and 6).

C. A person who intends to install and operate a carbon dioxide capture facility shall report to the Minister of Trade, Industry and Energy, and a person who intends to engage in a carbon dioxide transport business shall obtain approval from the Minister of Trade, Industry and Energy (Articles 7 and 8).

D. A person who has received approval for a transport business and intends to install and operate a carbon dioxide transport pipeline shall establish safety management regulations and obtain approval from the Minister of Trade, Industry and Energy, appoint a safety manager for safety management of carbon dioxide transport pipelines, and undergo a safety inspection of carbon dioxide transport pipelines (Articles 9 through 11).

E. It regulates matters related to the approval of exploration to excavate storage sites on land or at sea, selection and cancellation of selection as a candidate site for carbon dioxide storage, and closure of storage sites (Articles 13 to 17).

F. It allows the Minister of Trade, Industry and Energy to designate carbon dioxide capture, storage, and utilization integrated complexes for the purpose of fostering carbon dioxide capture, storage and utilization industries; it provides support grounds for integrated complexes, and provides grounds for evaluating the operation of integrated complexes (Articles 29 to 32).

G. It provides various support grounds, such as special provisions for carbon dioxide supply, certification of carbon dioxide utilization technology and products, confirmation of carbon dioxide utilization specialized companies, and implementation and special provisions of demonstration projects (Articles 33 to 43).

H. It provides grounds for the establishment of a carbon dioxide capture, storage, and utilization promotion centre (Article 44).

17.6.3 Supplementary Provisions

Article 1 (Enforcement Date)
This Act shall enter into force one year after the date of its promulgation.
Article 2 (Transitional Measures on Marine Underground Storage Permit)
Any person who has received a permit for marine underground storage of carbon dioxide streams pursuant to the Marine Waste and Marine Pollution Sediment Management Act prior to the enforcement of this Act shall be deemed to have received a permit for storage business pursuant to this Act.

17.7 Conclusion

According to the IEA report, CCS is not only very useful as a means of massive reduction of greenhouse gases, but also a very important technology for achieving the INDC 40% target in South Korea. In this regard, CCS is central to South Korea's environmental policy. The South Korean government recognizes CCS technology as a key technology for sustainable development and enacted the CCUS Act in February 2024, which came into effect on February 7, 2025. Institutional support is needed so that such important CCUS technology can be recognized as a core element of ESG and play a greater role in creating a sustainable society.

In Germany, legislative adoption on CCS was delayed because of strong opposition from the public. In Korea, strong opposition, for example the Pohang earthquake episode, may hinder the implementation of CCS projects. With the increasing occurrence of natural disasters, including earthquakes, people are more aware of the risks that CCS plants can bring, including leakage of CO_2. In this context, policies and regulations should be regularly updated in a serious and specific manner in the revision of the next legislation, preventing any trials and errors that may lead to mistrust between the public and the government with regard to CCS and other clean energy facilities.

Without public understanding, technological development cannot be fully implemented by the public and private entities involved. The public acceptance provision originally included in the CCUS draft Act is missing from the current Act, but given its importance, it should be reflected in future amendments of the CCUS Act as soon as possible.

Notes

1 IEA, "Carbon Capture Utilisation and Storage" available at: https://www.iea.org/energy-system/carbon-capture-utilisation-and-storage (Last visited on January 24, 2025).
2 Kim Jae-Chang et al., *Carbon Dioxide Capture, Storage and Conversion Technology*, Cheongmoon-gak, 2013, p. 27.
3 IEA, "CO2 Capture and Utilisation" available at: https://www.iea.org/energy-system/carbon-capture-utilisation-and-storage/co2-capture-and-utilisation (Last visited on January 25, 2025).
4 IEA, "CO2 Capture and Utilisation" above note 3 (Last visited on January 25, 2025).
5 Barry Barton, Kimberley Jordan and Greg Severinsen, *Carbon Capture and Storage: Designing the Legal and Regulatory Framework for New Zealand*, Centre

for Environmental, Resources and Energy Law, The University of WAIKATO, 2013.9, p. 14.

6 Koh Moon-Hyun and Ahn Tae-Yong, "Legal Issues Related to CCS," *Soongsil University Law Research Institute, Journal of Law*, vol. 35, January 2016, p. 36.; Jo In-seong, "Legislative Measures for Improving Social Acceptance of Commercialization of Carbon Capture and Storage (CCS) – Focusing on Discussions in Japan, the United States, and Germany," *European Constitutional Studies*, vol. 17, April 2015, p. 726.

7 Ian Havercroft and Richard Macrory, *Legal Liability and Carbon Capture and Storage*, GCCSI, October 2014, p. 11; "Various potential types of damage from CO2 Leakage have been identified, including risk to human health and harm to flora and fauna, but the current evaluation is that with proper site selection such risks are likely to be minimal."

8 Koh Moon-Hyun and Ryu Kwon-Hong, *Energy and Resources Law for Climate Change Response*, Soongsil University Press, 2015, p. 207.

9 Moon-Hyun and Kwon-Hong, *Energy and Resources Law for Climate Change Response*, p. 211.

10 IEA, "Carbon Capture", above note 3. https://www.iea.org/energy-system/carbon-capture-utilisation-and-storage (Last visited on January 25, 2025).

11 United Nations, *United Nations Framework Convention on Climate Change*. The Paris Agreement, 2015 December 12.

12 Koh Moon-Hyun, "Necessity of Legislation of CCS and Its Major Contents to Cut Down Greenhouse Gas", *Public Land Law Review*, vol. 75, 2016 May, pp. 317–341.

13 Green Growth Committee, The Korean National CCS Mater Action Plan, 2010 July 13.

14 Korea Carbon Capture & Sequestration R&D Center (KCRC), http://www.kcrc.re.kr/html/company.html

15 G. B. Cho, H. Y. Cho and N. E. Park, "A Study on Implementation and Deriving Future Tasks of 'The Korean National CCS Master Action Plan'", *Journal of Climate Change Research*, vol. 7, no. 3, 2016, pp. 237–247.

16 See IEA, Net Zero by 2050 – A Roadmap for the Global Energy Sector-, 2021, pp. 53–56.

17 IEA, Net Zero by 2050 – A Roadmap for the Global Energy Sector-, 2021, p. 60. In particular, CCUS is reported to play the most important role in reducing CO2 in cement production, p. 125.

18 Moonhyun Koh, Eunhae Shin and Woongchan Seo, "Outline of Korean Integrated CCS Act Draft and Its Implication", *Applied Energy Symposium and Forum*, Carbon Capture, Utilization and Storage, CCUS 2018, 27–29 June 2018, Perth, Australia, ScienceDirect, Energy Procedia 154, 2018, 15–21.

19 Moonhyun Koh, Bola Ju and Woongchan Seo, *A Review on Public Understanding of Carbon Dioxide Capture and Storage (CCS) in South Korea, Applied Energy Symposium and Forum, Renewable Energy Integration with Mini/Microgrids*, REM 2018, 29–30 September 2018, Rhodes, Greece, Energy Procedia 159, 2019, 319.

20 Koh, Ju and Seo, *A Review on Public Understanding*, 317.

21 https://www.law.go.kr/lsInfoP.do?lsiSeq=259701#0000 (Last visited: 2025. 1. 27).

List of Cases

A.P. Pollution Control Board v. *Prof. M.V. Nayudu (Retd.) & Others*, 1999 (2) SCC 718.
Advocate Khagenedra Subedi and Others v. *Government of Nepal, Prime Minister and Council of Ministers*, NKP 2018–19, no. 9, p. 1626
Advocate Ram Chandra Simkhada v. *Government of Nepal, Prime Minister and Council of Ministers*, NKP 2019, no. 2, p. 367.
Advocate Shailendra Prasad Ambedkar and Others v. *Govt of Nepal, Prime Minister and Office of the Council of Ministers and Others*, Writ No. 077-wc-0099 and other 7 writs, decided on April 20, 2022 (unpublished).
Advocate Sunil Ranjan Singh v. *Government of Nepal, Prime Minister and Council of Minister*, NKP 2020–21, no. 9, p. 1727.
Advocate Tulasi Simkhada v. *Government of Nepal, Prime Minister*, NKP 2019, no. 10, p. 1818.
Ambica Quarry Works v. *State of Gujarat* , AIR 1987 SC 1073
Bangalore Medical Trust v. *BS Muddappa*, (1991) 4 SCC 54.
Banwasi Seva Ashram v. *State of UP*, AIR 1987 SC 374
Binay Kumar Dalei v. *State of Odisha*, [(2022) 3 SCC 33
Bombay Dyeing & Mfg Co Ltd v. *Bombay Environmental Group*, AIR 2006 SC 1489
Centre for Public Interest litigation v. *Union of India and Others*, (2013) 16 SCC 279
Chief Conservator of Forest Wild Life v. *Nisar Khan*, AIR 2003 SC 187
Client Earth v. *European Investment Bank*, [2021] Case T-393/19.
Common Cause v. *UOI*, (2017) 9 SCC 499
Dharam Bahadur Lamichhane and others v. *Ministry of Physical Infrastructure and Transport and others*, NKP 2018–19, no. 9, p. 1615
Fomento Resorts and Hotels Ltd v. *Minguel Martins*, (2009) 3 SCC 571
G Sundarrajan v. *Union of India*, (2013) 6 SCC 620
Gene Campaign and Others v. *Union of India*, 2024 SCC Online SC 1793
Goa Foundation v. *UOI*, (2014) 6 SCC 590
Haghshenas v. *Canada (Citizenship and Immigration)*, 2023 FC 464
Hinch Lal Tiwari v. *Kamala Devi*, (2001) 6 SCC 496,

Hindustan Zinc Ltd v. *Rajasthan Electricity Regulatory Commission*, MANU/SC/0641/2015
Hospitality Association of Mudumalai v. *In Defence of Environment and Animals*, (2020) 10 SCC 589.
Hospitality Association of Mudumalai v. *In Defence of Environment and Animals*, AIR Online 2020 SC 776
Indian Council for Enviro-Legal Action and Ors. v. *Union of India and Ors.* (1996) 3 SCC 212
Indian Council for Enviro-legal Action v. *Union of India and Ors* ,[1996] Supp. (1) S.C.R. 507
Indian Council for Enviro-Legal Action v. *UOI*, AIR 1996 SC 1446
Intellectual Forum, Tirupati v. *State of Andhra Pradesh*, AIR 2006 SC 1350
Juliana v. *United States* [9th Cir. 2020] 947 F.3d 1159
Juniper Hotels Private Limited v. *Delhi Electricity Regulatory Commission*, MANU/DE/7411/2023
Justice K.S. Puttaswamy v. *Union of India*, (2017) 10 SCC 1
Karnataka Industrial Area Development Board v. *Kenchappa*, AIR 2006 SC 2038
Kisan Trimbal Kothula v. *State of Maharashtra*, (1977) 1 SCC 300
M.C. Mehta v. *Kamal Nath*, (1997) 1 SCC 388.
M.C. Mehta v. *Kamal Nath*, (2000) 6 SCC 213
M.C. Mehta v. *Union of India and Ors.*, 2020 SCC OnLine SC29
M.C. Mehta v. *Union of India*, 1986 SCC (2) 176,
M.K. Ranjitsinh & Ors. v. *Union of India & Ors* 2024 INSC 280
M.K. Ranjitsinh & Ors. v. *Union of India & Ors.*, 2024 SCC Online SC 570
Mantri Technose Private Ltd v. *Forward Foundation*, (2019) 18 SCC 494.
Massachusetts v. *Environmental Protection Agency* [2007] 549 U.S. 497 38
MC Mehta v. *Union of India*, (2009) 6 SCC 142.
MC Mehta v. *UOI*, (2020) 7 SCC 571,
MC Mehta v. *UOI*, 2012 8 SCC 123
MC Mehta v. *UOI*, AIR 1997 SC 734
MC Mehta v. *UOI*, AIR 1997 SC 734
MC Mehta v. *UOI*, AIR 1999 SC 2367.
MC Mehta v. *UOI*, AIR 1999 SC 2367.
MC Mehta v. *UOI*, AIR 2004 SC 4016.
MI Builders Pvt Ltd v. *Radhey Shyam Sahu*, AIR 1999 SC 2468
Milieudefensie and Others v. *Royal Dutch Shell PLC and Others* [2021] C/09/
Narmada Bachao Andolan v. *Union of India* (2000) 10 SCC 664
Narmada Bachao Andolan v. *Union of India* AIR 2000 SC 3751
Narmada Bachao Andolan v. *UOI*, AIR 2000 SC 3751,
ND Jayal v. *UOI*, AIR 2004 SC 867
PUCL v. *Union of India*, (2001) Writ Petition (Civil) No. 196 (SC)
Rajeev Suri v. *Delhi Development Authority*, decided on 05-Jan-2021

Rato Kumari Bikash Sewa Samiti v. *Prime Minister and Council of Ministers*, NKP 2018, no. 4, p. 699.
Roncarelli v. *Duplessis* [1959] SCR 121.
Rumlal Pun and others v. *Ministry of Energy, Water Resources and Irrigation*, Case no. 079-wo-1520, decided on Jan. 28, 2024 (unpublished)
Rural Litigation and Kendra v. *State of Uttar Pradesh*, AIR 1985 SC 652
Rural Litigation and Kendra v. *State of Uttar Pradesh*, AIR 1988 SC 2187
S Jagannath v. *UOI* , (1997) 2 SCC 87
Sabu Mathew George v. *Union of India and Ors.*, MANU/SC/1674/2017.
Safarian v. *Canada (Citizenship and Immigration)*, 2023 FC 775.
Senior Advocate Prakash Mani Sharma Representing Pro Public and Others and Senior Advocate Krishna Prasad Bhandari v. *Govt. of Nepal*, Writ No. 076-wf-0006 and 076-wf-0007, decided on May 26, 2022 (unpublished).
Shreya Singhal v. *Union of India*, AIR 2015 SC 1523
State of Bihar v. *Murad Ali Khan*, (1988) 4 SCC 655
State of Tamil Nadu v. *M/s Hind Stone and others*, 1981 (2) SCC 205
Subash Kumar v. *State of Bihar*, (1991) 1 SCC 598.
Subhash Kumar v. *State of Bihar*, AIR 1991 SC 420
Susetha v. *State of Tamil Nadu*, AIR 2006 SC 2893
Swami Achutanand Tirth and Ors. v. *Union of India and Others*, (2014) 13 SCC 314
T.N. Godavarman Thirumulpad v. *Union of India,* (2022) 4 SCC 289
T.N. Godavarman Thirumulpad v. *Union of India*, 2022 SCC On Line SC, Decided on April 26, 2023
T.N. Godavarman Thirumulpad v. *UOI*, an order dated June 3, 2022
TN Godavarman Thirumulpad (104) v. *Union of India,* (2008) SCC 222
TN Godavarman Thirumulpad (87) v. *Union of India*, (2006) 1 SCC 1
TN Godavarman Thirumulpad v. *UOI*, (2006) 5 SCC 47
TN Godavarman Thirumulpad v. *UOI*, (2010) 6 SCC 710
TN Godavarman Thirumulpad v. *UOI*, (2012) 3 SCC 277
TN Godavarman Thirumulpad v. *UOI*, (2012) 4 SCC 362.
Urgenda Foundation v. *The Netherlands* [2015] HAZA C/09/00456689
Vellore Citizens Welfare Forum v. *Union of India and Ors.*, (1996) 5 SCC 647
Virender Gaur v. *State of Haryana*, (1995) 2 SCC 577

Index

www.ingramcontent.com/pod-product-compliance
Lightning Source LLC
LaVergne TN
LVHW010552300926
849162LV00002B/307
* 9 7 8 1 0 4 1 2 2 1 4 5 6 *